Nanotechnology
The Future is Tiny

Nanotechnology
The Future is Tiny

Michael Berger
Nanowerk LLC, Berlin, Germany
Email: michael@nanowerk.com

THE QUEEN'S AWARDS
FOR ENTERPRISE:
INTERNATIONAL TRADE
2013

Print ISBN: 978-1-78262-526-1
PDF eISBN: 978-1-78262-887-3
EPUB eISBN: 978-1-78262-888-0

A catalogue record for this book is available from the British Library

Published by The Royal Society of Chemistry,
Thomas Graham House, Science Park, Milton Road,
Cambridge CB4 0WF, UK

Registered Charity Number 207890

Visit our website at www.rsc.org/books

Printed in the United Kingdom by CPI Group (UK) Ltd, Croydon, CR0 4YY, UK

Preface

This book is a collection of essays about researchers involved in all facets of nanotechnologies. Nanoscience and nanotechnology research are truly multidisciplinary and international efforts, covering a wide range of scientific disciplines such as medicine, materials sciences, chemistry, biology and biotechnology, physics and electronics.

Each of the following stories is based on a scientific paper that has been published in a peer-reviewed journal. Although each story revolves around one or two scientists who were interviewed for this book, many, if not most, of the scientific accomplishments covered here are the result of collaborative efforts by several scientists and research groups, often from different organizations and from different countries.

These stories take you on a journey of scientific discovery into a world so small that it is not open to our direct experience. While our five senses are doing a reasonably good job at representing the world around us on a macro-scale, we have no existing intuitive representation of the nanoworld, ruled by laws entirely foreign to our experience. This is where molecules mingle to create proteins; where you wouldn't recognize water as a liquid; and where minute morphological changes would reveal how much "solid" things such as the ground or houses are constantly vibrating and moving.

You will catch a glimpse of how diverse, wide-ranging and intriguing this research field is and what kind of amazing and exciting materials and applications nanotechnologies have in store for us.

We will showcase 176 very specific research projects that are taking place in laboratories around the world and you will meet the scientists who developed the theories, conducted the experiments, and built the new materials and devices that each will take us one tiny step further into our nanotechnology-influenced future.

Nanotechnology: The Future is Tiny
By Michael Berger
© Michael Berger 2016
Published by the Royal Society of Chemistry, www.rsc.org

Some stories are more like an introduction to nanotechnology, some are about understanding current developments, and some are advanced technical discussions of leading edge research. Reading this book will shatter the monolithic term "nanotechnology" into the myriad of facets that it really is.

Major technology shifts don't happen overnight; and rarely are they the result of a single breakthrough discovery. Nowhere is this more true than for the vast set of capabilities that we have come to simply call nanotechnology.

Nanotechnology is not an industry; nor is it a single technology or a single field of research. What we call nanotechnology consists of sets of enabling technologies applicable to many traditional industries (therefore it is more appropriate to speak of nanotechnologies in the plural).

Rather than standing on the shoulders of a few intellectual giants, nanotechnologies get created by tens of thousands of researchers and scientists working on minute and sometimes arcane aspects of their fields of expertise in a multitude of areas; they come from different science backgrounds; live in different parts of the world; work for different organizations (government labs, industry labs, universities, private research facilities) and follow their own set of rules – get papers reviewed and published; achieve scientific recognition from their peers; struggle to get funding for new ideas; look to make that breakthrough discovery that leads to the ultimate résumé item, a Nobel prize; get pushed by their funders to secure patent rights and commercialize new discoveries.

The collection of stories in this book is barely scratching the surface of the vast and growing body of research that leads us into the nanotechnology age. The selection presented here is not meant to rank some labs and scientists higher than others, nor to imply that the work introduced in this book is more important or valuable than the vastly larger body of work that is not covered. The intention is to give the interested reader an idea of the incredibly diverse aspects that make up nanotechnology research and development – the results of which will bring about a new era of industrial and medical technologies.

The development of nanotechnologies is not based on a few big and bold discoveries or inventions. Rather it is a painstakingly slow journey of gradual development, a result of which will be some truly revolutionary products and applications.

Michael Berger

Contents

Nanotechnology: The Future is Tiny
By Michael Berger
© Michael Berger 2016
Published by the Royal Society of Chemistry, www.rsc.org

CHAPTER 1

Generating Energy Becomes Personal

In the not-too-distant future, the way we generate energy will change dramatically. This shift will not just mean the ongoing transition from fossil fuels to renewables; it will also concern the way our day-to-day gadgets are powered. There still will be large-scale power plants to provide energy for industry, infrastructure and households, but some part of power generation will become decentralized all the way to its point of use: down to a personal level where gadgets, textiles—even implants—will generate their own power.

There is an almost infinite number of mechanical energy sources all around us—basically, anything that moves can be harvested for energy. These environmental energy sources can be very large, like wave power in the oceans, or very small, like rain drops or biomechanical energy from the heartbeat, breathing, and blood flow. Engineering at the nanoscale allows researchers to find more and more ways to tap into these pretty much limitless sources of energy and to make energy harvesting and storage much more efficient.

If current research is an indicator, form and shape of future electronics will go far beyond very small and ultra-thin devices and wearable, flexible computers. Not only will these devices be embedded in textile substrates but an electronics device or system could ultimately become the fabric itself. Electronic textiles ("e-textiles") will allow the design and production of a new generation of garments with fully integrated sensors and electronic functions. Such e-textiles will have the revolutionary ability to sense, act, store, emit, and move—think biomedical monitoring functions or new man–machine interfaces—while ideally leveraging an existing low-cost textile manufacturing infrastructure.

Nanotechnology: The Future is Tiny
By Michael Berger
© Michael Berger 2016
Published by the Royal Society of Chemistry, www.rsc.org

All these wearable and potentially textile-embedded electronics will require power; and it wouldn't make sense to have to plug your sleek flexible sleeve display into a bulky lithium-ion battery brick. Researchers are therefore pushing the development of wearable energy storage. Especially supercapacitors with a cable-type architecture could lead to flexible energy storage devices that can achieve a subversive technology to open up a path for radical design innovations.

1.1   Forget Batteries, Let a T-Shirt Power Your Smartphone

The continued miniaturization of portable electronics is increasingly challenged by the reliance on conventional battery technology. More powerful processors and displays require more power. However, ever-sleeker form factors require this increased power to come from shrinking battery sizes.

Micro- and even nanoscale devices will become widely used in health monitoring; infrastructure and environmental monitoring; internet of things; and of course defense technologies. In these application areas, battery design will have to go way beyond today's typical lithium-ion batteries. Rather than relying on stored power, nanodevices will make use of novel—also nanoscale—power sources. Self-powered technology based on piezoelectric nanogenerators aims at powering nanodevices and nanosystems using the energy harvested from the environment in which these systems operate. This offers a completely new approach for harvesting mechanical energy using organic and inorganic materials.

1.1.1   Self-Powered Smartwear

Nano-sized generators that possess piezoelectric properties—meaning that they accumulate an electric charge when mechanical stress is applied to them—allow them to convert into electricity the energy created through mechanical stress, stretches and twists of fabrics. Such energy-scavenging textiles could eventually lead to wearable "smart" clothes that can power integrated electronics and sensors through ordinary body movements.

Already, researchers have demonstrated a new type of fully flexible, very robust and wearable triboelectric nanogenerator (WTNG) with high power-generating performance and mechanical robustness.

The scientists applied a bottom-up nanostructuring approach where they used a silver-coated textile and polydimethylsiloxane (PDMS, a silicon-based organic polymer) nanopatterns based on zinc oxide (ZnO) nanorod arrays as active triboelectric materials.

The nanopatterning was achieved by coating PDMS directly over vertical ZnO nanorods grown on the silver-coated textile substrate.

"This nanopatterning promotes the triboelectrification effect by increasing the effective contact area and friction for high electrical output power and very high mechanical robustness by bottom-up nanostructuring between textile and nanostructure," explains Sang-Woo Kim, a professor in the School of Advanced Materials Science and Engineering at Sungkyunkwan University in Korea.

Kim, together with collaborators from Sungkyunkwan University and the Australian Institute for Innovative Materials, successfully demonstrated the self-powered operation of light-emitting diodes (LEDs), a liquid-crystal display (LCD) and a keyless vehicle entry system only with the output power of their WTNG without any help of external power sources.

This triboelectric power is generated when mechanical stress creates an electrical charge, which is much larger than the power generated from textile-based piezoelectric power generators reported before. This stress can arise through stretching or twisting the textile.

When integrated into clothing, the WTNG relies on its triboelectric property to produce an electrical charge when pressed, and could potentially allow users to power mobile electronics such as smart watches simply by moving or walking around.

In previous work,[1] Kim and his team already have reported fully transparent, flexible, and stretchable nanogenerators but they were not robust enough as wearable devices.

By contrast, the researchers tested a four-layer-stacked WTNG over 12 000 cycles and found no significant differences in the output voltages they measured.

With the previous devices, the problem was the very weak adhesion between textile and nanostructures, which caused mechanical durability issues. In this subsequent work, the researchers overcame these durability problems by bottom-up nanostructuring using ZnO nanorods and PDMS nanopattern coating.

"Triboelectric generation is one of the promising new energy harvesting methods with extremely high output voltage and efficiency, low cost, high versatility and simplicity in structural design and fabrication, stability and robustness, as well as environmental friend," concludes Kim. "We have been looking to find new materials for huge triboelectrification effects which have never been reported before. Furthermore, there is a need for developing highly efficient power management systems to effectively store electric power generated from triboelectric nanogenerators into textile-based energy storage platforms such as textile batteries or supercapacitors."

Featured scientists: The Sang-Woo Kim Group (http://nesel.skku.edu/index.html)

Organization: School of Advanced Materials Science and Engineering, Sungkyunkwan University (Republic of Korea)

Relevant publications: W. Seung, M. Gupta, K. Lee, K. Shin, J. Lee and T. Kim *et al.*, Nanopatterned Textile-Based Wearable Triboelectric Nanogenerator, *ACS Nano*, 2015, **9**(4), 3501, DOI: 10.1021/nn507221f

1.1.2 Cotton T-Shirts As Batteries

Generating power at the source of demand is one option. Another one is to provide energy storage, *i.e.* batteries, that possess the same size scale, flexibility, and ease of integration as the electronic circuitry they are intended to power.

Textile yarns are an obvious choice. Addressing the power source issue with natural materials, researchers have now found a simple way to provide cotton with a new function—storing energy.

Inspired by traditional activation of cellulose fibers into activated carbon fibers with a simple chemical route, a team of scientists has come up with a way to convert cotton textiles into activated carbon textiles for energy storage applications.

After this functionalization, the textile features were well preserved and the obtained activated carbon fibers were highly conductive and flexible, enabling an ideal electric double layer capacitor performance.

Xiaodong Li, CEC Distinguished Professor at the University of South Carolina, and Lihong Bao, a postdoctoral researcher in Li's group, demonstrated how cotton textiles can be functionalized into activated carbon textiles (ACTs) by a simple chemical activation route—a traditional dipping, drying and curing process. During this process, the cellulose fibers are converted into activated carbon fibers.

The researchers explain that the coupled superior capacitive characteristic and high porosity make the ACTs unique supporting backbones for controlled deposition of nanostructured pseudo-capacitive manganese dioxide (MnO_2) to construct a hybrid composite for enhanced electrochemical performance: "The key point of the design lies in activation of cotton T-shirt textiles into porous, mechanically flexible, and highly conductive ACTs with high electrolytically accessible surface area."

Performing cyclic voltammetry tests, the scientists explored the electrochemical performance of their hybrid MnO_2/ACT composite.

"The specific capacitances calculated from the cyclic voltammetry curves at different scan rates indicate remarkably improved capacitance performance with almost three times increase in specific capacitance compared to that of ACTs," says Li. "We think that the energy storage capacity can be further improved by depositing a nanometer-thick transition metal oxide film or graphene layer. This may open up a new field—biomaterial-enabled energy storage."

The key finding of this work is that cheap and environmentally friendly textiles such as cotton can be a low-cost and green solution for textile-based energy-storage devices.

"Cotton is an environmentally friendly, renewable, and easily recyclable material—unlike oil and engineered chemicals," says Li. "It is a promising candidate as a power source for future electronic devices."

In future research, the scientists may extend their method to process other cellulosic materials such as bamboo and pine straws into activated carbon for flexible high performance supercapacitor and battery electrodes.

Featured scientists: Xiaodong Li, CEC Distinguished Professor, Nanostructures and Reliability Laboratory (http://www.me.sc.edu/research/nano/index.htm)
Organization: Department of Mechanical Engineering, University of South Carolina, Columbia, SC (USA)
Relevant publication: L. Bao L, X. Li. Towards Textile Energy Storage from Cotton T-Shirts, *Adv Mater.*, 2012, **24**(24), 3246, DOI: 10.1002/adma.201200246

1.1.3 Graphene Yarns Turn Textiles into Supercapacitors

Textile yarns, such as the above-mentioned conductive cotton fibers, are an obvious choice for e-textiles.

On the other hand, high-performance multifunctional synthetic fibers, although of interest from a mechanical and electrical point of view, suffer from low electrochemical performance, which is crucial to the realization of multifunctional textiles required for the advancement of smart electronic devices.

To overcome this challenge, researchers at the University of Wollongong's Institute for Superconducting & Electronic Materials have fabricated flexible, durable, and self-assembled graphene textile electrodes for supercapacitors using a novel wet-spinning approach of ultra large graphene oxide (GO) liquid crystals followed by heat treatment to obtain graphene fibers.

These graphene textile electrodes exhibit unrivalled electrochemical capacitance properties of as high as 410 F g^{-1} (the theoretical value is 550 F g^{-1} whereas the best value achieved to date has been 265 F g^{-1}).[†]

As Seyed Hamed Aboutalebi, first author of a paper on this work, points out, careful, rational nanoarchitectonic design and spacing of individual graphene layers is crucial for high-performance energy storage devices. The key to producing such fibers and yarns is to preserve the large sheet size even after the reduction of GO while simultaneously maintaining a high interlayer spacing in-between graphene sheets.

"This in conjunction with maximizing the number of covalently bonded carbon atoms per unit volume or mass and significantly reducing the number of other atoms present at the system and attached to graphene sheets resulted in exceptional electrochemical performance," says Aboutalebi.

The team's wet-spinning process allows the fabrication of unlimited lengths of highly porous yet mechanically strong (Young's modulus in excess of 29 GPa), robust and flexible multifunctional fibers and yarns by taking advantage of the intrinsic soft self-assembly behavior of ultra-large graphene oxide liquid crystalline dispersions.

"Both fibers and yarns exhibited outstanding tensile strength and could be easily weaved into conductive textiles, opening up opportunities for the application of graphene in wearable electronic gadgets," notes Aboutalebi.

[†]F is the symbol for farad, the SI unit of capacitance.

These graphene yarns could lead the way to the realization of powerful next-generation multifunctional renewable wearable energy storage systems. The simplicity of the method used here and the abundance of graphene oxide precursors make these materials interesting and highly promising candidates for a range of applications such as wearable, lightweight multifunctional textiles and electronic gadgets and flexible energy storage devices to meet the demands of real-world energy storage systems.

Furthermore, the method introduced here along with the team's previous publications[2,3] can serve as a platform to process these materials at industrially highly-scalable levels for a whole range of both novel and existing applications such as coatings, fillers, molecular electronics, wearables, smart garments, RFID (radio-frequency identification) devices, printed electronics, organic field effective transistors and 3D bionic scaffolds.

Following up on these results, the team is working on the self-assembly process and lyotropic liquid crystallinity as an enabling platform to tailor-make processable self-assembled, self-oriented graphene-based hybrids with large-area molecular ordering.

Featured scientist: Hamed Aboutalebi, PhD
Organization: Institute for Superconducting & Electronic Materials, University of Wollongong (Australia)
Relevant publication: S. Aboutalebi, R. Jalili, D. Esrafilzadeh, M. Salari, Z. Gholamvand and S. Aminorroaya Yamini, *et al.*, High-Performance Multifunctional Graphene Yarns: Toward Wearable All-Carbon Energy Storage Textiles, *ACS Nano*, 2014, 8(3), 2456, DOI: 10.1021/nn406026z

1.1.4   Silky Substrate Makes Flexible Solar Cells Biocompatible

Another elegant solution for wearable electronic textiles would be if they could charge themselves with integrated, equally wearable, solar cells.

The most common flexible substrates used for flexible solar cells so far have been synthetic polymers such as polyethylene terephthalate (commonly known as PET) and polyethylene naphthalate (PEN). However, if organic solar cells are to be applied onto clothes and other soft surfaces—some of which come into direct contact with skin—they are required to be human-compatible, non-toxic and non-irritable.

One possible solution for a suitable substrate could be silk.

"The natural silk fibroin—extracted from the silkworm (*Bombyx mori*) cocoon—is a promising alternative material due to its good biocompatibility, biodegradability, non-toxicity, non-irritability and advantageous mechanical properties, as well as high optical transmittance (90–95%) of films," explains Baoquan Sun, a professor in materials science at Soochow University in Suzhou, China. "Furthermore, the biodegradable and mechanical properties of silk fibroin substrates can be tailored by controlling the fabrication process, such that they match the desired requirements for some specific application."

Sun and his team, together with researchers from the National Engineering Laboratory for Modern Silk, also at Soochow University, integrated a biocompatible silk fibroin with a mesh of silver nanowires to achieve a flexible, transparent, and biodegradable substrate for efficient plastic solar cells (Figure 1.1).

"Our flexible substrate can achieve a conductivity of ~11.0 Ω sq^{-1} and a transmittance of ~80% in the visible light range, which is much better than commercialized flexible substrates such as indium tin oxide (ITO) coated PET and indium tin oxide coated polyethylene naphthalate," says Sun. "The power conversion efficiency of 6.6% is relatively high on the silk fibroin substrate."

He points out that, even after extremely rigid bending, the devices retain a stable conductivity that is superior to traditional flexible ITO–PEN substrates. He also notes that the conductivity of bent silver nanowire silk fibroin substrates can be recovered *via* a "self-healing" process.

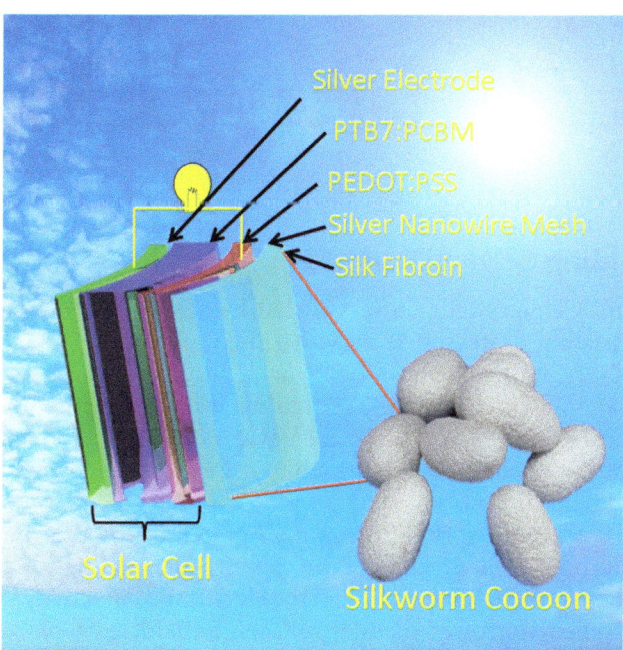

Figure 1.1   The figure shows the silkworm cocoon picture and the scheme of plastic solar cell with silk fibroin film as the substrate. The device can be flexible and compatible with human skin. Here, PEDOT:PSS, PTB7 and PCBM stand for poly(3,4-ethylenedioxythiophene), polystyrene sulfonate thieno[3,4-*b*]thiophene/benzodithiophene and [6,6]-phenyl-C71-butyric acid methyl ester, respectively. (Image: Sun group, Soochow University.)

In order to form a continuous conductive film on transparent films, silver nanowire mesh is usually fabricated by directly spin- or spray-coating onto a transparent substrate. The resulting film is rough due to the random distribution of silver nanowire piling on the substrate.

"This roughness is a disadvantage because substrate flatness is critical to fabricating the ~100-nm-thick active layer of the plastic solar cell," explains Sun. "Additionally, the deposited silver nanowire displayed poor adhesion properties on the substrate."

To resolve this problem, the researchers first deposited silver nanowires on a flat model substrate instead of being directly deposited onto the substrate. Then, the aqueous silk fibroin solution was coated onto the silver nanowire-covered substrate.

This work illustrates another step towards fully biocompatible plastic solar cells that one day might be integrated with objects and devices of everyday use and even with living tissue for some futuristic bionic applications.

Featured scientists: Prof. Baoquan Sun's Lab of Advanced Energy Materials (http://baoquansun.weebly.com/)
Organization: Institute of Functional Nano & Soft Materials, Soochow University (PR China)
Relevant publication: Y. Liu, N. Qi, T. Song, M. Jia, Z. Xia and Z. Yuan, *et al.*, Highly Flexible and Lightweight Organic Solar Cells on Biocompatible Silk Fibroin, *ACS Appl. Mater. Interfaces*, 2014, **6**(23),20670, DOI: 10.1021/am504163r

1.1.5   Folding Origami Batteries

A team from Arizona State University, led by associate professor Hanqing Jiang and assistant professors Hongyu Yu and Candace K. Chan, has demonstrated the fabrication of a highly deformable lithium-ion battery using standard electrodes and commercially standard packaging technologies.

The origami design concept of these devices enables Li-ion batteries with unprecedented mechanical deformability including folding, unfolding, twisting and bending.

The team's previous work[4] of using folding to improve the energy density helped them to implement some of the experiments leading to these results, such as using nanotube ink to maintain the electrical conductivity after folding.

The researchers borrowed the idea for their foldable battery from origami. One of the main characteristics of origami is compactness and deformability of the folded structures. Here, for example, the team used the Miura technique, which is a rigid origami folding pattern.

"The deformability, particularly the stretchability of this pattern, is achieved by the folding and unfolding of the creases while maintaining the paper rigid. This is ideal to develop highly deformable electronics and motivates us to conduct this work," Jiang explains.

Thanks to using the origami concept, the researchers were able to achieve significant system-level linear and areal deformability for their battery (over 1000%), large twistability and bendability, and up to 74% areal coverage.

Moreover, this origami lithium-ion battery uses commercially standard materials and packaging technologies.

"This work represents the fusion of the art of origami, materials science, and functional energy storage devices, and could provide a paradigm shift for architecture and design of flexible and curvilinear electronics with exceptional mechanical characteristics and functionalities," says Jiang.

According to the team, this origami battery can be incorporated with other deformable electronics components by two means:

The first approach is to build a functional system that may include energy harvesting devices (*e.g.*, solar cells), energy storage devices (*e.g.*, Li-ion batteries) and other devices (*e.g.*, sensor arrays and microprocessors) in the same origami platform to achieve a highly deformable system.

The second approach is to build a stand-alone Li-ion battery by encapsulating the origami battery with elastomeric materials and then integrating it with other functional devices.

Going forward, the team's plan is to develop a highly deformable system with all the necessary components—including energy harvesting, energy storage, sensors, processors, wireless communications, *etc.*, using the concept of origami.

They have already demonstrated locking an origami structure using phase change materials[5] and they also described the fabrication of origami-enabled deformable silicon solar cells.[6]

Featured scientists: Prof. Hanqing Jiang (http://jiang.lab.asu.edu/); Prof. Hongyu Yu (http://hongyuyu.asu.edu/); Prof. Candace Chan (http://faculty.engineering.asu.edu/chan/)
Organization: Arizona State University, Tempe, AZ (USA)
Relevant publication: Z. Song, T. Ma, R. Tang, Q. Cheng, X. Wang and D. Krishnaraju, *et al.*, Origami lithium-ion batteries, *Nat. Commun.*, 2014, 5, 3140, DOI: 10.1038/ncomms4140

1.1.6 Towards Self-Powered Electronic Papers

Paper as you know it is going to change. Various research efforts are about to make paper "smart". Take the European ROPAS project[7] which aims to develop a wireless sensor device on a paper surface which can be manufactured using high-end and low-cost printing techniques. Researchers have also already demonstrated various types of devices on paper—batteries, solar cells, RFID tags, or transistors.

Taking these developments a step further, an international research team—led by Jun Zhou, a professor at the Wuhan National Laboratory for Optoelectronics, and School of Optical and Electronic Information, Huazhong

University of Science and Technology, and Liangbing Hu, an assistant professor in the Department of Materials Science and Engineering and Energy Research Center at the University of Maryland—has designed and demonstrated novel self-powered, human-interactive, transparent nanopaper systems, utilizing transparent nanopaper as a base material.

This nanopaper system is based on an electrostatic induction mechanism and a dielectric material. That makes it self-powered, *i.e.* able to operate without the need for external power.

These so-called *electret* materials—the word is a combination of *electr* from "*electr*icity" and *et* from "magn*et*"—have the ability to store residual charges for rather a long time. This new approach to energy harvesting is already used in microphones and in MEMS (microelectromechanical system) devices, but these usually are solid devices. A recent example for a flexible application has been a paper generator.

The team's transparent paper-based flexible generator (TPFG) basically consists of two transparent electrode components: two sheets of nanopaper coated with carbon nanotubes (CNTs) where one of the sheets additionally is covered with a 30 µm polyethylene (PE) film, a transparent and non-fluorine electret. The two sheets are then joined together on their coated sides (the PE facing the CNTs).

The basic working mechanisms of the resulting devices are electrostatic induction effects caused by the retaining charges.

"The variation of the air gap between the two components plays a key role in the generation of the alternating electric currents," Zhou explains. "When the TPFG is in the original—equilibrium—state, no free electrons will flow in the external circuit. However, when a vertical compression is applied to the TPFG, the air gap becomes smaller and more positive charges will be induced in the bottom CNT electrode. As a result, the equilibrium state is broken and current will flow from the top CNT electrode to the bottom CNT electrode."

Charges will accumulate in the bottom electrode until a new equilibrium is reached and the current stops flowing. When the pressure is released, the process reverses: the air gap increases and the current starts to flow again in the other direction until again equilibrium has been reached.

"Our transparent nanopaper system could potentially be manufactured at a large scale for practical applications," adds Hu. "To demonstrate this, we have fabricated TPFG anti-theft and anti-fake systems."

He explains that the anti-theft system will not affect the appearance of the art and is also sensitive for detecting external pressure. Additionally, the system is compatible with paper arts and can be added or removed freely.

In a similar vein, the anti-fake system based on transparent nanopaper displays coded information—a date, ID number, *etc.*—which may have potential applications in the smart packaging of important documents, such as wills and birth certificates, or as product labels.

The researchers have begun working on improving their device, especially with regard to long-term stability and device integration (*e.g.* RFID). Furthermore, currently the output stability of the generator can be affected when the device works in a high humidity environment. Also, the carbon nanotube electrodes could pose a potential health concern if these devices are integrated into electronic skin applications and therefore the team is thinking about using other types of electret materials and conductive layers composed of silver nanowires or conductive polymers.

Featured scientists: Prof. Jun Zhou; Prof. Liangbing Hu
Organizations: Wuhan National Laboratory for Optoelectronics, Huazhong University of Science and Technology (PR China) (http://en.wnlo.cn/index.htm); Department of Materials Science and Engineering and Energy Research Center, University of Maryland (USA) (http://www.mse.umd.edu/)
Relevant publication: J. Zhong, H. Zhu, Q. Zhong, J. Dai, W. Li and S. Jang, *et al.*, Self-Powered Human-Interactive Transparent Nanopaper Systems, *ACS Nano*, 2015, **9**(7), 7399, DOI: 10.1021/acsnano.5b02414

1.1.7 Light-Driven Bioelectronic Implants Don't Need Batteries

Benefitting from the miniaturization enabled by nanotechnologies, bioelectronics is a growing research field that is concerned with the convergence of biology and electronics: the application of biological materials and processes in electronics; and the use of electronic devices in living systems.

Among the latter, implantable bioelectronic devices wirelessly powered by different stimuli provide electrical impulses to precisely modulate the body's neural circuits—although wireless powering and remote manipulation still remain a major challenge for the practical use of these devices, which include retinal and cochlear implants; deep brain stimulators for epilepsy and Parkinson's disease; pacemakers; and brain–machine interfaces.

Adding to the options for wirelessly powering implants from outside the body, researchers in China have proposed a light-driven powering device using near infrared rays (nIR). Flashing light impulses, which are absorbed by the device, induce temperature fluctuation, thereby generating voltage/current pulses which can be used for charging a battery or biological stimulations.

"Compared to the wireless power transport by electromagnetic coupling, near-infrared light with a wavelength of 760–1500 nm—known for its heating and medical physical therapy effects—provides an alternative wireless power that can penetrate into human tissue up to a depth of 4–10 cm," explains Prof. Hongzhong Liu and Dr Weitao Jiang, from the State Key Laboratory for Manufacturing Systems Engineering at Xi'an Jiaotong University.

Inspired by the photothermal effect of nIR in biomedical applications, Liu's team fabricated a remotely/wirelessly controlled battery-less implantable device driven by nIR.

"What has motivated us to add to the existing devices for wireless powering systems is that a miniature power supply without the need for a battery is greatly desired for bioelectronics," says Jiang. "Besides, for bioelectronic intervention inside the body, device flexibility and the controllability of stimuli are also the greatest challenges."

"Our flexible and compact device can generate electrical pulses with controllable amplitude and width when remotely irradiated by nIR," notes Liu. "Not only can it supply power to implantable bioelectronics, it also provides adjustable electrical pulses for nerve stimulation."

"In nerve stimulation, it is important to modulate the impulses that flow through the stimulated nerves," he elaborates. "The stimulus waveforms—amplitude; pulse width; monophasic *versus* biphasic; and the delay between the two phases of the biphasic pulse—provide the greatest excitability differences for different nerve fibers."

The scientists point out that, in contrast to wireless powering systems driven by electromagnetic coupling, nIR-driven systems can not only realize far-field energy transfer but also be fabricated as a metal-free system; which is a great advantage in *in vivo* applications.

This wireless powering system combines PVDF—a specialty polymer in the fluoropolymer family—as the active pyroelectric material with graphene as the electrode material.

"PVDF exhibits lightweight, mechanical flexibility and biocompatibility, which are particularly interesting attributes for wearable or implantable devices," says Jiang. "In addition, the strong infrared absorption of PVDF makes it highly suitable for nIR-driven wireless powering systems."

Due to its excellent electrical and thermal conductivity, high surface area, and high flexibility, graphene has attracted much attention in recent years. While graphene possesses low absorption inherently in the infrared, it exhibits a transparency of 97.7% in visible wavelengths and even higher transparency in the infrared.

Each cell in the team's implantable power device is composed of a laminated graphene–PVDF–graphene sandwich, combining the high transparency of graphene and the strong infrared absorption of a PVDF thin-film. This serves to enhance its electric properties while reducing the device temperature to avoid damage to intervening and surrounding normal tissue.

To demonstrate the practical use of their device, the researchers implanted their power generation device as a stimulator for real-time functional electrical stimulation of a sciatic nerve of a frog and a rat heart, by remote control and nIR irradiation.

Apart from pacemakers and nerve stimulation, direct electrical activation has been widely used to recover the function of neurons. Going forward, the team will explore their device's application in the area of neural recovery by remote control *in vitro* and integrate some functional devices, *i.e.*, a compact camera and stimuli tips on the developed internal stimuli systems.

Featured scientists: Prof. Hongzhong Liu, Dr Weitao Jiang
Organization: State Key Laboratory for Manufacturing Systems Engineering, Xi'an Jiaotong University (PR China) (http://www.sei.xjtu.edu.cn/)
Relevant publication: H. Liu, T. Zhao, W. Jiang, R. Jia, D. Niu and G. Qiu, *et al.*, Flexible Battery-Less Bioelectronic Implants: Wireless Powering and Manipulation by Near-Infrared Light, *Adv. Funct. Mater.*, 2015, **25**(45), 7071, DOI: 10.1002/adfm.201502752

1.1.8 A Stretchable Far-Field Communication Antenna for Wearable Electronics

Apart from electronic circuits, displays, and power supplies, wearable electronics, especially sensor devices, also need to become fully integrated into sophisticated monitoring systems. For that they require wireless interfaces to external communication devices such as smartphones. This necessitates far-field communication systems that, like the sensor systems, perform even under extreme deformations and during extended periods of normal daily activities.

"While the transistors used in radio frequency (RF) circuits can be made flexible and stretchable using several techniques already demonstrated, the main component of the communication circuit, the antenna for far-field communication, is still a challenge," says Muhammad Mustafa Hussain, an associate professor of electrical engineering at the King Abdullah University of Science and Technology (KAUST).

To complement existing designs for stretchable antenna systems—which usually radiate at different resonant frequencies and are expensive due to the complex processing involved or the exotic materials used—an international team led by Hussain has demonstrated a stretchable and wearable antenna that can provide a single frequency operation while flexing or stretching. The group of researchers included KAUST assistant professor Atif Shamim and Swanlund Chair Professor John Rogers of the University of Illinois at Urbana Champaign.

The team's flexible and stretchable metal (copper) thin-film antenna for far-field communication—up to 80 meters while mounted on a stretchable fabric and worn by a person—maintains its properties during stretching, bending and strain cycles.

"We fabricated our antenna using a metal/polymer bilayer process—the resulting structure combines the conductivity of the metal and the elasticity of the polymer—and the stretchability is imparted using a lateral spring structure," explains Aftab Hussain, a PhD candidate in Hussain's lab. "The key reason the antenna needs to be fabricated as a metal/polymer bilayer is that standalone metal thin films are very malleable, and deform plastically under application of stress."

That means that a metal thin film lateral spring structure cannot be used as a stretchable antenna, since it will only be able to undergo one stretch cycle.

The solution to this problem was to use a polymer backing that provides the restoration force which helps the spring return to its original shape after the release of the applied lateral force.

As a result, the key performance parameters of the antenna do not change with bending, stretching, flexing and twisting—hence the antenna can continuously communicate information in the WiFi frequency band while it is worn.

In their tests, the researchers found that their antenna retains all its essential properties such as gain, radiation pattern, directionality, operation frequency and bandwidth for up to 30% strain and for 2000 stretching cycles.

As a next step, the team will integrate their stretchable antenna into a fully integrated, flexible, stretchable and wearable sensor array for real-time communication of sensor information.

Featured scientists: Prof. Muhammad Mustafa Hussain, Aftab Hussain
Organization: Integrated Nanotechnology Laboratory, King Abdullah University of Science and Technology, Thuwal (Saudi Arabia) (https://nanotechnology.kaust.edu.sa/Pages/Home.aspx)
Relevant publication: A. Hussain, F. Ghaffar, S. Park, J. Rogers, A. Shamim and M. Hussain, Metal/Polymer Based Stretchable Antenna for Constant Frequency Far-Field Communication in Wearable Electronics, *Adv Funct Mater.*, 2015, 25(42), 6565, DOI: 10.1002/adfm.201503277

1.1.9 Reversibly Bistable Materials Could Revolutionize Flexible Electronics

Expect future generations of smartphones to be flexible. Already, Apple has been granted a patent for "flexible electronic devices"[8] that comprise a flexible housing and a flexible display screen and Samsung has a similar patent application pending.[9]

Rollable displays and other flexible, stretchable electronic systems are often enabled by the successful integration of nanostructured materials. Most commercially available flexible electronic circuits and devices are fabricated on flexible plastic substrates, such as polymeric amides, polyether ether ketone (PEEK) polymers, or transparent conductive polyester films. Although these substrates can be easily bent and rolled up, they cannot be used to fabricate rollable display-integrated gadgets that are fixed at a rigid perpendicular position on their own.

This is where the need for a mechanically flexible and reversibly bistable material arises—a material that can assume two stable states: a flexibly rolled state and an unbent state.

For the first time, the same research team at KAUST mentioned earlier has used a reversibly bistable material to demonstrate flexible electronics. In addition, the scientists introduced a performance metric—the *cumulative impact budget*—which takes into account the impact force imparting an impulse on the silicon fabric during the mechanical deformation of the substrate.

"We investigated the mechanical and electronic aspects of flexible, inorganic field-effect transistors physically supported by a mechanically bistable metallic substrate," Prof. Muhammad Mustafa Hussain summarizes the effort. "Our work combines basic and applied studies with findings that are supported by experimental data—semiconductor device analysis, scanning electron microscopy, energy-dispersive X-ray, and high-speed imaging—and theoretical discussion: impulse-momentum theory and approximation, and prediction of the kinetic energy losses and magnitude of impulsive forces with respect to impact speeds."

"This is the first demonstration and discussion of the effectiveness of a reversibly bistable material for free-form electronics," says Nasir Alfaraj, a PhD student in Hussain's group. "The material we used to form the bistable substrate is a porous iron-carbon metal alloy, which is inexpensive and commonly used in the fabrication of commercially available cycling safety wristbands and a variety of ankle bracelets for orthopedic health care. This motivated us to work with a low-cost, commonly available material and extend its functionality through the integration of logic and control components in order to create practical flexible display devices."

To fabricate their device, the team attached a flexible silicon-based metal-oxide-semiconductor field-effect transistor (MOSFET) on a mechanically flexible and optically semitransparent porous silicon onto a metallic bracelet.

This material platform has two stable and reversible mechanical states: stretched and rolled. Surface, cross-sectional, and elemental composition nanoscale examinations of the thin metallic structure, along with electrical measurements of the transistors, show that the distribution of nanopores throughout the structure allows the metal alloy to internally absorb strain energy and hence achieve flexibility. Nevertheless, the transistor devices on the thin silicon fabric maintained their integrity after accumulating an impulsive force budget about 300 times higher than the force that average adults experience as a result of their weight.

This work could have a significant impact on the electronics industry and open the door to commercializing flexible, large electronic devices. Reversibly bistable flexible transistors can be used in a variety of applications, including optoelectronic devices in which LEDs are controlled by reversibly bistable flexible transistors. As such transistors can handle high drive currents, they can be used to realize foldable display devices.

The researchers note that reversibly bistable electronics can also aid in the development of practical orthopedic tools and technologies that employ electronic devices required to handle high physical force loads.

As a next step, the team plans to design and demonstrate a naturally flexible and reversibly bistable polymer to be integrated with state-of-the-art logic and radio-frequency electronic devices.

"Several challenges need to be overcome before we can realize a fully functional, mechanically bistable electronic system," cautions Hussain. "They include the integration of modern and more complicated electronic devices based on complementary metal–oxide–semiconductor (CMOS) technology and the development of an easily influenced, bistable substrate to house such devices."

Featured scientists: Prof. Muhammad Mustafa Hussain, Nasir Alfaraj
Organization: Integrated Nanotechnology Laboratory, King Abdullah University of Science and Technology, Thuwal (Saudi Arabia) (https://nanotechnology.kaust.edu.sa/Pages/Home.aspx)
Relevant publication: N. Alfaraj, A. Hussain, G. Torres Sevilla, M. Ghoneim, J. Rojas and A. Aljedaani, *et al.*, Functional integrity of flexible *n*-channel metal–oxide–semiconductor field-effect transistors on a reversibly bistable platform, *Appl. Phys. Lett.*, 2015, **107**(17), 174101, DOI: 10.1063/1.4934355

1.1.10   Nanogenerators for Large-Scale Energy Harvesting

So far we have looked at how nanotechnologies enable the fabrication of miniscule energy sources that can be used to power tiny devices, even implants.

In an intriguing demonstration, Zhong Lin Wang, distinguished professor and director, Center for Nanostructure Characterization at Georgia Tech, and his team, have demonstrated that the technology offered by nanogenerators can also be used for energy harvesting on a much larger scale.

This work represents a mechanical energy-harvesting technique—a triboelectric nanogenerator—which effectively converts ambient mechanical energy into electrical energy.

"The working mechanism of our novel nanogenerator does not fall into the category of established conventional transduction mechanisms for mechanical energy harvesting, *i.e.* electrostatic, piezoelectric, and electromagnetic," says Wang. "Instead, energy generation is realized through the coupling between triboelectric effect and electrostatic effect."

He explains that, unlike electrostatic generators such as dielectric elastomer generators, the triboelectric nanogenerator does not require to be "jump-started" with an initial voltage before it can produce power. Instead, the triboelectric effect gives rise to contact surface charges that act as an inner "driving force" for the motion of electrons through the external circuit. This characteristic offers significant advantages including low cost, simple fabrication, and small volume.

The triboelectric nanogenerator device fabricated by Wang's team has a layered sandwich structure with two substrates of poly(methyl methacrylate) (PMMA) on the outside. On one side, a layer of contact electrode—which plays the dual role of electrode and contact surface—is prepared, consisting of a gold thin film and gold nanoparticles coated on the surface. On the other side, a thin film of gold is laminated between the substrate and a layer of polydimethylsiloxane (PDMS). The two substrates are connected by four springs installed at the corners, leaving a narrow spacing between the contact electrode and the PDMS.

According to Wang, the triboelectric nanogenerator represents an extremely cost-effective design in mainly three aspects:

"Firstly, the cost of materials is insignificant in comparison with other conventional mechanical energy-harvesting techniques. It does not involve synthesis of crystalline materials nor require certain materials that generate magnetic field. All components of the nanogenerator are commonly commercial products on the market. Although it uses gold in the design, extremely little amount is applied, only 50 nm in thickness. Besides, other non-precious metal can also be used as replacements.

"Secondly, very little amount of materials is consumed. The nanogenerator has a thin-film based structure. Besides the two metal electrodes (both 50 nm in thickness), the other material, PDMS layer, has a thickness as small as 10 μm. The small size not only further saves the cost of materials but also indicates the excellent scalability of this technique.

"Last but not the least, the fabrication is straightforward without sophisticated tools and complicated process flow. Therefore, in terms of cost, there stands little hurdle for widespread use of the nanogenerator."

The improvement over other types of nanogenerators is not limited to the simplified and low-cost fabrication, though. Wang's novel nanogenerator also delivers an unprecedented high electric output efficiency and power density.

"The proper selection of materials enables excellent tolerance on defects from materials and fabrication process," explains Wang. "The elastic property of the PDMS layer ensures that the entire surface area is fully utilized. Moreover, we applied nanoparticle-based surface modification on contacting surfaces of nanogenerators for the first time. It functions both physically and chemically, giving electric enhancement of at least 25 times compared to nanogenerators without modification."

Triggered by commonly available ambient mechanical energy such as human footfalls, a nanogenerator smaller than a human palm can generate a maximum short-circuit current of 2 mA, delivering an instantaneous power output of 1.2 W to an external load.

Wang points out that, at device level, the overall efficiency of the nanogenerator reaches 9.8%. If only energy forms that participate in conversion process are taken into consideration, the direct energy conversion efficiency can reach as high as 14.9%.

"We attribute the record high power output of our nanogenerator to optimized structure, proper materials selection, and nanoscale surface modification," says Wang.

The team demonstrates an application of the nanogenerator on a large scale, simultaneously lighting up as many as 600 commercial LED bulbs at an estimated output open circuit voltage of ~1200 V. In this case, the current pulse output was produced by human footfall, *i.e.* stepping on the nanogenerator.

According to the team, the mechanism demonstrated in this work can be utilized on a large scale to harvest mechanical energy from human footfalls, rolling wheels, wind power, and even ocean waves.

Featured scientist: Prof. Zhong Lin Wang
Organization: State Center for Nanostructure Characterization, Georgia Institute of Technology, Atlanta, GA (USA) (http://cncf.nanoscience. gatech.edu/)
Relevant publication: G. Zhu, Z. Lin, Q. Jing, P. Bai, C. Pan and Y. Yang, *et al.*, Toward Large-Scale Energy Harvesting by a Nanoparticle-Enhanced Triboelectric Nanogenerator, *Nano Lett.*, 2013, **13**(2), 847, DOI: 10.1021/ nl4001053

1.2 A Much More Sophisticated Way to Tap into the Sun's Energy

When you hear about solar cells, you probably think of the large bulky panels installed on rooftops. Tomorrow's solar cells will be very, very different: thanks to nanomaterials such as graphene and quantum dots, as well as innovative fabrication technologies, they can be sprayed onto walls; woven into fabrics; printed onto packaging materials; or painted onto car roofs. This will make energy harvesting devices from sunlight as ubiquitous and versatile as the gadgets that they will power.

1.2.1 Solar Cell Textiles

The future will see a major role for e-textiles that allow the design and production of a new generation of garments with built-in unobtrusive sensors and a variety of electronic functions. These e-textiles will have the revolutionary ability to sense, act, store, emit, and move—think biomedical monitoring functions or new man-machine interfaces—while ideally leveraging an existing low-cost textile manufacturing infrastructure.

Researchers in China have developed novel miniature polymer solar cell wires that, when woven into textiles, can serve as a power source.

These wire-shaped polymer solar cells are fabricated by incorporating a thin layer of titania nanoparticles between the photoactive material and electrode. The aligned carbon nanotube fiber enables high flexibility and stability of the resulting polymer solar cell.

Figure 1.2 schematically shows the structure of the wire-shaped polymer solar cell (PSC) with a titanium (Ti) wire and an aligned multiwalled carbon nanotube (MWCNT) fiber as cathode and anode, respectively. In a typical fabrication, a Ti wire is modified by growing aligned titania nanotubes on the surface by electrochemical anodization, followed by coating of a layer of

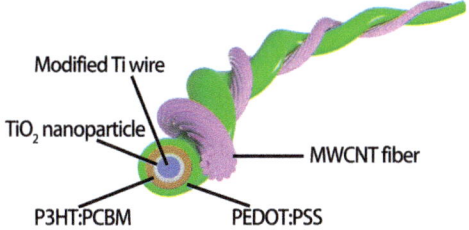

Figure 1.2 Schematic illustration of the wire-shaped polymer solar cell. (Image: The Peng Research Group, Fudan University.)

titania nanoparticles. Two polymer layers are then dip-coated onto the modified Ti wire. Finally, the resulting Ti wire is wound with an aligned MWCNT fiber to produce the wire-shaped polymer solar cell.

"We found that the titania nanoparticles enhance the adsorption of photoactive materials and charge transport, which increased the energy conversion efficiency by 36% compared with the wire-shaped polymer solar cell without the titania nanoparticle under the same condition," says Zhitao Zhang, a PhD candidate in Prof. Huisheng Peng's research group in the Department of Macromolecular Science at Fudan University.

According to the researchers, the nanocrystalline semiconductor oxide layer played a crucial role in providing pathways for charge transport. Here the incorporation of semiconducting nanoparticles at the tips of aligned TiO_2 nanotubes after titanium tetrachloride ($TiCl_4$) treatment helped to make the outer surface uniform.

In addition, the nanoparticle layer effectively increases the polymer load, decreases the electrical resistance for charge transport, and enhances light scattering.

"We discovered that the diameters of MWCNT fibers affect the wire-shaped PSCs to a large degree," explains Zhang. "The energy conversion efficiencies were increased from 0.72% to 1.78% with the increasing diameters from 18 to 32 μm and then decreased with the further increasing diameters (*e.g.*, 1.47% and 1.18% at 60 and 74 μm, respectively)."

The maximal efficiency occurred at approximately 32 μm. A smaller fiber produced a higher electric resistance with a lower current density, while a bigger fiber shaded the incident sunlight also to decrease the current density.

The team tested the wire-shaped PSCs for flexibility and bendability and, even after 1000 bending cycles, found no obvious damages or decreased energy conversion efficiencies (Figure 1.3).

The as-prepared wire-shaped PSC could be easily woven into various flexible structures such as textiles without the necessity for sealing that had been required by the widely studied wire-shaped dye-sensitized solar cells.

Going forward, the researchers will work on increasing the performance of their wire-shaped polymer solar cells and also, in addition to bendability, make them stretchable.

Featured scientists: Prof. Huisheng Peng's research group (http://www.polymer.fudan.edu.cn/polymer/research/Penghs/main_en.htm)
Organization: Department of Macromolecular Science, Fudan University (People's Republic of China)
Relevant publication: Z. Zhang, Z. Yang, Z. Wu, G. Guan, S. Pan and Y. Zhang, *et al.*, Weaving Efficient Polymer Solar Cell Wires into Flexible Power Textiles, *Adv. Energy Mater.*, 2014, **4**(11), 1301750, DOI: 10.1002/aenm.201301750

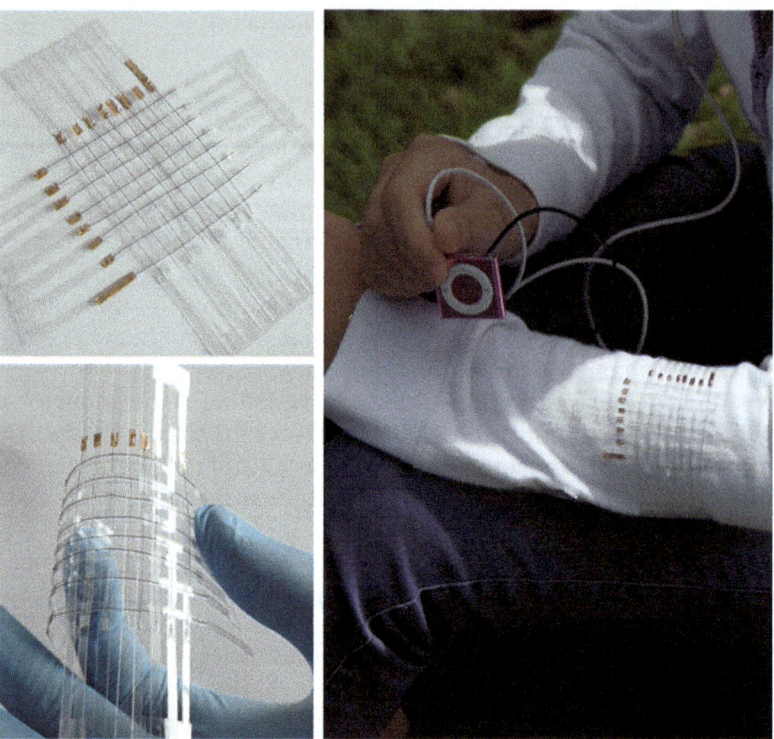

Figure 1.3   Left panels: PSC textile laid out flat and being bent. Right panel: the PSC applied to a shirt sleeve powers an iPod nano. (Image: The Peng Research Group, Fudan University.)

1.2.2   Complete Solar Cells Printed by Inkjet

Organic photovoltaic (OPV) technologies have the potential to become a thin-film alternative to inorganic silicon photovoltaics due to their intrinsic potential for low-cost print processing from solution: high-speed and at low temperature. Organic solar cells can be integrated into building facades and windows because they are optically translucent and can be manufactured on large areas at high throughput.

"Despite impressive progress over the past decade, OPVs are still some way behind other thin-film technologies harvesting solar energy because of low efficiencies and short lifetimes," says Sungjune Jung, an assistant professor in the Department of Creative IT Engineering at Pohang University of Science and Technology (POSTECH) in South Korea. "To date, no route is available for monolithically integrating solar cells into a system in which other components such as transistors, sensors, or displays are already fabricated."

Jung and an international team of researchers were the first to report the fabrication and measurement of all-inkjet-printed, all-air-processed organic solar cells.

Inkjet printing has already proven effective in the fabrication of organic light-emitting diodes (OLEDs), printed electronics, and photodetectors.

The researchers demonstrated a high-efficiency solar cell with a homogeneous inkjet-printed donor–acceptor thin-film, which was achieved by engineering the semiconducting blend ink with a tailored ternary solvent. "The results show that our inkjet-printed blend layer exhibits similar nanoscale structure and excited state dynamics to its spin-coated counterparts," says Jung.

Notably, all four layers of the solar cell are printed at ambient temperatures. The team achieved a power conversion efficiency (PCE) of about 2% with all-inkjet-printed organic solar cells. They achieved a 5% PCE with a printed three-layer structure with the cathode as the only non-printed layer (made by evaporation).

"Overall, our results show that inkjet-printing is an attractive digital printing technology for cost-effective, environmentally-friendly integration of photovoltaic cells onto plastic substrates," says Jung. "With this inkjet printing technique, we could integrate the energy harvesting units into a system where other components such as transistors, sensors or displays are already fabricated."

This means that these inkjet-printed solar cells could allow for full integration into the manufacturing process of portable flexible electronic systems.

Jung cautions, though, that in order to make this happen, the efficiency of the all-inkjet-printed all-air-processed solar cells need to be increased by further developing the processes and materials optimized for inkjet-printing.

Featured scientist: Prof. Sungjune Jung
Organization: Department of Creative IT Engineering, Pohang University of Science and Technology (Republic of Korea) (http://cite.postech.ac.kr/)
Relevant publication: S. Jung, A. Sou, K. Banger, D. Ko, P. Chow and C. McNeill, *et al.*, All-Inkjet-Printed, All-Air-Processed Solar Cells, *Adv. Energy Mater.*, 2014, **4**(14), 1400432, DOI: 10.1002/aenm.201400432

1.2.3 Solar Paint Paves the Way for Low-Cost Photovoltaics

Using quantum dots as the basis for solar cells is not a new idea, but attempts to make such devices have not yet achieved sufficiently high efficiency in converting sunlight to power. So far, advances in quantum dot photovoltaics have resulted in solar cell power conversion efficiencies slightly above 7%. Although these performance levels are promising, all high-performing device results to date have relied on a multiple-layer-by-layer strategy for film fabrication rather than employing a single-layer deposition process.

The attractiveness of using quantum dots for making solar cells lies in several advantages over other approaches: they can be manufactured in an energy-saving room-temperature process; they can be made from abundant, inexpensive materials that do not require extensive purification, as silicon does; and they can be applied to a variety of inexpensive and even flexible substrate materials, such as lightweight plastics.

A research team from the University of Toronto and KAUST developed a semiconductor ink with the goal of enabling the coating of large areas of solar cell substrates in a single deposition step and thereby eliminating tens of deposition steps necessary with the previous layer-by-layer method.

The team's "solar paint" is composed of semiconductor nanoparticles synthesized in solution—so-called colloidal quantum dots (CQDs). They can be used to harvest electricity from the entire solar spectrum because their energy levels can be tuned by simply changing the size of the particles.

"We sought an approach that would achieve highly efficient utilization of CQD materials," says Professor Ted Sargent from the University of Toronto, who, together with Osman Bakr, an associate professor in the Solar & Photovoltaics Engineering Research Center at KAUST, led the work. "To achieve this, we made a solar cell ink that can be deposited in a single step, which makes it an excellent material for high-throughput commercial fabrication."

Previously, films made from these nanoparticles were built up in a layer-by-layer fashion where each of the thin CQD film deposition steps is followed by curing and washing steps to densify the film and form the final semiconducting material. These additional steps are required to exchange the long ligands that keep the CQDs stable in solution for short ligands that allow efficient charge transport. However, this means that many steps are required to build a thick enough film to absorb enough sunlight.

"We simplified this process by engineering the CQD surfaces with short organic molecules in the solution phase to enable a stable colloidal solution and reduce the film formation to a single step," Bakr explains. "At the same time, the post processing steps are reduced significantly, since the semiconducting material is formed in solution. This means that CQD films can be deposited quickly and at low cost, similar to a paint or ink."

Besides the reduction in processing steps, the new process is also much more efficient in terms of materials usage. While the layer-by-layer, solid-state treatment approach provides less than 0.1% yield in its application of CQD materials from their solution phase onto the substrate, the new approach achieves almost 100% use of available CQDs.

"This means that for the same amount of CQD material, we could make a 1000-fold larger area of solar cells compared with conventional methods," Bakr points out. "Our technology paves the way for low-cost photovoltaics that can be fabricated on flexible substrates using roll-to-roll manufacturing, similar to a printing press," adds Lisa Rollny, a PhD candidate in Sarget's group. "Our ink is also useful in biological applications, *e.g.* in biosensors and tracing agents with an infrared response."

"In previous work, we found new routes of passivating the CQD surface using a combination of organic and inorganic compounds in a solid state approach with large improvements in efficiency," says Rollny. "We intend to integrate this knowledge with our solar CQD ink to further improve the performance of this material, especially in terms of how much solar energy is converted into usable electrical energy."

Although the team has developed an effective method for producing a CQD film in a single step, the electronic properties of the resulting films are not yet optimized. This is due to the very small imperfections on the CQD surface that reduce the usable electricity output of a solar cell. Through careful engineering of CQD surfaces in solution, the researchers plan to eliminate these unwanted surface sites in order to make higher quality, higher efficiency CQD solar cells using their single step process.

Featured scientists: Prof. Osman Bakr, Prof. Ted Sargent
Organization: Solar & Photovoltaics Engineering Research Center, King Abdullah University of Science & Technology, Thuwal (Saudi Arabia) (https://sperc.kaust.edu.sa/Pages/Home.aspx); The Sargent Group, University of Toronto (Canada) (http://www.light.utoronto.ca/)
Relevant publication: A. Fischer, L. Rollny, J. Pan, G. Carey, S. Thon and S. Hoogland, *et al.*, Directly Deposited Quantum Dot Solids Using a Colloidally Stable Nanoparticle Ink, *Adv Mater.*, 2013, **25**(40), 5742, DOI: 10.1002/adma.201302147

1.2.4   Paper Solar Cells

Transparent and flexible substrates are widely explored for flexible electronics and researchers have been working on techniques to develop thermally stable and biodegradable materials that are as easily printable as paper.

Liangbin Hu's research group at the University of Maryland has reported a novel transparent paper substrate design optimized for solar cells. This paper is made of earth-abundant wood fibers that simultaneously achieve an ultrahigh transmittance (~96%) and ultrahigh optical haze (~60%). The work was a collaboration with Prof. Jinsong Huang's team from the University of Nebraska-Lincoln, which led the demonstration of solar cell applications.

"We demonstrated that simple lamination of transparent paper on solar cell devices can increase the devices' efficiency by 10–20% due to its high optical transmittance and light scattering," says Hu.

"Whereas displays and touch screens need high clarity and low optical haze, this is a property preferably maximized in transparent substrates integrated into solar devices," he explains. "This means that current commercial substrates are best suited for displays but are not optimized for solar cell devices. We developed a new paper-based on wood cellulose material, which has both high optical transmittance and high optical haze."

Transmission haze is an important optical property for optoelectronic devices and refers to the percentage of light diffusely scattered through a transparent surface from the total light transmitted. Higher transmission haze improves the light absorption efficiency of solar cells.

Hu notes that the paper material's novel optical properties allow a simple light-management strategy for improving solar cell performances. Transmission haze tends to increase with an increase in paper thickness while the optical transmittance increases slightly with a decrease in paper thickness. Combining the optical haze and transmittance for substrates toward different applications allows fine-tuning of the desired results.

"We demonstrated this with an organic solar cell by simply laminating a piece of transparent paper and observed that its power conversion efficiency (PCE) increased from 5.34 to 5.88%," says Hu.

Additionally, the mechanical properties of paper—such as toughness and strength—are important for various applications.

The team fabricated their transparent paper by creating a specially treated pulp from pine wood. This pulp was diluted with water and then dried as thin coatings.

Hu points out that this transparent paper demonstrates a much higher optical transmittance than nanopaper made of nanoscale fibers while using much less energy and time to process paper with a similar thickness.

"Such low-cost, highly transparent, and high haze paper can be utilized as an excellent film to enhance light-trapping properties for photovoltaic applications such as solar panel, solar roof, or solar windows," he adds. "We could also see uses in microfluidic devices as well as displays operating in a bright environment like outdoor GPS navigation devices."

The next steps for the team include integrating devices in transparent paper through scalable methods, and addressing the stability issues.

Featured scientists: Prof. Liangbin Hu's research group (http://bingnano. com/); Prof. Jinsong Huang's research group (http://www.huanggroup. unl.edu/)
Organizations: A. James Clark School of Engineering, University of Maryland, College Park, MD (USA); Department of Mechanical Engineering, University of Nebraska-Lincoln, Lincoln, NE (USA)
Relevant publication: Z. Fang, H. Zhu, Y. Yuan, D. Ha, S. Zhu and C. Preston, *et al.*, Novel Nanostructured Paper with Ultrahigh Transparency and Ultrahigh Haze for Solar Cells, *Nano Lett.*, 2014, **14**(2), 765, DOI: 10.1021/nl404101p

1.2.5   Recharging Wearable Textile Battery by Sunlight

As we have seen in Section 1.1, going hand in hand with the development of wearable electronic textiles, researchers are also pushing the development of wearable and flexible energy storage to power e-textiles. For instance, scientists demonstrated that flexible cotton threads can be used as a platform to fabricate a cable-type supercapacitor.

Taking this one step further, a research team in Korea has developed wearable textile batteries that can be integrated with flexible solar cells and thus be recharged by solar energy.

The scientists demonstrated a fully functional wearable textile battery by finding unconventional materials for all of the key battery components and integrating them systemically: nickel-coated polyester yarn as a current collector for efficient stress release; polyurethane binder for strong adhesion of active materials; and a polyurethane separator with superior mechanical, electrochemical, and thermal properties.

"In developing wearable rechargeable batteries, the component requiring the most significant alteration from the conventional cell configuration is the current collector because it largely dictates the mechanical properties of the entire cell," explains Jang Wook Choi, an associate professor at the Korea Advanced Institute of Science and Technology (KAIST), who led the work. "Along this direction, one of the most natural approaches would be to use textiles as current collectors after integration with conductive materials."

Choi points out that the use of carbon nanomaterials for the fabrication of textile supercapacitors and paper batteries has encountered limitations in increasing the cell size and rate performance. Also, the investigation of metal-incorporating current collectors, such as silver-coated textiles, runs into cost issues as well as mechanical challenges with regard to fatigue failure.

"In contrast, our battery cells endure extremely severe mechanical tests while delivering comparable electrochemical properties to those of the conventional foil-based counterparts," says Choi. "It is important that, for the wearable capability of the rechargeable battery, the binder and separator also support the mechanical endurance of the overall system."

To this end, the researchers decided to investigate unconventional materials rather than modify existing ones—which have been developed for conventional flat cells that do not need to withstand the demands of wearables. After searching a wide range of material candidates, they found that polyurethane— due to its chemical and polymeric structure—possesses various material properties that are suitable for both binder and separator in a wearable battery.

Furthermore, flexible and lightweight solar cells based on plastic substrates were integrated onto the outer surface of the textile battery for recharging the textile battery without physical connection to a power outlet.

While changing or recharging the batteries in a watch or cell phone is a relatively simple operation, doing this with wearable batteries integrated into textiles might not be such an easy task. The team has therefore integrated a series of flexible polymer solar cells with its wearable textile battery (Figure 1.4).

Choi notes that the wearable textile battery exhibits decent cycling and rate performance. "First, without folding–unfolding motions, our battery showed comparable performance to that of its foil-based counterpart. More importantly, the wearable textile battery showed good cycling performance even during severe folding–unfolding repetitions. After 40 cycles equivalent to 5500 deep folding–unfolding cycles, the textile battery retained 91.8% of the original capacity."

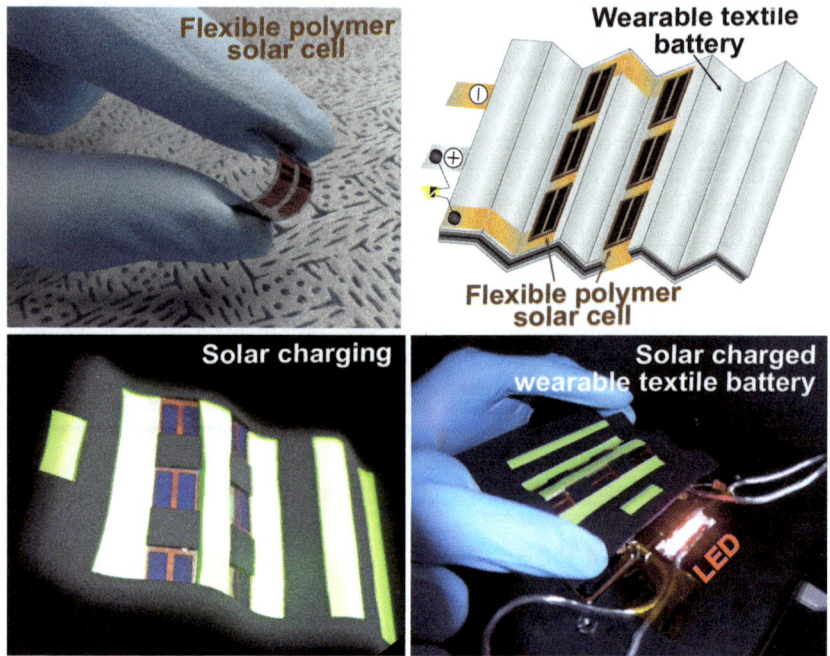

Figure 1.4   Integration with flexible polymer solar cells. (Image: Dr Jang Wook Choi, KAIST.)

The team attributes the relatively lower capacities during the folding–unfolding motions to the fact that such mechanical motions make ionic transport less efficient at various points of lithium diffusion including through the separator, at the electrode–electrolyte interface, and within the electrode films.

"Our investigation demonstrates that materials and fabrication processes can be systemically united to realize a wearable textile battery with exceptional mechanical stability particularly in the forms of clothes and watch-straps," Choi sums up this work. "It is also reasonable to expect that further tuning of the cell dimensions will find various future applications beyond what can be conceptualized now, especially with the aid of solar-charging capabilities."

Featured scientist: Prof. Jang Wook Choi (http://nest.kaist.ac.kr/p/prof-choi.html)
Organization: Korea Advanced Institute of Science and Technology (KAIST), Daejeon (Republic of Korea)
Relevant publication: Y. Lee, J. Kim, J. Noh, I. Lee, H. Kim and S. Choi, *et al.*, Wearable Textile Battery Rechargeable by Solar Energy, *Nano Lett.*, 2013, **13**(11), 5753, DOI: 10.1021/nl403860k

References

1. D. Choi, M. Choi, W. Choi, H. Shin, H. Park and J. Seo, *et al.*, Fully Rollable Transparent Nanogenerators Based on Graphene Electrodes, *Adv. Mater.*, 2010, **22**(19), 2187, DOI: 10.1002/adma.200903815.
2. S. Naficy, R. Jalili, S. Aboutalebi, R. Gorkin III, K. Konstantinov and P. Innis, *et al.*, Graphene oxide dispersions: tuning rheology to enable fabrication, *Mater. Horiz.*, 2014, **1**(3), 326, DOI: 10.1039/C3MH00144J.
3. R. Jalili, S. Aboutalebi, D. Esrafilzadeh, K. Konstantinov, J. Razal and S. Moulton, *et al.*, Formation and processability of liquid crystalline dispersions of graphene oxide, *Mater. Horiz.*, 2014, **1**(1), 87, DOI: 10.1039/C3MH00050H.
4. Q. Cheng, Z. Song, T. Ma, B. Smith, R. Tang and H. Yu, *et al.*, Folding Paper-Based Lithium-Ion Batteries for Higher Areal Energy Densities, *Nano Lett.*, 2013, **13**(10), 4969, DOI: 10.1021/nl4030374.
5. E. Kim, H. Tu, C. Lv, H. Jiang, H. Yu and Y. Xu, A robust polymer microcable structure for flexible devices, *Appl. Phys. Lett.*, 2013, **102**(3), 033506, DOI: 10.1063/1.4788917.
6. R. Tang, H. Huang, H. Tu, H. Liang, M. Liang and Z. Song, *et al.*, Origami-enabled deformable silicon solar cells, *Appl. Phys. Lett.*, 2014, **104**(8), 083501, DOI: 10.1063/1.4866145.
7. http://cordis.europa.eu/project/rcn/98854_en.html.
8. Apple Inc., *U.S. Patent*, 8,929,085, 2015.
9. Samsung Electronics Co., Ltd, *U.S. Patent Appl.*, US20150062840 A1, 2015.

CHAPTER 2

No More Rigid Boxes—Fully Flexible and Transparent Electronics

Unlike today's electronics based on rigid silicon technologies, stretchable devices can conform to almost any kind of surface shape and provide unique functionalities, which are unreachable with a simple extension of conventional technologies.

Flexible electronics are mechanically robust electronics that can be bent, folded, crumpled and stretched and are a major research focus of nanotechnology scientists towards next-generation wearable and implantable devices.

This promises an entirely new design factor for novel devices: ultra-thin smartphones that wrap around the wrist; flexible displays that fold out as newspapers or as large as a television; photovoltaic cells and reconfigurable antennas that conform to the body parts of our cars; adhesive sensor patches for the skin; or flexible implants that can monitor and treat cancer.

To get electronics out of their rigid casings and integrated into flexible materials such as textiles or even stretchable biomedical devices that interface directly with organs such as the skin, heart and brain, researchers have come up with solutions ranging from hydrogel electrodes in dielectric elastomers to stretchable supercapacitors using buckled carbon nanotube (CNT) macrofilms, CNT-coated cotton yarns and sponges.

In order to fabricate entirely flexible electronic devices, the components that power them—such as batteries—not only need to be fully flexible as well,

Nanotechnology: The Future is Tiny
By Michael Berger
© Michael Berger 2016
Published by the Royal Society of Chemistry, www.rsc.org

but they have to be compatible with commercially available manufacturing technologies. This would require achieving a high degree of deformability without using elastomeric materials.

2.1 Ultra-Stretchable Silicon

The flexibility required to fabricate flexible electronic components has led to the use of plastic substrates, carbon nanomaterials, and different transfer techniques. One of the biggest obstacles to mass adoption of flexible electronics has been the incompatibility of most of these solutions with industry's state-of-the-art silicon-based CMOS (*complementary metal-oxide-semiconductor*) processes—which still produce about 90% of today's electronics.

However, researchers have found a way to make monolithic single-crystal silicon stretchable. The design is based on an all silicon-based network of hexagonal islands connected through spiral springs.

"With this structure, we have been able to achieve a remarkable stretch ratio of about 1000% using a brittle material such as silicon," says Muhammad Mustafa Hussain, an associate professor of electrical engineering at King Abdullah University of Science and Technology (KAUST).

The fabrication process is based on conventional microfabrication techniques consisting of five basic steps: starting with a silicon-on-insulator wafer (50 μm).

(1) a gold hard mask is first deposited on silicon-on-insulator and then patterned using
(2) photolithography and
(3) reactive ion etching (RIE). Next,
(4) the silicon is deeply, anisotropically etched (DRIE) until the buried oxide layer is reached and then the hard mask is removed. Finally,
(5) the silicon structure is released with vapor hydrofluoric acid, which removes the underlying oxide layer.

Hussain points out that device implementation can be achieved through CMOS-compatible fabrication prior to the five-step release process described.

The resulting single-spiral structures can be stretched to a ratio more than 1000%, while remaining below a 1.2% strain. Moreover, these network structures have demonstrated area expansions as high as 30-fold in arrays.

"While handling still remains a challenge, our method can provide ultra-stretchable and adaptable electronic systems for distributed network of high-performance macro-electronics especially useful for wearable electronics and bio-integrated devices," says Hussain.

The researchers are planning to demonstrate electronic devices implemented on top of their hexagon islands and interconnected through the spiral springs to form stretchable sensor networks with outstanding electrical performance and mechanical robustness.

Featured scientist: Prof. Muhammad Mustafa Hussain
Organization: Integrated Nanotechnology Laboratory, King Abdullah University of Science and Technology, Saudi Arabia (https://nanotechnology.kaust.edu.sa/Pages/Home.aspx)
Relevant publication: J. Rojas, A. Arevalo, I. Foulds and M. Hussain, Design and characterization of ultra-stretchable monolithic silicon fabric, *Appl. Phys. Lett.*, 2014, **105**(15), 154101, DOI: 10.1063/1.4898128

2.2 Rewritable, Transferable and Flexible Sticker-Type Organic Memory

The advanced electronic systems of the future will be fabricated on soft substrates by integrating multiple crucial components such as logic and memory devices as well as their power supply. Organic memory with a simple sandwich structure has been considered a promising information storage element in future soft electronics for its exceptional merits such as low production cost, mechanical elasticity, flexibility, low-temperature processing and roll-to-roll printability. So far, rewritable organic memories have been fabricated mostly on rigid, flat, and smooth substrates, such as metal, glass, plastic and silicon, which greatly limit their soft and flexible merits.

The fact that organic films are often fabricated by solution processes—*i.e.*, spin-coating or ink-jet printing—presents a great hurdle in constructing flexible organic memory in stack or 3D architecture. The reason for this is that the use of solution processes can result in severe damage to the bottom soft organic device or substrate due to the solvent used for the vertical integration of the subsequent devices.

Therefore, for future applications, it is essential to establish a proper fabrication strategy for rewritable organic memory onto diverse substrates or devices.

A team of researchers in Taiwan has reported a methodology to overcome these challenges. They have, for the first time, successfully demonstrated a rewritable, transferable, and flexible sticker-type organic memory on arbitrary non-conventional substrates through a simple, low-temperature and cost-effective one-step methodology.

The researchers demonstrated that this re-writable organic memory can be simply stuck on various desired substrates, including rigid, flexible, non-planar, and rough substrates, promising that the information storage devices can be greatly broadened in diversified future applications.

"Compared to traditional bottom-up solution processes for organic bistable memory, the advantages of our transferable and flexible memory device include several unique aspects," Yang-Fang Chen, a professor in the Department of Physics at National Taiwan University, explains. "Most importantly, the transferable and self-adhered features of the organic memory pave an easy route to vertically integrate digital organic memories with other flexible organic devices with minimal solvent issue; hence harsh synthesis and unaccustomed fabrication steps on non-conventional substrates can be avoided."

Furthermore, as Chen notes, the combination of organic memory and protective layer for transferring graphene electrodes eliminated the need for chemical treatment processes for the graphene's protective layer. With the flexible and adhesive graphene–electrode underlay, the presented memory can be simply molded and functioned on desired non-conventional substrates, including non-planar and soft ones. This versatile substrate selection advantage might greatly broaden memory applications.

Finally, considering the cost-effective production and because both the organic materials and the bottom CVD-graphene possess the capability for roll-to-roll processes, the resulting memory is suitable for industrial large-area printing manufacture.

CVD-grown graphene with a unique ultra-thin feature is a crucial conductive material in this novel, label-like memory. Due to its good conductivity, mechanical flexibility together with low-cost and capability for roll-to-roll fabricating features, the researchers chose it as their flexible conductive layer for the organic memory.

"In particular," says Chen, "utilization of its astonishing mechanical properties and strong attractive interfacial adhesion force make CVD-grown graphene an ideal interfacial electrode for adhering to non-conventional surfaces and flexible electronic applications. Moreover, its unique ultra-thin characteristic can significantly reduce device thickness, which is beneficial for adhesion and bending operations."

Thanks to its combination of low fabrication cost, facile processes, and mechanical flexibility combined with high-speed electrical access capability, application areas for this flexible memory could potentially be found in low-cost digital information storage such as smart labels, e-tags, and portable disks. By introducing an antenna circuit, it could be accessed wirelessly. Also, it is nonvolatile, which means it can keep the data without a battery.

Featured scientist: Prof. Yang-Fang Chen (http://web.phys.ntu.edu.tw/semiconductor/)
Organization: Department of Physics, National Taiwan University.
Relevant publication: Y. Lai, Y. Wang, Y. Huang, T. Lin, Y. Hsieh and Y. Yang, *et al.*, Rewritable, Moldable, and Flexible Sticker-Type Organic Memory on Arbitrary Substrates, *Adv. Funct. Mater.*, 2014, **24**(10), 1430, DOI: 10.1002/adfm.201302246

2.3 Roll-to-Roll Production of Carbon Nanotube-Based Supercapacitors

Ultra- or supercapacitors are emerging as a key enabling storage technology for use in fuel-efficient transport as well as in renewable energy systems (for instance as a power grid buffer). These devices combine the advantages of conventional capacitors (they can rapidly deliver high current densities on demand) and batteries (they can store a large amount of electrical energy).

Supercapacitors offer an alternative source of energy to replace recharge-able batteries in applications ranging from power tools to mobile electronics and electric vehicles. Although the energy density of capacitors is quite low compared to batteries, their power density is much higher, allowing them to provide bursts of electric energy that for instance can help electric cars to accelerate at comparable or better rates than traditional petrol-only engine vehicles, while achieving a significantly reduced fuel consumption.

"Among the various types of supercapacitors, carbon nanotube (CNT) based devices have shown an order of magnitude higher performance in terms of energy and power densities," explains Ramakrishna Podila, an assistant professor in the Department of Physics and Astronomy at Clemson University. "The bottleneck for transferring this technology to the marketplace, however, is the lack of efficient and scalable nanomanufacturing methods."

To address this issue, Podila's team, in collaboration with Professor Apparao Rao's lab, also at Clemson University, has developed a scalable method to directly spraycoat CNT-based supercapacitor electrodes.

"Much like painting a car or a wall in your home, we can spray CNT solutions on flexible electrodes, porous aluminum foils in our case, to achieve high energy density supercapacitor electrodes without the need of any binder," explains Podila.

The resulting supercapacitors have an energy density 10 times higher than state-of-the-art supercapacitors on the market.

Theoretically, CNTs offer an ultra-high surface area; in practice, though, the net capacitance of the CNT electrodes is smaller than the predicted values—based on surface area—due to the presence of a so-called small quantum capacitance in series.

In their work, the Clemson researchers together with Cornell Dubilier, Inc (a leading capacitor manufacturer in Liberty, SC) and Sai Global Technologies (a newly founded manufacturer of tailored nanomaterials in San Antonio, TX), have demonstrated that nitrogen-doped CNT electrodes overcome the quantum capacitance limitations and exhibit high power density along with high energy density on par with thin-film Li-ion batteries.

"The quantum capacitance must be increased, ideally to infinity, for realizing the true potential of nanocarbons in energy storage," remarks Rao, who is director of the Clemson Nanomaterials Center.

"Heteroatomic doping, which has been a valuable tool in the semiconductor industry, can provide a solution," adds Podila. "Here we showed that doping provides a handle to control the energy states where electrons could reside in supercapacitor electrode materials and thereby increase the quantum capacitance."

The team points out that their supercapacitors show excellent cycle stability with very little degradation over at least 10 000 cycles.

"At the end point of the electrode lifetime, the CNTs from the used electrode could be recycled to make another new electrode," Rao notes. "The recycled electrode could perform as efficiently as 60% of the original method—a great advantage in terms of sustainability."

Another advantage of the roll-to-roll spray-coating process is significantly lower cost. As the researchers report, the final price of the spray-coated CNT electrodes could be reduced by almost 17%, which includes material and production cost.

Featured scientists: Prof. Ramakrishna Podila, Prof. Apparao Rao
Organizations: Nano-Bio Lab, Clemson University (https://clemsonnanobio.wordpress.com/); Clemson Nanomaterials Center (http://www.raonanolab.net/); Clemson, SC (USA)
Relevant publication: M. Karakaya, J. Zhu, A. Raghavendra, R. Podila, S. Parler and J. Kaplan, *et al.*, Roll-to-roll production of spray coated N-doped carbon nanotube electrodes for supercapacitors, *Appl. Phys. Lett.*, 2014, **105**(26), 263103, DOI: 10.1063/1.4905153

2.4 Foldable Capacitive Touch Pad Printed with Nanowire Ink

Paper has become a popular substrate for fabricating electronics—it is a cheap, abundant material and easy to print on. Electronics printed on paper are inexpensive, flexible, and recyclable, and could lead to applications such as smart labels on foods and pharmaceuticals or wearable medical sensors. Paper has been used for printed memory, as gas sensors or bioactive sensors to detect neurotoxins.

Most printed electronics applications rely on some kind of ink formulated with conductive nanomaterials.

"Paper electronics have been extensively studied in the past but a printing protocol has yet to be developed," according to Anming Hu, an assistant professor at the University of Tennessee, Knoxville. "We wanted to develop a way to print it directly on a variety of paper to make a sensor that could respond to touch or specific molecules, such as glucose."

"We also proposed a simplified theoretical model to elucidate the capacitive operation of touch pads, which closely approximates empirical data," notes Hu.

Hu and his collaborators developed a technique that uses a programmed printing machine with postdeposition sintering using a camera flashlight to harden the deposited silver nanowire ink. This is a rapid and facile method to fabricate a foldable capacitive touch pad.

The researchers point out that the resulting paper-based touchpads produced by direct writing with silver nanowire inks offer several distinct advantages over existing counterparts including:

- low-cost and disposable;
- rapid sintering of nanowires through surface plasmonic excitation, typically requiring three flash pulses and less than 20 seconds using a commercial camera flash;
- ultra-thin and ultra-light: less than 0.1 mm thickness with printing and inkjet paper substrates and less than 60 mg for a single keypad on printing paper; and

- flexible and robust: the device responded to touch even when curved, folded and unfolded 15 times, and rolled and unrolled 5000 times.

The team is now working on printable biosensors and energy devices with paper-based or polymer-based substrates. "We hope that we can integrate micro-sized batteries into a sensor and form a stand-alone microsystem," says Hu.

Featured scientist: Prof. Anming Hu (http://mabe.utk.edu/peopletwo/anming-hu/)
Organization: Department of Mechanical, Aerospace and Biomedical Engineering, University of Tennessee, Knoxville, TN (USA)
Relevant publication: R. Li, A. Hu, T. Zhang and K. Oakes, Direct Writing on Paper of Foldable Capacitive Touch Pads with Silver Nanowire Inks, *ACS Appl. Mater. Interfaces*, 2014, **6**(23), 21721, DOI: 10.1021/am506987w

2.5   Computer Memory Printed on Paper

It seems that computer memory technology is coming full circle. Pioneers in the early 19th century, *e.g.* Charles Babbage, first proposed the use of paper memory (albeit non-electronic), where a bit was stored as the presence or absence of a hole in a paper card. State-of-the-art research today again is pro-posing the use of paper as a memory device. This time, although the paper may be very similar, the bits are not crudely punched holes but nanofabri-cated device structures.

Traditionally, electronic devices are mainly manufactured by photolithog-raphy, vacuum deposition, and electroless plating processes. In contrast to these multi-staged, expensive, and wasteful methods, inkjet printing offers a rapid and cheap way of printing electrical circuits with commodity inkjet printers and off-the-shelf materials.

Although transparent plastic substrates are widely explored for flexible electronics, they have intrinsic problems: most of them are not thermally stable; are not based on green—*i.e.*, renewable and biodegradable—materi-als; and are not as easily printable as paper.

Researchers from National Taiwan University added to this list by demon-strating a paper-based, nonvolatile memory device. To do that, the team used a combination of inkjet and screen printing to fabricate resistive RAM mem-ory cells on commercial printing paper.

This is the first nonvolatile memory built on paper. Resistive random access memory (RRAM), however, is not a new concept. The combination of RRAM and paper is ideal because the structure of RRAM is simple—only one insulator and two electrodes are required for a bit. Operation is also sim-ple; such memory is operated by changing the resistances of the insulator material, whose resistive states—0 and 1—vary greatly as different voltages impose across it. In addition, RRAM is nonvolatile so an embedded power source is not required.

"One challenge of using regular cellulose paper as a base for electronic memory is that, because it is made of fibers, it is very rough and porous on a microscopic level, making it difficult to lay down the thin, uniform layers of materials that typical memory technologies require," says Jr-Hau He, who, at the time, led the work as an associate professor at the Institute of Photonics and Optoelectronics & Department of Electrical Engineering, National Taiwan University.

To get around this problem, He's team decided to fabricate RRAM, a relatively new type of memory with a structure simple enough to cope with such surface variations.

In a RRAM device, an insulator can be set to different levels of electrical resistance by applying a voltage across it; one level of resistance corresponds to the 1s of digital logic, the other to the 0s. So each bit in RRAM consists of an insulator sandwiched by two electrodes.

This device is fabricated by printing techniques—inkjet printing and screen printing—so the memory can be printed by a commercial printer or roll-to-roll techniques. In the first fabrication step, commercial printing paper is coated several times—until the surface roughness of the paper is smoothed out—with a layer of carbon paste that will serve as the bottom electrode. After curing, patterns of titanium oxide nanoparticles are inkjet-printed onto the paper and dried. Finally, silver nanoparticle ink is inkjet-printed on top of the TiO_2 patterns in order to form the top electrode.

He notes that, if combined with other printable devices, it will become possible for consumers to design a functional circuit on their own computer, or download one from the internet, and directly fabricate it at home instead of in a lab or factory setting.

"And then a lot of applications will become possible," says He. "For example, integrated with RFIDs, a printed RRAM device can be used for ticketing; dot arrays can serve as QR code and provide better security; or it can just be memory embedded in your books, paper cups, or any other printable object".

Going forward, one of the issues the researchers will focus on is to increase the memory density of their printed RRAM. The current work uses a regular inkjet printer and has a dot size of ~50 μm with pitch resolution of 25 μm. Thus, one bit will occupy a square of about 100 μm on each side, or 10^4 bits per cm^2. With this density, a fully printed A4 paper can hold 6.237×10^6 bits or roughly 780 KB of data.

With a higher resolution printer, it will be possible to achieve a dot resolution of 1 μm and nearly the same pitch width. That means the density could be enhanced by about 2500 times, and 1.56×10^{10} bits, or almost 2 GB, could be obtained on a single A4 paper.

According to He, the team is currently exploring RRAM crossbar architectures and trying to build 1D-1R or 2D-1R structures. It is possible to stack the cross-bar architecture, so that 3D arrays can be achieved, and the density can be increased.

They are also exploring the use of arrays of single memory dots, for instance in arrangements such as QR codes. "The dots can be painted, so combining the pigments (dark and light) and resistive states (high and low) the stored information can be doubled and secured," says He.

Featured scientist: Prof. Jr-Hau He
Organization: Institute of Photonics and Optoelectronics & Department of Electrical Engineering, National Taiwan University (http://gipo.ntu.edu.tw/eng/e_index.php)
Relevant publication: D. Lien, Z. Kao, T. Huang, Y. Liao, S. Lee and J. He, All-Printed Paper Memory, *ACS Nano*, 2014, **8**(8), 7613, DOI: 10.1021/nn501231z

2.6   Nanopaper Transistors

Cellulose papers have been explored to replace plastic substrates as a lightweight substrate for low-cost, versatile, and roll-to-roll printed electronics. Researchers have already demonstrated various types of devices on paper: batteries, solar cells, RFID tags, and even transistors.

For these applications, transparent nanopaper has many advantages over regular paper as well as plastic substrates. Nanopaper, made from cellulose like traditional paper, shows much lower surface roughness and much higher transparency than traditional paper. This is due to the nanoscale dimensions of the cellulose fibers used for its production.

With regard to fabricating electronics, and compared to plastic substrates, the crucial advantages of nanopapers are their better thermal stability and the fact that they tolerate a much higher processing temperature than plastic.

A research group, led by Liangbin Hu, an assistant professor in the Department of Materials Science and Engineering at the University of Maryland, has fabricated transistors on specially designed nanopaper. They show that flexible organic field-effect transistors (OFETs) with high transparency and excellent mechanical properties can be fabricated on tailored nanopapers.

Nanopaper is transparent and, in order to keep the high transparency of the device, the semiconductor materials also need to be transparent. To that end, the researchers used a highly conductive single-walled carbon nanotube film to serve as the transparent gate electrode of the transistor.

"Instead of a transparent conductive oxide (TCO) film we used a carbon nanotube film because TCO is brittle and can crack during the fabrication process," Hu explains. "Furthermore, carbon nanotube film can be deposited by various low-cost methods such as rod coating and simple drawing methods, while the deposition of high quality TCO film usually requires expensive methods such as vacuum deposition or high temperature annealing."

"The nanopaper OFETs exhibit good transistor electrical characteristics," says Hu. "To demonstrate the flexibility of nanopaper OFETs, we measured devices before and during bending. We observed only a 10.2% and a 9.8% decrease in mobility when we bent the device in the direction parallel to the conduction channel direction and vertical to the conduction channel direction, respectively."

He points out that these excellent optical, mechanical, and electrical properties suggest the great potential of nanopaper FETs in next-generation flexible and transparent electronics and in a broad range of other cost-efficient and practical applications.

Featured scientists: Prof. Liangbin Hu's research group (http://bingnano.com/)
Organizations: University of Maryland, A. James Clark School of Engineering.
Relevant publication: J. Huang, H. Zhu, Y. Chen, C. Preston, K. Rohrbach and J. Cumings, *et al.*, Highly Transparent and Flexible Nanopaper Transistors, *ACS Nano*, 2013, 7(3), 2106, DOI: 10.1021/nn304407r

2.7 Approaching the Limits of Transparency and Conductivity with Nanomaterials

Following up on his group's work on nanopaper transistors, Hu's group began working on transparent conductive coatings—a group of materials that pervades modern technology. They are a critical component of optoelectronic devices such as smartphone and tablet displays as well as solar cells. The most widely used standard coating is indium tin oxide (ITO), although the use of carbon nanomaterials—carbon nanotubes and graphene—is on the rise.

The search for novel transparent electrode materials with good stability, high transparency and excellent conductivity is driven by the required trade-off between transparency and conductivity: metals are very conductive but not transparent; plastics are quite transparent but not conductive. However, many optoelectronic applications ideally require electrodes with both high transparency and high conductivity.

Simultaneously increasing the conductivity and transparency of ultra-thin graphite—ranging from 3–60 graphene layers in thickness—by lithium intercalation, Hu's group, who collaborated with the research group of Prof. Michael Fuhrer at Monash University, has achieved the highest combined performance of sheet resistance and transmittance so far reported among all continuous thin-films.

The researchers designed a methodology *via* a planar nanobattery that allowed them to conduct *in situ* studies of the electrical and optical properties of few-layer graphene sheets during electrochemical intercalation and deintercalation of lithium.

"We doped few-layer graphene by inserting lithium inbetween the graphene layers," explained Jiayu Wan, a PhD student in Hu's lab. "As a result of this electrochemical intercalation, the Fermi level is upshifted by the doping effect, resulting in a more transparent and conductive material."

In previous studies, researchers already tried to achieve a combined increase of both transmittance and conductivity. Some of the best results have been achieved by acid doping and $FeCl_3$ intercalation.

"However, these previous doping/intercalation efforts just increased the conductivity compared to pristine few-layer graphene; the transmittance in the visible range at best stayed unchanged or even decreased," notes Wan.

Hu points out that, due to the unusual band structure of graphene, lithium intercalation can simultaneously increase the DC electrical conductivity and increase optical transmission in the visible, allowing Li-intercalated few-layer graphene to achieve an unprecedented Figure of Merit $\sigma_{dc}/\sigma_{opt} = 920$, significantly higher than any other material and approaching the ultimate limit expected for doped graphene systems.

As the lithium intercalated graphene material is not very stable in air, the team improved the stability of the intercalation compound with an air-tight sealing.

"In addition to elucidating the limits of conductivity and transparency in ultra-thin graphite, we expect that the experimental techniques developed here will be broadly useful for studying the intercalation dynamics and correlated optoelectronic properties of other 2D nanomaterials that can be intercalated electrochemically," concludes Hu.

Featured scientists: Prof. Liangbin Hu's research group (http://bingnano. com/); Fuhrer Research Group (http://fuhrerlab.physics.monash.edu.au/) *Organizations*: University of Maryland, A. James Clark School of Engineering; Monash University. *Relevant publication*: W. Bao, J. Wan, X. Han, X. Cai, H. Zhu and D. Kim *et al.*, Approaching the limits of transparency and conductivity in graphitic materials through lithium intercalation, *Nat. Commun.*, 2014, 5, 4224, DOI: 10.1038/ncomms5224

2.8   Adaptive Electronics for Implants

For years, scientists and engineers have worked to design electronics that can interface with the body. However, typical silicon wafer-based electronics, which are planar and stiff, are not suited to interface with the soft, curvilinear, and dynamic environment that biology presents.

By exploiting the features of shape-memory polymer (SMP) substrates, an international team of researchers from the University of Tokyo and the University of Texas, Dallas, has demonstrated a unique form of adaptive electronics, which softly conform or deploy into 3D shapes after exposure to a stimulus. The resulting organic thin-film transistors (OTFTs) can change their mechanical properties from rigid and planar, to soft and compliant, in order to enable soft and conformal wrapping around 3D objects, including biological tissues.

This work is the continuation of previous efforts, one from the Someya lab[1] and one from the Voit lab.[2]

"With this work, we are demonstrating the first electronic device which is fabricated while flat and rigid using adapted microelectronics techniques, but which softens and adapts to the morphology of soft tissue after exposure to physiological conditions," explains Jonathan Reeder, a PhD student in Prof. Walter Voit's Advanced Polymer Research Lab at UT Dallas. "Additionally, these transistors can be made to deploy into 3D shapes in response to heat, which could have useful applications in electronics which can grip tissue such as nerves, blood vessels, or muscles (Figure 2.1)."

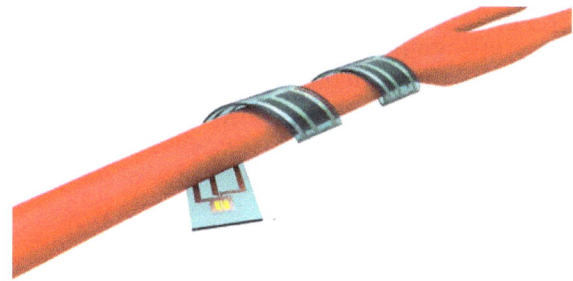

Figure 2.1 Illustration of an organic thin-film transistor on a shape-memory polymer substrate conforming to a warm surface. (Image: Voit lab, University of Texas, Dallas.)

Flexible electronics have been fabricated on many types of plastic substrates, but typically the softening temperature of the plastic is well above or well below body temperature. As a result, these devices exhibit the same mechanical properties (stiffness and shape) no matter their environment.

"By specifically engineering the electronic substrate to soften in response to body temperature and humidity, we allow electronics to adapt to the dynamic and curvilinear environment inside the body, all while maintaining excellent electrical properties," notes Reeder.

Shape memory polymers are "smart" materials capable of storing a metastable shape and returning in a controlled manner to a globally stable shape upon response to a stimulus such as heat or light. However, as the researchers point out, in order to fabricate organic thin-film transistors on SMPs for implantable electronics, it is critical to simultaneously achieve both high mobility and low voltage operation of OTFTs on shape changing substrates that adapt at physiologically relevant temperatures; this will enable the detection of small body signals and limit undesirable physiological responses.

Another crucial criteria for OTFTs on SMP substrates for implants is the reliability and stability of the devices inside the body.

The team fabricated their mechanically adaptive OTFTs by two methods: the first type is manufactured directly onto SMP substrates. The second type is manufactured on 1.4 μm PET foils, encapsulated by 1 μm of Parylene, and subsequently coated with a layer of SMP using a transfer-by-polymerization process. For both processing methods, the SMP layer can be partially polymerized, deformed into a complex 3D shape, and then fully cured, allowing for the fabrication of 3D electronics.

"After the initial shape changes in the device due to softening after insertion, the device is able to withstand the natural motion of the surrounding soft tissue," says Reeder.

The researchers also demonstrated that their OTFTs not only can conform to complex shapes in physiological environments, but also deploy into a 3D shape to actively grip a target site.[3]

This research lays the fabrication groundwork for future high-performance, robust bioelectronics, which will have specific applications in biomedical devices. In particular, devices that can stimulate and interrogate the cerebral cortex, spinal cord, gastrointestinal system, and peripheral nervous system are all areas of ongoing work in the Voit lab. Many of these devices are investigated in collaboration with the Texas Biomedical Device Center, which is working to pair scientists, engineers, surgeons, and neuroscientists in the Dallas area to create research and clinical devices.

To date, a main obstacle to flexible electronics becoming commonplace and effective as implants is the long-term biocompatibility of the electronic components. Rigid electronics implanted in soft tissue ultimately fail for the same reason splinters are eventually ejected from under your skin: the body's immune response. This, of course, limits the types of treatments that current devices can be used for, due to the limited timeline of effectiveness.

Alternatively, electronics have been fabricated on soft plastics in the past such as PDMS and elastomeric polyurethanes. However, these devices are not able to penetrate soft tissue, which is required for brain applications in particular.

Another major issue with implantable electronics is the transfer of power and data to and from the device, as connectors are common failure points. Because of this, wireless technology, which can eliminate wires, is highly desired but has been demonstrated in only a few devices yet.

Featured scientists: Prof. Walter Voigt's Advanced Polymer Research Lab (http://voitlab.com/)
Organization: The University of Texas at Dallas (USA)
Relevant publication: J. Reeder, M. Kaltenbrunner, T. Ware, D. Arreaga-Salas, A. Avendano-Bolivar and T. Yokota, *et al.*, Mechanically Adaptive Organic Transistors for Implantable Electronics, *Adv Mater.*, 2014, 26(29), 4967, DOI: 10.1002/adma.201400420

2.9 Integrating Nanoelectronic Devices onto Living Plants and Insects

As we have seen, there are basically two ways of achieving flexible electronic devices: develop a low-cost generic batch process using a state-of-the-art CMOS process to transform conventional silicon electronics into flexible and transparent electronics, while retaining high-performance, ultra-large-scale-integration density and cost.

Or develop new substrates and techniques such as inkjet-printing of graphene or other semiconductor inks on flexible substrates.

Taking the approach of flexible electronics one step further, researchers in Korea have integrated all-carbon based electronic devices onto live plants and insects.

"Technologies to interface electronic circuits, especially sensor networks that have capabilities of transferring information and power wirelessly with living flora and fauna, can monitor the conditions of the environment, including the detection of chemical weapons, pollution, and infections," Kyongsoo Lee, a postdoctoral fellow in Prof. Jang-ung Park's Flexible Nanoelectronics and Biotechnology Lab at Ulsan National Institute of Science and Technology (UNIST), explains. "In addition, the attached devices can function consistently as sensors even after the *in vivo* activities of animals and plants have stopped."

Lee and a team of fellow researchers from UNIST, developed an unconventional approach for the *in situ* synthesis of monolithically integrated electronic devices based on single-walled carbon nanotube (SWCNT) channels and graphitic electrodes. The highly flexible transistors were formed directly by the *in situ* synthesis using patterned metal catalyst films and subsequently could be transferred to both planar and nonplanar substrates, including papers, clothes, and fingernails (Figure 2.2).

Carbon nanotubes have high sensitivity to a large number of different gases and vapors, which are important in areas as diverse as environmental monitoring, process monitoring in industry, agriculture, personal safety, medicine, or security screening. Gas sensors often operate by detecting the subtle changes that deposited gas molecules make in the way electricity moves through a surface layer. One advantage that CNTs offer for gas sensors is their fast response time and the fact that they react with gases at lower temperatures, sometimes even as low as room temperature.

"On the basis of these capabilities, we developed a sensor platform which can be interfaced with inherent life forms in nature for monitoring environmental conditions wirelessly," says Lee. "To demonstrate our sensor technology, we conducted experiments with real-time gas sensor arrays on a leaf of a live plant (*Dracaena sanderiana cv. Virens*) and on the epidermis of a live insect (*Lucanus maculifemoratus dybowskyi parry*) for the detection of stimulants of sarin nerve agent (DMMP)."

He adds that this all-carbon electronic device demonstrated superb mechanical flexibility and good adhesion to the nonplanar surfaces of biomaterials.

The researchers are confident that their device technology may also find uses in a variety of other applications such as artificial skin that includes sensors and displays for collecting information, threat detection (toxins or pollutants), or as a component of wearable displays.

The team plans to develop artificial skin for electronics with unconventional geometries.

Figure 2.2   Left: SWCNTs-graphite arrays transferred onto the surface of a live leaf and (right) onto a live beetle. (Images: Dr Lee, UNIST.)

Lee points out that remaining technical challenges arise from requirements for commercial applications, such as the need for long-distance detection, which can be improved by circuit design and using different nanomaterials.

Featured scientists: Prof. Jang-ung Park's Flexible Nanoelectronics and Biotechnology Lab (http://wearable.unist.ac.kr/main/main.php)
Organization: Ulsan National Institute of Science and Technology (UNIST), Ulsan Metropolitan City, (Republic of Korea)
Relevant publication: K. Lee, J. Park, M. Lee, J. Kim, B. Hyun and D. Kang, *et al.*, In-situ Synthesis of Carbon Nanotube–Graphite Electronic Devices and Their Integrations onto Surfaces of Live Plants and Insects, *Nano Lett.*, 2014, **14**(5), 2647, DOI: 10.1021/nl500513n

2.10   Nanoelectronics on Textiles, Paper, Wood and Stone

Nanofabrication techniques often depend on creating a structure on one substrate and then transferring it *via* various processes onto another, desired, substrate. Nanoimprinting lithography (NIL) is such a pattern transfer process, as is poly(methyl methacrylate) (PMMA)-mediated peeling, or transfer printing with a polydimethylsiloxane (PDMS) stamp.

In the field of flexible electronics, transfer-printing techniques also have been successfully used to deploy very small and ultra-thin silicon nanomembranes or nanoribbons on top of polymeric materials. Unfortunately, these methods rely on unconventional and expensive substrates and are limited in terms of integration density.

To get around these limitations, researchers have developed a simple double-transfer printing technique that allows them to integrate high-performing electronic devices—featuring state-of-the-art, non-planar, sub-20 nm Fin Field Effect Transistor (FinFET) devices—fabricated on novel flexible thin silicon sheets with several kinds of materials exhibiting complex, asymmetric surfaces including textiles, paper, wood, stone and vinyl.

"Our simple double-transfer process utilizes soft materials to integrate non-planar FinFET and planar traditional MOSFET devices onto various wavy, curvilinear, irregular, or asymmetric surfaces, which helps to reduce the applied stress to the devices, and more importantly, assists to preserve performance with insignificant deterioration even at a bending state," says Muhammad Mustafa Hussain, an associate professor of electrical engineering at King Abdullah University of Science and Technology (KAUST), who led the team. "The use of a soft polymeric material gives us a way to not only provide a strong adhesion layer to reduce the strain and achieve more compliant systems but also encapsulate ultra-thin, silicon fabric-based, flexible electronics."

Jhonathan P. Rojas, first author of a paper on this work, points out that this novel technique offers the opportunity for large-area, full-die and even full-wafer transfer with ultra large-scale integration (ULSI) density capability.

"This is a great example of how hetero-integration of diverse materials with different mechanical properties can work together to improve deployment and compatibility to all kinds of shapes, while retaining performance," he says.

The team used a thin layer of PDMS as the flexible substrate to provide final adhesion and isolation from applied strains. PDMS has favorable mechanical and insulating properties and is biocompatible. Moreover, it is inexpensive, transparent, and stretchable.

Currently, the KAUST team is working on demonstrating more complex and faster circuitry to build high performing applications such as microprocessors and memory arrays, as well as system level implementations that include sensing, data processing, storage and communication as well as energy harvesting/management units.

"We are also focusing on effective and automated ways to develop 3D integration of multiple silicon fabrics for multifunctional capabilities with extended area efficiency," notes Hussain. "There might be several scenarios where limited space and complex shapes are requiring electronics to have the ability to conformably adapt to such asymmetric forms in an area-efficient way, without compromising on electric performance. One advantage of our technique is that it can potentially enable 3D integration by allowing the stack of thin silicon sheets with devices, which remain high-performance and could be vertically interconnected."

Featured scientist: Prof. Muhammad Mustafa Hussain's Integrated Nanotechnology Laboratory (https://nanotechnology.kaust.edu.sa/Pages/Home.aspx)
Organization: King Abdullah University of Science and Technology, Saudi Arabia
Relevant publication: J. Rojas, G. Torres Sevilla, N. Alfaraj, M. Ghoneim, A. Kutbee and A. Sridharan, *et al.*, Nonplanar Nanoscale Fin Field Effect Transistors on Textile, Paper, Wood, Stone, and Vinyl via Soft Material-Enabled Double-Transfer Printing, *ACS Nano*, 2015, **9**(5), 5255, DOI: 10.1021/acsnano.5b00686

References

1. M. Kaltenbrunner, T. Sekitani, J. Reeder, T. Yokota, K. Kuribara and T. Tokuhara, *et al.*, An ultra-lightweight design for imperceptible plastic electronics, *Nature*, 2013, **499**(7459), 458, DOI: 10.1038/nature12314.
2. T. Ware, D. Simon, K. Hearon, C. Liu, S. Shah and J. Reeder, *et al.*, Three-Dimensional Flexible Electronics Enabled by Shape Memory Polymer Substrates for Responsive Neural Interfaces, *Macromol. Mater. Eng.*, 2012, **297**(12), 1193, DOI: 10.1002/mame.201200241.
3. See video: https://youtu.be/EpOE2AlDuEI.

CHAPTER 3

Nanofabrication

All the cool functionalities described in the previous chapters require sophisticated engineering skills and fabrication technologies operating on a scale smaller than the size of bacteria.[†]

Proponents of "atomically precise manufacturing" and "molecular manufacturing" love to talk about the mind-boggling possibilities that these technologies could offer one day. These visions range from the modest, such as improved materials and more efficient production methods for chemicals (already on the horizon), to the outrageous, such as molecular desktop fabs to make any desired product (far, far out).

Articles about revolutionary nanotechnology almost always skip the hard part, *i.e.* the tremendous amount of research breakthroughs that are required to get from where we are today to the promised land.

Nanotechnology researchers today are still struggling with very basic problems such as being able to fully control the synthesis of nanoparticles or 2D materials such as graphene in order to consistently obtain materials with predictable, precisely defined parameters. Not to mention the challenge of scaling up to industrial scale production.

Nevertheless, the fabrication techniques that are being developed today are very impressive, sophisticated tools and applications that offer a glimpse of the changes that are in store for industrial manufacturing.

3.1 Fabricating Complex Micro- and Nanostructures

Complex surfaces with precisely fabricated three-dimensional (3D) micro- and nanosized features are needed in applications in electronics, metamaterials, substrates for cell culture and tissue engineering, smart active surfaces, and

---

[†]Most bacteria range from 1000 to 5000 nanometers in size (1–5 microns).

Nanotechnology: The Future is Tiny
By Michael Berger
© Michael Berger 2016
Published by the Royal Society of Chemistry, www.rsc.org

lab-on-a-chip systems, just to name a few. Although there are several methods of fabricating 3D microstructures—lithography techniques, capillary forming, 3D-printing, or direct ink writing—all have their drawbacks, requiring trade-offs in feature geometry, heterogeneity, resolution, and throughput. The challenge for researchers is to find techniques for constructing controllable 3D self-assembled microstructures in a simple and convenient way.

3.1.1 Assembling Nanoparticles into 3D Structures with Microdroplets

According to Yanling Song, a professor at the Institute of Chemistry, Chinese Academy of Sciences, and Director of the Key Laboratory of Green Printing, "existing fabrication techniques for 3D microstructures usually suffer from complicated equipment, time-consuming processes, and insufficient controllability of precise structures."

Song and his team propose a facile strategy to directly assemble nanoparticles into controllable 3D structures from one microdroplet based on hydrophilic pinning patterns.

The scientific core of these findings is that "0D" patterns—hydrophilic pinning dots on a hydrophobic surface—are used to print 3D structures by controlling the dynamic asymmetric dewetting of the three phase contact line (TCL) based on energy difference.

This novel fabrication technique utilizes asymmetric dewetting of the TCL of hydrophilic patterns to fabricate 3D structures with controllable morphologies for the first time.

"The preparation of fine 3D microstructures is an attractive issue, however, it is of limited use for large-area fabrication processes and intricate morphology manipulation," explains Song. "Here, our strategy to fabricate controllable 3D structures and morphologies can be done from one droplet *via* ink-jet printing. The whole process can be easily handled and the morphologies can be well manipulated."

Previous approaches fabricated structures only from replication: they mainly focused on restricting structures at the hydrophilic region, whose morphologies rely on the shape of hydrophilic patterns. So far, there has been no report using patterned hydrophilic pinning dots to direct the fabrication process of 3D structures.

Compared to these previous results, in this work, the hydrophilic dots were used as pinning dots to control the dynamic asymmetric dewetting of the TCL. This process leads to the assembly of nanoparticles into various 3D morphologies with closely packed structures.

"Through the cooperative regulation of liquid properties and the pinning points' parameters, we are able to precisely design and control the 3D morphology," notes Song.

He elaborates: "Generally, nanoparticles pack tightly on a hydrophobic surface, on which the droplet has a large surface receding contact angle (θ_R), while pack loosely on a hydrophilic surface with a small θ_R."

Closely-packed morphologies are assumed to have superior properties for practical applications, and numerous efforts have been made to achieve a uniform deposition by adjusting particle shapes and regulating capillary flow.

By manipulating the retracting TCL by 0D hydrophilic patterns, the team controlled the assembly of nanoparticles on the hydrophobic region with large θ_R as well as on the hydrophilic region with small θ_R to achieve a hexagonal closely-packed structure.

In their experiment, the researchers directly printed 3D colloidal crystals from a single microdroplet on manually designed hydrophilic pinning patterns. By designing the 0D hydrophilic pinning patterns, they prepared different 3D microcolloidal crystals with precise morphologies, including line, triangle, square, star, hexagon, and octagon shapes.

"Encouraged by the controllable morphologies and photonic properties of the 3D microcolloidal crystals, we designed a macro/microhierarchical structure color pattern, which is of great significance for the fabrication of multi-information carriers or anticounterfeiting materials," says Song. "The printed 3D array displays bright structure color even with the existence of reflective rainbow color. The composition, morphology, and location of each single 3D structure as well as their permutation and combination could be well manipulated, which is of potential use as multiprotocol optical codes."

Since well-defined 3D architectures can lead to collective functional properties, the as-prepared hierarchical pattern will be promising for the development of novel 3D photonics and other functional devices.

"In the future, our strategy can be applied in guiding the morphology of bio-materials, which will be encouraging for 3D biological manufacture and further applications in bio-engineering," says Song. "Here, we have utilized surface energy difference to manipulate TCL of droplets. Likewise, surface energy differences can be utilized to control adhesion of proteins, cells, or bacteria. Take cells for example: firstly, this method could direct cell assembly onto predetermined locations and into different morphologies. Secondly, the morphology of a single cell itself can be directed by a predesigned pattern."

Song's group is focused on developing new technologies to print nanomaterials, especially nanoparticles, to assemble microstructures from 0D to 3D. Another recent example of their work is a controllable strategy to print precise 3D microstructures *via* 2D interface manipulation of droplets on surfaces with tunable dynamic dewetting properties under magnetic guidance.[1]

Song points out that the challenge facing future research in this area could be the substrate fabrication process, which should be bio-friendly. Also, the regulation of adhesion between the substrate and the bio-materials or the nutrient substance culturing the bio-materials is a difficult problem to manipulate.

Featured scientist: Prof. Yanling Song (http://159.226.64.162/web/29070/home)
Organization: Institute of Chemistry, Chinese Academy of Sciences
Relevant publication: L. Wu, Z. Dong, M. Kuang, Y. Li, F. Li and L. Jiang, *et al.*, Printing Patterned Fine 3D Structures by Manipulating the Three Phase Contact Line, *Adv. Funct. Mater.*, 2015, 25(15), 2237, DOI: 10.1002/adfm.201404559

3.1.2 A Design Guide to Self-Assemble Nanoparticles into Exotic Superstructures

Magnetite is the most abundant magnetic mineral of the Earth and magnetite particles are used in numerous products and applications: inks, magnetic liquids, and medical contrast agents, but also as memory elements in data storage media. Nanoscale magnetite has been extensively investigated by scientists during the past decades, also in the context of self-assembly.

For millions of years, magnetite nanoparticles have been synthesized by magnetotactic bacteria—as well as by a variety of other species: birds, fishes *etc.*—where they self-assemble into fine needles which serve as an internal "compass", allowing them to orient themselves along the lines of the magnetic field of the Earth.

Researchers have shown that, by varying the shape of magnetite nanoparticles, they can control the nature of the structures as the nanoparticles assemble. This work provides guidelines for the design of new self-assembled materials.

Magnetite nanoparticles are essentially very tiny magnets, which can interact with each other by magnetic dipole–dipole interactions—in addition to other types of forces.

"We had envisioned that cubic nanoparticles of magnetite would self-assemble differently from other shapes. The reason is that the properties of magnetite—specifically, so-called magnetocrystalline anisotropy, that is, preferential magnetization along given crystallographic orientation—tell us that when two cubes interact by magnetic forces, they could align in a corner-to-corner fashion," Rafal Klajn, an assistant professor in the Department of Organic Chemistry at the Weizmann Institute of Science, explains. "On the other hand, when particles assemble, they also want to maximize contact (van der Waals interactions) which entails a side-to-side arrangement of two cubes. Therefore we have a 'competition' of two types of forces that want to align nanoparticles in different ways. This competition can lead to the emergence of exotic shapes: in our case, helices (Figure 3.1)."

Klajn's group reported their observation on the emergence of helical nanoparticle superstructures during the self-assembly of superparamagnetic nanocrystals in collaboration with a theory group led by Petr Král at the University of Illinois at Chicago.

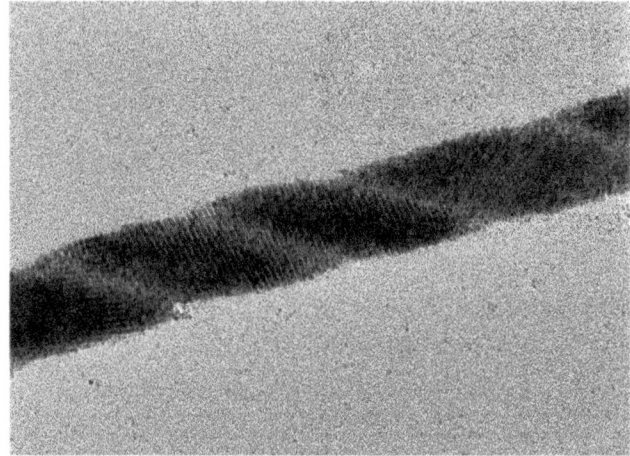

Figure 3.1   Transmission electron micrograph of an individual self-assembled
helix. (Image: Gurvinder Singh, Department of Organic Chemistry,
Weizmann Institute of Science.)

The team is broadly interested in the self-assembly of nanoparticles as
a route to obtain novel functional materials. They note that materials self-
assembled from nanoparticles often have fascinating optical, magnetic,
catalytic, and other properties, in particular when the nanoparticles are
assembled into non-close-packed lattices.

"Self-assembly of nanoparticles driven by competing forces can result
in truly unique structures, the diversity and complexity of which could be
particularly striking if the building blocks were simultaneously coupled by
short- and long-range forces of different symmetries," says Klajn.

The team shows that it is possible to rationally design such "frustration"
during self-assembly of nanoparticles, thereby producing exotic superstruc-
tures/assemblies. This is probably the first method to self-assemble nanoparticles into helical assemblies without the use of any (*i.e.* helical) templates.

At the same time, it is a way to assemble achiral building blocks into
chiral assemblies, whereby symmetry breaking and chirality amplification is
observed.

Klajn points out that it actually is quite striking that, given all the previous research on magnetite nanoparticles, helical assemblies of magnetite
nanoparticles—which form spontaneously when the solvent is evaporated in
the presence of the magnetic field—are reported only now.

Although this work is about a fundamental discovery and, at this point, the
research is quite remote from any practical applications, the scientists think
these materials could interact with light in very interesting ways; they should
also have interesting magnetic properties.

"In addition," says Klajn, "the fact that they are helical and chiral, and can
be prepared very quickly and cheaply, makes them excellent candidates for

templates for the assembly of other materials. They also could be coupled with achiral catalysts to produce catalysts that could potentially be chiral: that is, potential application in asymmetric catalysis."

Having shown that the competition between magnetic and van der Waals interactions leads to unprecedented helical assemblies, the next step for the group is to add more types of interactions to the overall balance of forces in an effort to produce even more exotic assemblies.

Another direction is the use of different materials. "Up to now we only worked with magnetite; there are a host of other magnetic nanoparticles—*e.g.* cobalt, nickel—that can behave similarly," says Klajn. "But the main focus in the future will be on novel properties/functions/applications enabled by these self-assembled materials."

Featured scientists: Prof. Rafal Klajn Research Group (http://www.weizmann.ac.il/Organic_Chemistry/Rafal/index.html); Petr Král Research Group (http://www2.chem.uic.edu/pkral/)
Organizations: Department of Organic Chemistry, Weizmann Institute of Science, Rehovot (Israel); Department of Organic Chemistry, University of Illinois at Chicago (USA)
Relevant publication: G. Singh, H. Chan, A. Baskin, E. Gelman, N. Repnin and P. Kral, *et al.*, Self-assembly of magnetite nanocubes into helical superstructures, *Science*, 2014, **345**(6201), 1149, DOI: 10.1126/science.1254132

3.1.3   3D Nanopatterning with Memory-Based, Sequential Wrinkling

Wrinkling and buckling can occur at all length scales in materials composed of a stiff thin film on a strained supporting layer. When the strain is removed, either by thermal or mechanical stimuli, different surface patterns can form.

This phenomenon—now starting to be realized at nanometer-length scales—is emerging as a powerful bottom-up nanopatterning method to program surfaces with unique properties. It has many applications in the design and fabrication of flexible electronics and devices, micro-cell arrays, optical gratings, and so on.

"Strain-induced surface wrinkling of a stiff skin on a soft base layer is emerging as a powerful method to realize ordered and disordered patterns over large areas," explains Teri W. Odom, Board of Lady Managers of the Columbian Exposition Professor of Chemistry and Professor of Materials Science and Engineering at Northwestern University. "One of the next grand challenges is to create bio-mimetic, three-dimensional hierarchical architectures that display a diverse range of applications including reversible dry adhesion, selective filtration, and self-cleaning surfaces."

She explains that two critical issues are important to design such functional surfaces: (1) control over length scale patterns; and (2) the ability to order features within hierarchy architectures.

However, the realization of 3D hierarchy with independent control over wrinkle wavelength and orientation has not yet been achieved using existing wrinkling systems, which limits wrinkling as a bottom-up route to 3D functional surfaces.

Now though, Odom and her team describe how a memory-based, sequential wrinkling process can transform flat polystyrene sheets into multi-scale, 3D hierarchical textures over large areas (>100 cm^2).

In the fabrication process devised by the team, multiple cycles of plasma-mediated skin growth followed by directional strain relief of the substrate resulted in hierarchical architectures with characteristic generational (G) features.

Independent control over wrinkle wavelength and wrinkle orientation for each G was achieved by tuning plasma treatment time and strain-relief direction for each cycle.

The researchers demonstrated Lotus-type superhydrophobicity on three-dimensional G1-G2-G3 hierarchical wrinkles as well as tunable superhydrophilicity on these same substrates after oxygen plasma.

"Our materials system provides a general approach for nanomanufacturing based on bottom-up sequential wrinkling that will benefit a diverse range of applications, and especially those that require large area, multi-scale, three-dimensional patterns," notes Odom.

A distinct feature of this process is that smaller features are preserved even as larger ones are formed; this the basis for memory-based processing to create hierarchical 3D architectures.

"Memory-based sequential nanowrinkling can generate 3D hierarchical wrinkles with up to three different generations of wrinkles," explains Odom. "Multiple cycles of plasma-mediated skin growth followed by directional strain-relief enabled independent control over wavelength and orientation for each generation of wrinkles. Because of a very small Young's modulus[†] ratio of the skin layer to polystyrene substrate (<10) as well as fine control over skin thickness, we achieved tunable wavelengths at scales from tens of nanometers to tens of microns."

Access to such 3D, hierarchical wrinkles enabled superhydrophobicity as well as tunable superhydrophilicity using oxygen plasma to treat the same substrate.

"Our memory-based hierarchical nanowrinkling provides a general approach to construct polymeric, 3D patterns that have potential in unidirectional liquid transport, anti-biofouling substrates, and omniphobic surfaces," concludes Odom.

[†]Young's modulus is a numerical constant, named for the 18th-century English physician and physicist Thomas Young, that describes the elastic properties of a solid undergoing tension or compression in only one direction, as in the case of a metal rod that after being stretched or compressed lengthwise returns to its original length. Young's modulus is a measure of the ability of a material to withstand changes in length when under lengthwise tension or compression. (Source: *Encyclopædia Britannica*.)

Featured scientist: Prof. Teri W. Odom Research Group (http://chemgroups.northwestern.edu/odom/)
Organization: Department of Chemistry, Northwestern University, Evanston, IL (USA)
Relevant publication: W. Lee, C. Engel, M. Huntington, J. Hu and T. Odom, Controlled Three-Dimensional Hierarchical Structuring by Memory-Based, Sequential Wrinkling, *Nano Lett.*, 2015, **15**(8), 5624, DOI: 10.1021/acs.nanolett.5b02394

3.1.4   Spraying Light—the Fabrication of Light-Emitting 3D Objects

OLED (organic light-emitting diode) technology is based on the phenomenon that certain organic materials emit light when fed by an electric current. What makes OLEDs so attractive is that they do not require a backlight to function and therefore require less power to operate; also, since they are thinner than comparable LEDs, they can be printed onto almost any substrate. The current commercial fabrication of OLEDs is a clean room technology that depends on time- and cost-driving process steps under high vacuum and/or an inert atmosphere. This has limited OLED displays to smaller electronic devices such as phones, cameras and small TV screens.

"The light-emitting electrochemical cell (LEC) shares several external attributes with the OLED, notably the opportunity for soft areal emission from thin-film devices, but its unique electrochemical operation eliminates the principal requirement on inert-atmosphere/vacuum processing as it can comprise solely air-stable materials," says Ludvig Edman, a professor at Umeå University in Sweden, who heads the Organic Photonics and Electronics Group there. "This important intrinsic advantage has inspired recent work on an ambient-air fabrication of LEC devices using scalable means; but as-of-yet, a fault-tolerant and cost-effective fabrication of large-area and uniformly-emitting devices is lacking."

Back in 2010, the group introduced "a unifying model for the operation of light-emitting electrochemical cells".[2] Now, in cooperation with LunaLEC, a Swedish company that was formed in 2012 to commercialize the technology developed by Edman's group and Nathaniel Robinson's team at Linköping University, they have developed a spray-sintering method for the fabrication of LECs.

"Spray-sintering is a new process of depositing material that works uniquely for LEC-fabrication," says Patric Stafshede, Managing Director of LunaLEC. "The common knowledge for fabrication of light-emitting devices has been that you need to deposit a wet, homogeneous film of material in order to get a smooth surface without any pinholes in it that would create black spots or uneven emission. By contrast, the properties of the LEC allow for spraying layers that are inhomogeneous and that allows for spraying of individual droplets that dry one at a time."

This has not been proven to work before and allows for moving some of the LEC-work out of the lab into commercialization.

Introducing a new, purpose-designed spray-sintering deposition technique, Edman's group reports that it is possible to spray liquid inks onto essentially any surface for the achievement of light emission.

"Through sequential spraying of three thin layers of different inks onto the substrate-of-choice, we are able to realize uniform large-area light-emission at a low drive voltage of 3–5 V, and we are working toward the realization of any desired color," says Dr Andreas Sandström, previously a researcher in Edman's group, now CTO of LunaLEC, and first author of a paper reporting the work.

He points out that the entire fabrication process can be executed under ambient air using a simple airbrush.

The team also demonstrates that the inclusion of an additional sprayed layer allows for the creation of multi-colored light-emission patterns, and that light-emitting devices can be fabricated directly onto arbitrarily shaped surfaces.

Stafshede explains that the motivation behind the work has been the potential to increase the size of the objects the team works with. "In the lab, we regularly have made 10 mm by 10 mm squares, but when we want to show the potential in the LEC-technology, homogeneous light-emission over large areas, we need to fabricate larger area samples. The need for a reliable deposition technique over large areas—currently up to 20 cm by 30 cm— prompted this development."

He notes that, initially, to develop the technology further, the technology will be used for marketing items, decorations or design applications where the unique form factors of the LEC are important. That means that to start with, it will be used for 2D objects such as small panels.

"When spraying 3D-objects will become commercially viable, we will see both decorative and illumination applications," he adds. "A 3D-object can utilize already existing surfaces, for instance a door knob, or a child's night light, without the need to have additional light fixtures."

Featured scientist: Prof. Ludvig Edman' Organic Photonics and Electronics Group (http://www.physics.umu.se/english/research/photonics/organic-electronics/)
Organization: Umeå University (Sweden)
Relevant publication: A. Sandström, A. Asadpoordarvish, J. Enevold and L. Edman, Spraying Light: Ambient-Air Fabrication of Large-Area Emissive Devices on Complex-Shaped Surfaces, *Adv. Mater.*, 2014, **26**(29), 4975, DOI: 10.1002/adma.201401286

3.1.5 Microfabrication Inspired by LEGO™

Most tools used for the pick-and-place of microobjects are either single-ended microprobes or double-ended microgrippers. For these microprobes, both the pick-up and placement are challenging due to the adherent forces. For microgrippers, the pick-up is easier and secure due to the gripping motion, but the placement is still difficult. When a microgripper opens its gripping fingers, the microobject still adheres to one of the fingers by strong adhesion forces.

Owing to force scaling laws, the adhesion forces at the microscale—*i.e.*, capillary forces, van der Waals forces, and electrostatic forces—dominate gravity. These adhesion forces make both pick-up step and placement step of a microobject challenging. Specifically, during the pick-up, the adhesion forces between the microobject and the substrate must be overcome. During the placement, the adhesion forces between the microobject and the manipulation tool must be overcome.

These challenges are inherent to the tools, motivating the development of more versatile tools that modulate the adhesion forces drastically. A number of techniques and tools have been proposed to facilitate the pick-and-place process, such as using a microgripper for secure grasping and vibrating the tool to release the adhered microobject. Nevertheless, few of them can achieve deterministic assembly of planar LEGO™ block or brick-like structures without relying on mating interfaces and adhesives.

In 2012, a team of researchers from the University of Illinois at Urbana-Champaign demonstrated a manufacturing route to three-dimensional silicon microsystems—which they termed "micro-masonry"—based on individual manipulation.[3]

"Our approach is inspired by LEGO work and its procedure can be compared to masonry in a construction site," says Seok Kim, an assistant professor in the Department of Mechanical Science & Engineering. "The microobjects made of silicon, silicon dioxide, or gold are fabricated in a donor substrate as bricks are manufactured in a brickyard. A specially designed soft polymeric stamp transfers microobjects from the donor substrate to target areas on a receiver substrate as a mason carries and places bricks. Finally, placed microobjects are permanently bonded with rapid thermal annealing as bricks are bound with mortar."

The team's approach makes use of the adhesion forces between a polymeric stamp and a microobject for the pick-up. After the pick-up, the adhesion forces between those drastically decrease to enable the placement of the microobject on a target location.

"This adhesion force modulation allows us to realize LEGO-like microassembly," says Kim.

This work presents a new paradigm for the pick-and-place assembly process. A microtipped polymeric stamp is specially designed to modulate the adhesion forces between the stamp and the microobject. In this way, the microobject can be picked up with ease and placed on a target location deterministically.

Hoyun Keum, the first author of a paper on this work, points out that this manufacturing strategy provides many attractive features such as cost-effectiveness through fabrication of highly dense individual units on a single donor substrate that can be assembled on multiple foreign substrates in a sparse manner. Its manufacturing throughput can be increased when combined with automation. Moreover, it is highly fault tolerant since one unit assembly failure does not affect other neighboring units' assembly.

Making use of this micro-masonry technique and advancing it further, Kim and his team have demonstrated in a follow-up report the assembly of

MEMS mechanical sensors and actuators. Here they report the microfabrication processes for retrievable, complex MEMS components and the microassembly processes for integrating those components into MEMS sensors and actuators. Furthermore, they demonstrate the integration of gold films onto the assembled silicon device *via* micro-masonry to form metal contacts and to facilitate a subsequent wire bonding process.

"In comparison with our group's previous assembly of 3D silicon microstructures, MEMS device assembly requires more complex and fragile structures such as combs and suspended flexure beams to be fabricated as retrievable components on a donor substrate and to be subsequently transferred to a receiver substrate," explains Yong Zhang, a member of the team.

This work is demonstrating the micro-masonry technique's capability of constructing 3D microdevices that are impossible or difficult to realize with previous monolithic microfabrication.

"We envision the assembly of high-performance microscale weight sensors, microscanners, and vibration-driven energy harvesters in the future," says Kim.

Other opportunities include the exploration of similar assembly concepts with not only silicon but also other materials including metals, dielectrics, and polymers.

Kim notes that, like other pick-and-place assembly methods, the micro-masonry technique is performed serially. "Thus, we think micro-masonry is currently more appropriate for multi small batch rapid fabrication. To increase its manufacturing scalability, a reel-to-reel process with multiple stamps can be adopted for continuous and parallel operation. In addition, automating the process is appealing since it will speed up the process and reduce manual work."

Featured scientist: Prof. Seok Kim's Lab (http://skm.mechse.illinois.edu/)
Organization: Department of Mechanical Science & Engineering, University of Illinois at Urbana-Champaign (USA)
Relevant publication: Y. Zhang, H. Keum, K. Park, R. Bashir and S. Kim, Micro-Masonry of MEMS Sensors and Actuators, *J. Microelectromech. Syst.*, 2013, 23(2), 308, DOI: 10.1109/JMEMS.2013.2273439

3.1.6   Atomic Calligraphy

The difficulties associated with precisely manipulating nanomaterials to turn nanoscale structures into reliable functional devices—at a reasonable cost—is one of the key challenges that needs to be overcome in mass-manufacturing nanodevices (other than computer chips, the fabrication of which requires massive amounts of capital investment).

One of the most restricting parameters in nanofabrication is the difficulty involved with controllably patterning materials at precise locations in a repeatable manner over relatively large areas. The traditional process of randomly placing nanomaterials on a substrate typically leads to highly variable performance of the resultant functionalized devices.

Conventional lithography methods that are used in computer chip manufacturing are not only very expensive and wasteful, they also are reaching physical limitations. To overcome these issues, researchers have been developing a range of alternative, resist-free nanopatterning techniques, among them dip pen nanolithography, oxidation nanolithography, and colloidal self-assembly.

A novel microelectromechanical system (MEMS)-based mask writer has been developed by a team of researchers at Boston University, led by David Bishop, a professor of physics and Head of the Division of Materials Science and Engineering. The device allows the direct writing of structures at the nanoscale without the need to use photoresist, lift-off techniques or other complex and expensive approaches. The technique uses a MEMS plate with apertures drilled into it and a shutter so that one can, in effect, spray paint with atoms. With the shutter, the process can be turned on and off.

"Our results extend and build upon previous work," says Bishop. "Other techniques have used static stencils, AFMs with apertures on them, or dip pen lithography. Our approach helps mitigate some of the limitations of earlier approaches by allowing for arrays of devices, with more complex patterns using a wider range of materials than could be previously accommodated."

The researchers are confident that their approach opens the door to being able to build atomic-scale devices using a cost-effective manufacturing process.

To fabricate their MEMS writers, the team uses a standard lithography batch method used in foundries. The chips are inexpensive single use devices enabling high flexibility and turnover.

The individual writers, consisting of electrostatic comb actuators, folded springs and a central plate, sit on a 2.5×2.5 mm^2 die stack of 600 nm silicon nitride and doped silicon handle 675 μm thick. The central plate is suspended over the substrate by four doubly folded flexure springs and tethers. The springs and tethers can be combined into a single device that can move laterally >10 μm in all four quadrants; using the substrate as an additional electrode enables z-axis pull in.

"Our design is similar to MEMS based nanopositioners described in the literature using both comb or piezoelectric actuation methods," notes Bishop. "The smallest feature or dot that can be patterned is defined by the aperture dimension. We use a focused ion beam (FIB) to mill an aperture in the plate of the writer. By leveraging the strengths of the scalable MEMSCAP PolyMUMPs process with the nanoscale resolution of a FIB we obtain MEMS devices 2 mm across with customized feature sizes below 50 nm."

Bishop points out that this technology can be used to build electronic circuits and structures *in situ* out of materials typically not used in nanolithography.

"The integrated MEMS shutter makes it possible to control stochastically the number of atoms passing through the aperture down to of order one," he says. "Integrating this writer together with a MEMS based evaporator, as well as resonant sensors for deposition rate and temperature creates a cheap and versatile 'Fab on a Chip'. This will enable new mesoscopic experiments of quench condensed films, quantum dots, and single atom effects."

Featured scientist: Prof. David Bishop (http://www.bu.edu/ece/people/faculty/a-g/david-bishop/)
Organization: Division of Materials Science and Engineering, Boston University (USA)
Relevant publication: M. Imboden, H. Han, J. Chang, F. Pardo, C. Bolle and E. Lowell, *et al.*, Atomic Calligraphy: The Direct Writing of Nanoscale Structures Using a Microelectromechanical System, *Nano Lett.*, 2013, **13**(7), 3379, DOI: 10.1021/nl401699w

3.1.7 Complex Assemblies Based on Micelle-Like Nanostructures

One major challenge in contemporary science is to accomplish with synthetic building blocks what nature does so well, that is, creating complex and functional structures through multiple levels of assembly of biomolecules. Bottom-up engineered nanostructures that assemble themselves from polymer molecules are bound to become useful tools in chemistry, energy, and medicine. To that end, researchers are using block copolymer based micellar architectures to form hierarchical superstructures with defined shape and geometry.

"In spite of tremendous achievement in the past decade, a model platform to describe the behavior of these nanoscale building blocks and how they organize to form higher ordered structures is missing," says Zhihong Nie, an assistant professor in the Department of Chemistry and Biochemistry at the University of Maryland. "Ideally, such building blocks should illustrate the well-defined chemistry and architectures of block copolymer micelles."

Nie and his group have demonstrated that nanoparticles tethered with block copolymers resemble micelles that can assemble into well-ordered higher-level mesostructures. These assembly sub-units are not only much easier to create—compared to micellar building blocks purely based on block copolymers—but also have precisely defined dimension, chemistry, shape, and composition.

This work is a systematic study of how to design complex hierarchical structures based on micelle-like nanostructures. Potential applications for these assemblies are drug delivery, cancer imaging and treatment, and the design of the next generation of optoelectronic materials.

"Our system serves as a simple model to elucidate the behavior both in terms of chemistry and spatial length scales of higher level of self-assembly in block copolymer based complex nanostructures," explains Jie He, a post-doctoral researcher in Nie's group. "Basically, what we did was to establish a simple but effective model system comprising of nanoparticle core and grafted block copolymers to understand the design rules that control this multiple level self-assembly, and to design functional assembled materials with controlled properties for various applications."

The scientists used gold nanoparticles as scaffolds to bring multiple block copolymers together in a controlled manner, resulting in a micelle-like structure—but without multistep synthesis and self-assembly.

"This presence of gold nanoparticles as the 'micelle' cores endows the localized surface plasmon resonance properties to such building blocks," says He. "The controlled assembly of these amphiphilic plasmonic micelle-like nanoparticles enables the modulation of the plasmon coupling between individual building blocks, thus the collective properties of assemblies."

In another report, the team demonstrates an application of their plasmonic nanostructures as a theranostic platform for cancer imaging and treatment.[4]

Although it is a useful model to understand the hierarchical structures, the team points out that a more quantitative model is required to predict and design more complex nanostructures.

"In our model, we interpreted that deformability of the micelle-like nanoparticles plays a crucial role but we cannot quantitatively predict the assemblies and correlate various factors to the final products of assembly yet," says Nie.

He notes that there is still no effective means of controlling and predicting the multiple-level assembly behaviors of molecules and particles.

"In future research, it will be critical to design effective model systems to quantitatively evaluate the kinetics and thermodynamics of assembly, and to assess the roles of various factors governing assembly."

Featured scientist: Prof. Zhihong Nie Research Group (http://www2.chem. umd.edu/groups/znie/index.html)

Organization: Department of Chemistry and Biochemistry, University of Maryland, College Park, MD (USA)

Relevant publication: J. He, X. Huang, Y. Li, Y. Liu, T. Babu and M. Aronova, *et al.*, Self-Assembly of Amphiphilic Plasmonic Micelle-Like Nanoparticles in Selective Solvents, *J. Am. Chem. Soc.*, 2013, **135**(21), 7974, DOI: 10.1021/ja402015s

3.1.8 Precise Manipulation of Single Nanoparticles with E-Beam Tweezers

The idea of using laser light to trap or levitate small particles goes back to the pioneering work by Arthur Ashkin of Bell Laboratories in the 1970s and 1980s. Ashkin found that radiation pressure—the ability of light to exert pressure to move small objects—could be harnessed to constrain small particles.[5] This discovery has since formed the basis for scientific advances such as the development of optical tweezers, which are frequently used to control the motion of small biological objects. These techniques provide an extremely versatile toolbox for manipulating small particles and the systems being studied by optical trapping extending from large particles to biological molecules and even single atoms. However, optical trapping of nanoparticles remains a challenging task because the forces are often too small when the sizes of the objects are reduced to the nanometer scale.

Findings from scientists at Lawrence Berkeley National Laboratory and National University of Singapore fill a gap and also open the door to new discoveries by demonstrating trapping and manipulating nanometer-size particles using an electron beam instead of optical forces. It could also lead to new force spectroscopy where nanostructures can be assembled one nanoparticle at a time.

In their work, the team trapped gold nanoparticles inside an electron beam and showed that they can direct their movements over a membrane surface inside an environmental cell.

"Since an electron beam, such as the beam source of a transmission electron microscope (TEM) or a scanning electron microscope (SEM), can be focused into subnanometer sizes and easily operated to scan a surface up to millimeters, the ability to manipulate nanoparticles using an electron beam opens the opportunity to create a versatile tool for nanotechnology," Dr Haimei Zheng, a staff scientist in the Materials Sciences Division at Lawrence Berkeley National Laboratory, explains.

"Using the electron beam, we can also trap multiple nanoparticles and assemble nanoparticles on the surface," says Zheng. "Manipulation of the nanoparticles using the electron beam can be easily combined with the conventional electron microscopy techniques, so that both manipulation and directly imaging the nanoparticles of interest can be achieved."

A key tool leading to the team's findings arises from the microfabricated environmental cell used for this study: it contains two liquid reservoirs for liquid sample loading and an electron transparent silicon nitride window allowing the liquid sample to be examined under a TEM. Nanoparticles can be placed in the cell and subsequently moved in fluids with the beam without encountering the issues that other electron microscopy samples have, such as nanoparticles stuck on a TEM grid and exposed to vacuum.

The team controlled the movements of the electron beam at approximately 10 nm per second—faster or slower movement of the beam is possible—and the single or multiple gold nanoparticles trapped inside the beam are dragged along with the beam while moving chaotically inside the beam.

Utkur Mirsaidov, Assistant Professor of Biological Sciences, Department of Physics and Graphene Research Center at National University of Singapore, notes that the interaction between nanoparticles can be complex and diverse and although there is an attractive force between gold nanoparticles inside the electron beam, nanoparticles are bouncing vigorously and are not aggregated during an extended period of time.

Using electron beam tweezers instead of optical tweezers makes it possible to precisely manipulate nanoparticles and to fabricate devices using nanoparticles as building blocks. It also is a great tool to probe the interaction forces between nanoparticles. This could potentially revolutionize nanoscale device fabrication—nanoparticles can be collected and fused together, allowing the creation of new materials architectures.

Going forward, the team would like to position nanoparticles precisely and rapidly in a programmed fashion. This involves overcoming the random motion of nanoparticles and working with a confined space between membranes.

Featured scientists: Dr Haimei Zheng (http://haimeizheng.lbl.gov/); Prof. Utkur Mirsaidov (http://cbis.nus.edu.sg/utkur-mirsaidov/)
Organizations: Materials Sciences Division, Lawrence Berkeley National Laboratory, Berkeley, CA (USA); Centre for BioImaging Sciences, National University of Singapore
Relevant publication: H. Zheng, U. Mirsaidov, L. Wang and P. Matsudaira, Electron Beam Manipulation of Nanoparticles, *Nano Lett.*, 2012, **12**(11), 5644, DOI: 10.1021/nl302788g

3.1.9 Trapping Individual Metal Nanoparticles in Air

While there is a great deal of knowledge on optical manipulation of metallic nanoparticles in liquids, aerosol trapping of metallic nanoparticles is essentially unexplored. In general, very little is known about optical manipulation of any type of particle in air, where the physics appear to be rather different than in water. For instance, the relation between laser power and trapping strength is found to be linear in water but not in air.

"The ability to manipulate and study individual metallic or semiconductor nanostructures in air or vacuum would open up many exciting opportunities," says Lene Broeng Oddershede, a professor at the Niels Bohr Institute, University of Copenhagen. "This includes, for example, the study of catalytic processes, of heat transfer at the solid–gas interface at the nanoscale, or of the construction of advanced nanostructures away from a surface where electron-beam lithography cannot be used."

"Our results show how to realize optical control of aerosol metallic nanoparticles and give the first hints on the physics involved, for instance by showing that the linear relation between laser power and trapping strength is reversed in a certain size regime," says Liselotte Jauffred, a postdoc in Oddershede's Optical Tweezer Group and first author of the paper that describes the work. "We also show how to obtain the positions visited by the metallic nanoparticle in the trap, and this information can be used to shed light on Brownian motion in air with temperature gradients."

The researchers' results were made possible because they carefully minimized turbulence in their custom-made trapping chamber and because they precisely counteracted spherical aberrations.

As they describe in their paper, they were able to demonstrate stable aerosol trapping of individual metallic nanoparticles with diameters from 80 to 200 nanometers and quantified the corresponding trapping strengths.

"This is exciting because it allows for the possibility to build nanostructures in air, away from any surfaces that might restrict the geometry of the structure or chemically affect the structure," notes Oddershede. "As metallic nanoparticles have plasmonic properties and as the thermal conductance of air is much lower than that of water, the heating of laser-trapped airborne metallic nanoparticles is significant and can easily exceed hundreds of degrees, maybe even a thousand degrees."

"This combined manipulation and heating of metallic nanoparticles makes it possible to perform aerotaxy, *i.e.* controlled growth of nanostructures by 'soldering' one nanoparticle to another in air," she adds.

Aerotaxy is expected to be a highly efficient method for mass-production of well-defined nanostructures that cannot be obtained by other means.

These results pave the way for fabricating nanoscale architectures in air, either by means of simply placing particles in a certain structure, or by means of aerotaxy.

Also, metallic nanoparticles are known to increase electromagnetic field strength and therefore can be used to amplify weak signals, for instance in different types of microscopy. With these results, electromagnetic signal amplification is also possible in air.

Furthermore, these findings can help in understanding the physical mechanisms behind aerotaxy. As the results provide the positions of the particle as a function of time, these time series shed light on how metallic nanoparticles move in air, and in particular for achieving a fundamental understanding of "hot Brownian motion", the science of how a particle moves in a thermal gradient.

Since it is not exactly known how much a trapped airborne metallic nanoparticle heats, it would not only be of scientific interest to measure the temperature profile of an irradiated airborne nanoparticle but also a requirement for potential technological applications, *e.g.* aerotaxy.

"It may be worth noticing," says Oddershede, "that the reason quantification of heating is not trivial—it cannot be directly theoretically predicted—is that the focal intensity distribution in air is highly aberrated at the nanoscale and the exact distribution *a priori* unknown. Hence, it is also of interest to map out the intensity distribution at the nanoscale of a focused laser beam in air."

The method could probably as well be used for optical manipulation of droplets, potentially from the atmosphere, with the goal of studying their size, composition, gas uptake, dynamics and coagulation. In addition, each droplet could be taken in a controlled fashion through a spectroscopic beam with the goal of analyzing the content.

The motivation for conducting this work came from Oddershede's previous work on optical manipulation of metallic nanoparticles in liquid[6] and from her work on how to minimize spherical aberration[7]—"which we knew would be crucial in order to realize aerosol trapping," she says.

The researchers conclude that the future scientific directions of the field of aerosol optical manipulation will probably go towards achieving control of other types of nanoparticles; controlling a larger range of particle sizes; and towards gaining control over particle orientation and possibly particle growth, too.

Featured scientists: Prof. Lene Broeng Oddershede's Optical Tweezers Group (http://tweezers.nbi.dk/)
Organization: Niels Bohr Institute, University of Copenhagen (Denmark)
Relevant publication: L. Jauffred, S. Taheri, R. Schmitt, H. Linke and L. Oddershede, Optical Trapping of Gold Nanoparticles in Air, *Nano Lett.*, 2015, **15**(7), 4713, DOI: 10.1021/acs.nanolett.5b01562

3.1.10   Plant Viruses Assist with Building Nanoscale Devices

Plant viruses have become the focus of intense research in the field of nanotechnology due to their promising applications as biotemplates for bottom-up nanofabrication. The tobacco mosaic virus (TMV) plays a special role: it was the first virus to be isolated and characterized, its structure is very simple, and it is chemically and physically unusually resilient. TMV is a classical model of self-assembled protein particles. The particle length is 300 nm and its coat proteins assemble to well-defined 18-nm-thick tubes with 4 nm-wide channels. TMV and other plant viruses have shown the capability to be filled and/or coated with metallic and magnetic materials to form wires or clusters.

"Perhaps the biggest advantage of plant viruses in nanotechnology is that we know exactly which chemical groups are located in which position," says José María Alonso, a former postdoctoral researcher at the Self-Assembly Group, led by Alexander Bittner, at CIC nanoGune in Donostia-San Sebastián, Spain (Alonso is now a researcher at the Laboratory of Organic Chemistry at Wageningen University). "We use the channel and outer surface for modifications with the aim of making new nanoscale devices such as conductive wires, magnetic tubes, and small containers for liquids."

In order to build nanoscale devices based on plant viruses, it is necessary to incorporate the virus with standard micro- and nanofabrication techniques—something that still remains a considerable challenge.

Bittner's group, together with researchers at CEMES-CNRS (Toulouse, France) have shown that TMV particles are compatible with electron beam lithography (EBL) processes and can be integrated in nanostructures made of positive and also of negative EBL tone resists.

"Two major challenges had to be taken into account when integrating TMV with EBL processes," Alonso explains. "First, although TMV tolerates many more solvents than other viruses, conventional EBL developers cause denaturation. This led to the investigation of alternative methods to remove the polymer masking layer based on less reactive solutions. The second challenge is the thermal sensitivity of the TMV. As a biological supramolecular assembly, TMV loses its structural integrity, with an apparent collapse to irregular particles, above 90 °C. Again, this makes TMV much more tolerant than most other viruses, but is still not sufficient for conventional processing in EBL."

The novelty in this fabrication technique is that the researchers used a lithography scheme that relies on extremely low processing temperatures of 50 °C—the temperature of the polymer resist used to cover TMV—and on development of tone resists with organic solvents that were chosen for their biocompatibility. As a result, viral particles maintain their biochemical functionality after all fabrication steps, which was verified through selective immunocoating of the TMV.

As a proof of concept to demonstrate the post-lithography biochemical functionality of TMV, the scientists performed selective immunocoating of the viral particles with primary and with secondary gold-labelled antibodies, and used immobilized TMV as a direct immunosensor.

Alonso explains that this concept can be broadened by coating TMV with other peptides or proteins, and especially with various epitopes (the part of an antigen that is recognized by the immune system)—in fact, plant viruses are excellent vehicles for vaccine production, when they are coated with relevant epitopes.

"We believe that our fabrication methods should work also for other types of sensitive materials that are incompatible with standard EBL processing, *e.g.* DNA, RNA, protein fibers/tubes, or soft polymers," he says. "Moreover, taking into account the dimensions of TMV, our structures are ideal templates to study nanofluidic events. Such 'virus nanofluidics' is in fact very much basic science, operating close to the ultimate (molecular) scale, *i.e.,* below 5 nm."

This work is an example of the first steps that nanotechnology researchers are taking to integrate nanobiostructures into typical solid-state device nanofabrication techniques.

Featured scientists: Prof. Alexander Bittner's Self-Assembly Group (http://nanogune.eu/self-assembly)
Organization: CIC Nanogune, Donostia—San Sebastian (Spain)
Relevant publication: J. Alonso, T. Ondarçuhu and A. Bittner, Integration of plant viruses in electron beam lithography nanostructures, *Nanotechnology*, 2013, 24(10), 105305, DOI: 10.1088/0957-4484/24/10/105305

3.1.11 Sculpting 3D Silicon Structures at the Single Nanometer Scale

As the semiconductor industry has shrunk the size of transistors, they have also had to shrink the size of the masks that define them. Defining these tiny masks has been one of the most difficult and expensive parts of making smaller and smaller transistors. A novel nanofabrication approach uniquely sidesteps this problem.

Typical methods to create structures at this size rely on "bottom-up", self-assembled techniques. Unfortunately, bottom-up techniques are notoriously hard to predict as they rely heavily on surface-energies and other parameters for assembly that are difficult to maintain from wafer to wafer.

Researchers at the Kavli Nanoscience Institute at California Institute of Technology, led by Axel Scherer, the Neches Professor of Electrical Engineering, Applied Physics and Physics, have come up with a novel method to three-dimensionally sculpt silicon nanostructures that is easily integratable with existing massively parallel fabrication.

According to the team, their novel sculpting method allows them to create structures without having to define etch masks at nanoscopic sizes—rather they can define larger masks and shrink the pattern with the etch.

"Our plasma etching method allows us to use the alignment tools afforded by 'top-down' lithography to reliably create identical structures across several wafers," says Sameer Walavalkar, a postdoctoral scholar in Scherer's Nanofabrication Group. "Furthermore, our method allows us to fabricate devices in the vertical direction letting us fit transistors onto a chip with previously unheard-of densities."

The etching process employed by the team is based on the mixed-mode "pseudo-Bosch" process that utilizes simultaneous etching and passivation.

"We have seen that this etch chemistry is not orientation dependent and from our testing we have found that it etches (100), (110), and (111) silicon wafers identically," says Walavalkar. "The etching is conducted in an inductively coupled plasma reactive ion etcher. The use of the inductively coupled plasma (ICP) allows for independent control of the degree of gas species ionization (ICP power) from the forward power that accelerates ions towards the sample. This gives a unique capacity to tune the physical and chemical character of the etch."

He points out that, by shaping silicon at this size scale, it becomes possible to unlock new regimes of behavior in silicon previously unseen outside certain laboratory-based experimental methods.

"For example," he says, "silicon is normally a 'dark' material as it has poor luminescence properties, but at this size scale its energy band structure is rearranged and we find that it becomes an efficient light emitter that can emit throughout portions of the visible and near-IR spectrum."

Furthermore, the researchers can create heterostructures, which have been traditionally created from two or more types of materials out of a single piece of silicon.

This opens the door for the creation of room temperature quantum devices that could, in the near future, be incorporated with existing planar silicon transistors. The team has termed this material manipulation through geometry *geometric bandgap engineering*.

Importantly, this novel etching technique is CMOS compatible and can be directly integrated with current CMOS fabrication technologies. And since it is possible to manipulate silicon with sub-10 nanometer precision and create fully suspended structures without the use of sacrificial layers, its application is basically limitless.

The team is currently fabricating and testing devices with features that are in the single-digit nanometer range.

"Once we have a better idea of how they function we can port them directly to the semiconductor industry," says Walavalkar. "This includes room-temperature quantum tunneling devices and silicon LEDs."

Featured scientists: Prof. Axel Scherer Research Group (http://nanofab. caltech.edu/)
Organization: Kavli Nanoscience Institute, California Institute of Technology, Pasadena, CA (USA)
Relevant publication: S. Walavalkar, A. Homyk, M. Henry and A. Scherer, Three-dimensional etching of silicon for the fabrication of low-dimensional and suspended devices, *Nanoscale*, 2013, 5(3), 927, DOI: 10.1039/C2NR32981F

3.1.12 Probing the Resolution Limits of Electron-Beam Lithography

The boundaries of electron beam lithography (EBL), the workhorse of current nanofabrication processes, is constantly being pushed further down into the single nanometer range by researchers' efforts to overcome the various limitations of EBL resolution—spot size, electron scattering, secondary-electron range, resist development, and mechanical stability of the resist.

A team of scientists, led by Karl K. Berggren, an associate professor of electrical engineering who heads the Quantum Nanostructures and Nanofabrication Group at MIT, has now achieved the EBL fabrication of 2 nm feature size and 10 nm periodic dense structures, which are the highest resolution patterns ever achieved with common resists.

The minimum feature size, 2 nm, is roughly 10 atoms wide, and with just a few atoms of standard deviation. The researchers calculated that this lithography system has the potential to provide even higher resolution by using an optimized resist and they expect this technique to impact a wide array of fields that strive for sub-5-nm patterning, such as excitonics, plasmonics, nano-optics, and molecular electronics (Figure 3.2).

"We investigated the resolution limits of EBL using an aberration-corrected scanning transmission electron microscope (STEM) as the exposure tool," explains Vitor R. Manfrinato, a PhD student in Berggren's group. "The STEM provides high-energy electrons (200 keV), which reduce primary electron scattering, and provide the smallest spot size available (0.15 nm). We also used an e-beam resist with the highest reported resolution available (hydrogen silsesquioxane; HSQ) and performed patterning metrology with transmission electron microscopy (TEM)."

"To our knowledge," adds Berggren, "this is the first EBL study using aberration-corrected STEM for EBL. In addition, the TEM metrology provides

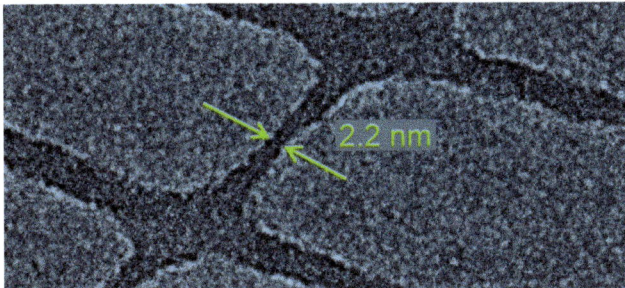

Figure 3.2   Top-down image obtained using a transmission electron microscope of a 2 nm feature (dark region) fabricated using 200 keV electrons in a scanning transmission electron microscope. (Image: Prof. Karl Berggren, MIT.)

an accurate method for quantifying the resolution limit. Furthermore, this study provides a good opportunity to investigate the energy loss mechanisms during the resist exposure."

As the researchers explain, these minimum features were only achieved by the combined use of subnanometer spot size, proper STEM stability, high-contrast development, high-resolution (and subnanometer line-width roughness) capabilities of HSQ resist, and subnanometer metrology obtained with TEM.

Manfrinato points out that the main challenges for sub-2-nm features and sub-10-nm periodic features are feature collapse due to capillary force during development, poor adhesion of the resist to the substrate, and mass-transport limitation during development.

"An optimized resist/development system should result in even higher resolution," he says.

Featured scientists: Prof. Karl K. Berggren's Quantum Nanostructures and Nanofabrication Group (http://www.rle.mit.edu/qnn/)
Organization: Massachusetts Institute of Technology, Cambridge, MA (USA)
Relevant publication: V. Manfrinato, L. Zhang, D. Su, H. Duan, R. Hobbs and E. Stach, *et al.*, Resolution Limits of Electron-Beam Lithography toward the Atomic Scale, *Nano Lett.*, 2013, **13**(4), 1555, DOI: 10.1021/nl304715p

3.1.13   Foldable Glass

Have you ever heard of foldable glass?

Exactly.

Glass is notorious for its brittleness. Although industry has developed ultra-thin (~0.1 mm), flexible glass (such as Corning's Willow® Glass) that can be bent for applications such as curved TVs and smartphone displays, fully foldable glass had not been demonstrated. Until now.

Usually, the thinner a substrate or device is, the more flexible it becomes. Folding can be considered an extreme case of bending. As we have seen in previous chapters, researchers have already made progress towards foldable electronics, origami-foldable solar cells, and foldable Li-ion batteries.

"The use of ultrathin substrates (usually with a thickness of 1–2 microns) is especially suitable for conformable, patch-like skin sensors or devices, whereas it is not so suitable for displays or other applications due to its mechanical instability," says Dahl-Young Khang, an associate professor in the Department of Materials Science and Engineering at Yonsei University. "Such ultrathin substrates—and devices fabricated on them—can easily be buckled or crumpled."

Khang and his group have demonstrated substrate platforms of glass and plastics, which can be reversibly and repeatedly folded at pre-designed location(s) without any mechanical failure or deterioration in device performances.

"We have engineered the substrates to have thinned parts on which the folding deformation should occur," says Moon Jong Han, at the time of this work a graduate student in Khang's lab. "This localizes the deformation strain on those thinned parts only."

He adds that this approach to engineering substrates has another advantage regarding device materials: "There is no need to adopt any novel materials such as nanowires, carbon nanotubes, graphene, *etc.* Rather, all the conventional materials that have been used for high-performance devices can be directly applied on our engineered substrates."

Intriguingly, even indium tin oxide (ITO), a very brittle transparent conducting oxide, can be used as an electrode on this novel foldable glass platform.

The design concept of this foldable platform comes from the methodology used in folding mats and hinge-less oriental folding screens, all of which commonly have pre-defined linear regions much thinner than the adjacent thick, rigid parts.

This selective thinning of substrates for foldability, contrary to ultra-thin substrates over the whole sample, is a novel approach. By applying this strategy, it has now become possible to fold even notoriously brittle glass substrates.

Furthermore, the whole sample can be folded twice (dual folding), folding along the x-direction first followed by another folding along the orthogonal y-direction, which reduces the sample size down to 1/4 of its original size.

The width of the thinned parts, the gap width, plays the key role in implementing dual foldability. The other key element is the asymmetric design of the gap width for the second folding.

The scientists suggest a simple design equation for the dual folding, which they found to agree very well with experimental data.

A dual fold might come especially handy for electronic devices that need portability and a large size at the same time. A large-area display, photovoltaics, sensors, or detectors can easily be transported in a small footprint by dual folding, then they can be expanded into a large size upon use.

"Although our approach relieves the mechanical instability a lot, it was not possible to completely overcome the instability by those thinned part(s) of the substrate," cautions Han. "This is a kind of fundamental dilemma. In other words, the foldability demands thin or ultrathin substrates, while mechanical stability requires the opposite: thick substrates. Although these contradicting requirements can be alleviated to some degree by devising the foldable platform, it is not possible to completely solve the contradiction."

Thus, in actual foldable hand-held devices, a proper foldable housing (the team has already filed a patent on a foldable housing design) would be necessary not only for mechanical stability but also for the integration of other elements required for electronic devices such as a battery, circuit board, *etc.*

Khang and his group are trying to apply this design principle to other devices, such as solar cells and sensors and, together with industrial partners, make prototype foldable devices.

Featured scientist: Prof. Dahl-Young Khang (http://mse.yonsei.ac.kr/xe/eng_faculty/entry/document_srl/732/sort_index/title/order_type/asc)
Organization: Department of Materials Science and Engineering, Yonsei University, Seoul (Republic of Korea)
Relevant publication: M. Han and D. Khang, Glass and Plastics Platforms for Foldable Electronics and Displays, *Adv. Mater.*, 2015, 27(34), 4969, DOI: 10.1002/adma.201501060

3.1.14   Plasmonic Biofoam Beats Conventional Plasmonic Surfaces

So far, most of the applications of plasmonic nanostructures have relied on solid substrates such as silicon, glass, plastic, or paper. These substrates offer a rather limited accessible surface area, thus severely limiting the volumetric density of the nanostructures.

To overcome these limitations, researchers have developed a novel 3D material with a high density of plasmonic nanostructures that are completely accessible.

"We have introduced an optically-active 3D material that can be treated as a frozen solution of metal nanostructures in which the nanostructures are completely accessible," says Dr Rajesh R. Naik of the Air Force Research Laboratory (AFRL). "We believe that this 3D plasmonic biofoam opens a lot of opportunities in using the unique optical properties of plasmonic nanostructures."

Naik, who is Chief Scientist, 711th Human Performance Wing, AFRL at Wright-Patterson Air Force Base in Dayton, Ohio, together with Professor Srikanth Singamaneni from the Washington University in St. Louis, and their teams, have demonstrated that the surface-enhanced Raman spectroscopy

(SERS) and photothermal performance of this novel 3D material is superior compared to that of conventional 2D plasmonic surfaces.

This optically active multifunctional material platform is based on biomaterial foam, which can be created from various biopolymers—in this case bacterial nanocellulose and regenerated silk fibroin. The nanocellulose material is processed into freeze-dried aerogel and then uniformly coated with a high density of plasmonic nanostructures (in this case gold nanorods).

"The highly open porous structure of the biomaterial foam combined with high volumetric density of nanostructures makes plasmonic foam a solid-state analogue of a highly concentrated solution of plasmonic nanostructures," notes Singamaneni.

This approach opens the door for using biocompatible materials as bioplasmonic foams that can be used in a variety of applications, mainly in biomedical and energy harvesting investigations.

Naik and his collaborators see three main application areas for the new material platform:

(1) Plasmonic biofoam, owing to the high density of plasmonic nanostructures, is ideally suited for chemical and biological sensing based on SERS;

(2) The large and tunable absorption of plasmonic nanostructures combined with high photothermal efficiency of these nanostructures make plasmonic biofoam an excellent candidate for solar energy harvesting and photothermal steam generation;

(3) The ability of loading and releasing cargo from these materials with external triggers, such as light, makes them excellent carriers for catalysts and drugs, for example optically triggered release of entrapped reagents in wound dressings.

In this work, the team has employed plasmonic nanostructures to achieve optically active biofoams. However, the functionality of the biofoams can be broadened with the incorporation of other functional nanomaterials such as graphene, carbon nanotubes and catalytically active nanomaterials.

"We would like to explore the use of other nanomaterials, making the aerogels more robust and creating other 3D hybrid functional materials using this approach," says Naik.

Also, in this study, the thickness of the plasmonic foam is higher than the light penetration depth. To minimize the amount of nanostructures employed, in the next step, the researchers will limit the nanostructures to the top portion of the aerogel while the bottom portion will serve as a support layer.

"We have employed gold nanorods as plasmonic materials in this study; considering that gold is an expensive material, it would be ideal to replace gold with other materials that are relatively inexpensive while preserving the unique optical properties of these functional aerogels," concludes Singamaneni. "A lot of effort has been made in recent years in developing inexpensive plasmonic materials. These developments would certainly broaden the applications of plasmonic materials and plasmonic biofoams."

Featured scientist: Dr Rajesh R. Naik (http://www.wpafb.af.mil/library/biographies/bio_print.asp?bioID=18809); Prof. Srikanth Singamaneni's Soft Nanomaterials Laboratory (http://softnano.wustl.edu/)
Organization: Air Force Research Laboratory, Dayton, OH; Dept. of Mechanical Engineering & Materials Science, Washington University in St. Louis, MO (USA)
Relevant publication: L. Tian, J. Luan, K. Liu, Q. Jiang, S. Tadepalli and M. Gupta, *et al.*, Plasmonic Biofoam: A Versatile Optically Active Material, *Nano Lett.*, 2016, **16**(1), 609, DOI: 10.1021/acs.nanolett.5b04320

3.1.15   Nanotechnology in a Bubble

Bubble-pen lithography (BPL) is a novel optically controlled nanofabrication technique that can be widely applied to pattern colloidal and biological particles on substrates in order to build functional optic, electronic, and magnetic devices.

"In BPL, an optically controlled microbubble is generated to capture and immobilize colloidal particles on plasmonic substrates through the coordinated actions of Marangoni convection, surface tension, gas pressure, and substrate adhesion in the substrate–bubble–solution system," explains Yuebing Zheng, Assistant Professor of Mechanical Engineering and Materials Science & Engineering at the University of Texas at Austin. "The irradiation of a plasmonic substrate with a focused laser beam at the plasmon resonance wavelength generates a microbubble at the substrate–solution interface. Due to the plasmon-enhanced photothermal effects, bubbles of variable sizes can be generated at a significantly reduced power."

With this new lithographic technique, the researchers can generate bubbles down to 1 μm in diameter, which are much smaller than the microbubbles (diameter in the range of 50–100 μm) commonly used in microfluidic devices for manipulation of particles. The smaller bubbles provide an enhanced patterning resolution.

The larger bubbles limited the patterning resolution while the high laser power that was used in previous techniques could change the particles' properties or even damage the particles.

"In BPL, both natural convection and Marangoni convection are responsible for the particle trapping at the microbubbles," explains Zheng. "The former is caused by the temperature gradient on the plasmonic substrate. The latter is induced by the surface-tension gradient along the microbubble surface. The convective flow drags the colloidal particles down to the plasmonic substrate and the in-plane drag force drives the particles towards the microbubble. The trapping occurs when a particle touches the microbubble."

The technique allows real-time and arbitrary patterning of colloidal particles at the micro- and nanoscale with simple and low-power optics. The low-power operation is crucial to retain the integrity of the patterned particles.

As a result, the patterned particles can lead to the development of functional devices with desired properties and, when biological particles are used, improve tissue engineering and cellular biology.

In their experiments, the team took advantage of gold nanoparticles as the plasmonic substrate to achieve continuous writing of the patterns of colloidal particles on the substrate by scanning the laser beam or translating the sample stage.[§]

Going forward, the team plans to achieve parallel patterning with multiple beams and microbubbles for high-throughput and high-volume production, and push the patterning resolution down to single nanoparticles and even single molecules.

"The challenges will be to achieve the ultimate resolution down to single-molecule level with non-invasive optical power and serve as a platform for fundamental research on colloidal nanoscience and life sciences using this lithographic technique," concludes Zheng.

Featured scientist: Prof. Yuebing Zheng's Research Group (http://zheng. engr.utexas.edu/)
Organization: Cockrell School of Engineering, The University of Texas at Austin, TX (USA)
Relevant publication: L. Lin, X. Peng, Z. Mao, W. Li, M. Yogeesh and B. Rajeeva, *et al.*, Bubble-Pen Lithography, *Nano Lett.*, 2016, **16**(1), 701, DOI: 10.1021/acs.nanolett.5b04524

3.1.16 Self-Assembly Machines—A Vision for the Future of Manufacturing

Industrial production processes, by and large, rely on robotic assembly lines that place, package, and connect a variety of disparate components. While the manufacturing world is dominated by robots, there are applications where the established processes of serial "pick-and-place" and manipulation of single objects reach scaling limits in terms of throughput, alignment precision, and the minimal component dimension they can handle effectively.

By contrast, the emerging methods of engineered self-assembly are massively parallel and have the potential to overcome these scaling limitations.

"So far, researchers have focused on small-area proof-of-concept demonstrations testing new self-assembly concepts—for example, testing a different driving force, agitation, or component handling mechanism—inside a small container to define the environment where self-assembly takes place,"

[§]This movie shows the real-time continuous writing process: https://youtu.be/9PS6ay3sHlU.

says Heiko Jacobs, a professor at Technische Universität Ilmenau and Chair of the university's Nanotechnology Group. "Over the past 10 years, tremendous knowledge has been gained. However, for many observers it lacks the proof that any of the processes can be scaled to enable high-volume production of real devices and products as promised at the outset."

"Personally, I am convinced that this is possible and we are now at a stage where we have gained sufficient experience that will allow us to put the knowledge together to begin building 'self-assembly machines'," he continues.

Jacobs's team, together with researchers from the University of Minnesota, has described the first example of a device to assemble and distribute LEDs over large areas for solid-state lighting applications.

It provides the blueprints and operational parameters of a reel-to-reel (RTR) fluidic self-assembly platform to assemble and electrically connect semiconductor chips with a yield clearly exceeding a 99% benchmark set by robotic pick-and-place machines.

This machine is based on surface-tension-directed self-assembly. As the team points out, their assembly process is no longer a discontinuous small-batch hand-operated process but resembles an automated machine-like process involving a conveyer belt and a RTR type assembly approach with automated agitation.

The researchers' current design achieves 15 000 chips per hour using a 2.5 cm wide assembly region. Jacobs notes that scaling to 150 000 chips per hour would be possible using a 25 cm wide web, which would be a factor of 20 better than one of the faster chip pick-and-place machines in use today in the semiconductor industry.

"In principle, scaling to any throughput should be possible considering the parallel nature of self-assembly and the fact that there is no requirement for any pick-and-place operations," he says. "This means that it is fairly straightforward to scale this technology to any desired assembly rate by increasing the substrate width and number of nozzles."

Under optimized operational conditions, the group achieved an assembly yield of 99.8% using their self-assembly process.

The team says that their approach can be extended using shape recognition concepts to enable unique angular alignment and contact pad registration or using sequential batch assembly processes to assemble more than one component type on the substrate if desired. Moreover, it is possible to transfer the chip onto other flexible or stretchable substrates.

As a demonstration, the researchers applied their self-assembly machine to the realization of area lighting panels incorporating distributed inorganic light emitting diodes (GaN LEDs). The lighting modules are fabricated in three steps: (1) assembly of LEDs on the bottom electrode using the RTR system; (2) isolation and passivation of assembled LEDs with a UV curable polymer; and (3) lamination of the top conductive layer to complete the electrical connections.

Although this is still basic research, Jacobs and his collaborators have set their eyes on the development of industrial-scale "self-assembly machines" that would allow the assembly and electrical interconnection of semiconductor chips over wide area substrates in large quantities with high speed and high precision.

"I am convinced that processes of self-assembly will eventually be used in the manufacturing industry of the future," says Jacobs. "Today it might appear as science fiction to some critical observers. However, researchers worldwide, including my group, continue to work out the details."

Featured scientists: Prof. Heiko Jacobs
Organization: Institut für Mikro- und Nanoelektronik, Technische Universität Ilmenau (Germany) (http://www.tu-ilmenau.de/mne-nano/)
Relevant publication: S. Park, J. Fang, S. Biswas, M. Mozafari, T. Stauden and H. Jacobs, A First Implementation of an Automated Reel-to-Reel Fluidic Self-Assembly Machine, *Adv. Mater.*, 2014, **26**(34), 5942, DOI: 10.1002/adma.201401573

3.2   Nanotechnology and 3D Printing

Fabrication of three-dimensional objects through direct deposition of functional materials—also called *additive manufacturing*—has been a subject of intense study in the area of macroscale manufacturing for several decades. This fabrication technology is becoming a viable alternative to conventional manufacturing processes in an increasing number of applications ranging from children's toys to cars, fashion, architecture, military, biomedical science, and aerospace, just to name a few.

3D printing techniques are reaching a stage where desired products and structures can be made independent of the complexity of their shapes—even bioprinting body tissues is now in the realm of the possible.

Applying 3D printing concepts to nanotechnology could bring similar advantages to nanofabrication—speed, less waste, economic viability—that they are expected to bring to manufacturing technologies.

In addition, pre-patterned micro- or nanostructures could be used as substrates, allowing researchers to realize unprecedented manufacturing flexibility, functionality and complexity at the nanoscale.

3.2.1   Getting Closer to 3D Nanoprinting

Researchers in Korea have shown that nanoscale 3D objects such as free-standing nanowalls can be constructed by an additive manufacturing scheme. Even without the motion of the substrate, nanojets are spontaneously laid down and piled to yield nanowalls.

"Electrospinning that produces polymer nanojets is a relatively simple and inexpensive method to yield nanoscale fibers, but the fiber streams are so chaotic that control of individual fibers has been considered almost impossible," explains Ho-Young Kim, a professor at Seoul National University. "In our work, we show that an electrospun polymer solution jet, which tends to become unstable as traveling in the air due to Coulombic repulsion, can be stably focused onto a thin metal electrode line."

Kim and his team also elucidated the fundamental electromechanical mechanism that enables the spontaneous stacking of a nanofiber onto itself to provide a physical basis behind this novel nanofabrication process.

In this method, a thin metal line on an insulating plate strongly focuses the electrical field, thus the whipping instability of the electrical nanojets is suppressed.

To stack the fibers in a controlled fashion, the researchers manipulate the fiber deposit into attracting the incoming nanojets rather than repelling them by draining the electrical charge quickly. Then they get a nanowall that lines the ground, implying that various free-standing structures can be created by patterning the microscale ground lines in a desired shape.

The construction of a free-standing nanowall is the most fundamental step to achieve 3D nanoprinting.

This process is so attractive because it needs only a power supply and a linear stage to build free-standing nanowalls after drawing metal microlines, all of which can be conducted under normal laboratory conditions.

Kim points out that this technique has a significant economic advantage as compared to conventional nanomanufacturing processes used to build nanowalls, such as DRIE (deep reactive ion etching).

The current scheme of repeatedly stacking nanofibers onto a conducting line is suited for fabricating nanoelectrodes consisting of straight walls and nanochannel field effect transistors (FETs) utilizing insulating nanowalls as gaps of metal patterns.

Further sophisticated methods to control the nanojets have the potential to realize rapid 3D printing of complicated shapes, which can be used for bio-scaffolds, nanofilters and even nanorobots.

However, further developments, such as lowering the nanojet speed and positioning the target place with high precision, are necessary to make the current focused electrojetting process fully capable of 3D printing of complicated shapes.

"Full 3D control of an electrospun nanojet would possibly revolutionize the current nanofabrication technology, which we aim to achieve in the long run," says Kim. "However, we believe that such great achievement cannot be made with a single step. Further development for the precise control of the nanojet could realize full 3D nanofabrication."

Featured scientist: Prof. Ho-Young Kim's Micro Fluid Mechanics Laboratory (http://fluids.snu.ac.kr/)
Organization: Department of Mechanical and Aerospace Engineering, Seoul National University (Republic of Korea)
Relevant publication: M. Lee and H. Kim, Toward Nanoscale Three-Dimensional Printing: Nanowalls Built of Electrospun Nanofibers, *Langmuir*, 2014, **30**(5), 1210, DOI: 10.1021/la404704z

3.2.2   The Emergence of 3D-Printed Nanostructures

Graphene has received a great deal of attention for its promising potential applications in electronics, biomedical and energy storage devices, sensors and other cutting-edge technological fields, mainly because of its fascinating properties such as an extremely high electron mobility, a good thermal conductivity and a high elasticity (read more about graphene in Chapter 4).

The successful implementation of graphene-based devices invariably requires the precise patterning of graphene sheets at both the micrometer and nanometer scale. Finding the ideal technique to achieve the desired graphene patterning remains a major challenge.

There are different methods to build 3D graphene monoliths—for example freeze casting or emulsion templating—but they are limited to building simple shapes such as cylinders or cubes.

For the first time, a research team led by Professor Seung Kwon Seol from Korea Electrotechnology Research Institute (KERI), has demonstrated 3D printed nanostructures composed entirely of graphene using a new 3D printing technique.

"We developed a nanoscale 3D printing approach that exploits a size-controllable liquid meniscus to fabricate 3D reduced graphene oxide (rGO) nanowires," Seol explains. "Different from typical 3D printing approaches that use filaments or powders as printing materials, our method uses the stretched liquid meniscus of ink. This enables us to realize finer printed structures than a nozzle aperture, resulting in the manufacturing of nanostructures."

The researchers note that their novel solution-based approach is quite effective in 3D printing of graphene nanostructures as well as in multiple-materials 3D nanoprinting.

"We are convinced that this approach will present a new paradigm for implementing 3D patterns in printed electronics," says Seol.

For their technique, the team grew graphene oxide (GO) wires at room temperature using the meniscus formed at the tip of a micropipette filled with a colloidal dispersion of GO sheets, then reduced it by thermal or chemical treatment (with hydrazine).

The deposition of GO was obtained by pulling the micropipette as the solvent rapidly evaporated, thus enabling the growth of GO wires. The researchers were able to accurately control the radius of the rGO wires by tuning the pulling rate of the pipette; they managed to reach a minimum value of 150 nm.

Using this technique, they were able to produce arrays of different freestanding rGO architectures, grown directly at chosen sites and in different directions: straight wires, bridges, suspended junctions, and woven structures.

"So far, to the best of our knowledge, nobody has reported 3D printed nanostructures composed entirely of graphene," says Seol. "Several results reported the 3D printing (millimeter- or centimeter-scale) of graphene or carbon nanotube/plastic composite materials by using a conventional 3D printer. In such composite system, the graphene (or carbon nanotube) plays an important role for improving the properties of plastic materials currently used in 3D printers. However, the plastic materials used for producing the composite structures deteriorate the intrinsic properties of graphene."

He points out that this 3D nanoprinting approach can be used for manufacturing 2D patterns and 3D architectures in diverse devices such as printed circuit boards, transistors, light emitting devices, solar cells, sensors and so on. Reducing the 3D printable size to below 10 nm and increasing the production yield remain challenges, though.

Featured scientist: Prof. Seung Kwon Seol
Organization: Korea Electrotechnology Research Institute (KERI), Seong-san-gu (Republic of Korea) (http://www.keri.re.kr/html/en/)
Relevant publication: J. Kim, W. Chang, D. Kim, J. Yang, J. Han and G. Lee, *et al.*, 3D Printing of Reduced Graphene Oxide Nanowires, *Adv. Mater.*, 2015, 27(1), 157, DOI: 10.1002/adma.201404380

3.2.3   Printing in Three Dimensions with Graphene

Using a different approach to graphene 3D printing, researchers have used flakes of chemically modified graphene together with very small amounts of a responsive polymer (a polymer that changes behavior and conformation when a "chemical switch" is activated), to formulate water-based ink or pastes.

"Our formulations have the flow and physical properties we need for the filament deposition process required in 3D printing: they need to flow through very small nozzles and set immediately after passing through them, retaining the shape and holding the layers on top," says Dr Esther García-Tuñon, a research associate at the Centre for Advanced Structural Ceramics at Imperial College London (ICL). "We use this two-dimensional material as a building block to create macroscopic 3D structures and a technique called direct ink writing (DIW), also known as direct write assembly (DWA), or Robocasting."

García-Tuñon is first author of a paper where a team of scientists from ICL, the University of Warwick, the University of Bath, and the Universidad de Santiago de Compostela, describe their technique.

This technique is based on the continuous deposition of a filament following a computer design. The 3D structures are built layer by layer from bottom to top.

"Our inks allow printing through nozzles as thin as 100 μm and their rheology could also be tailored for other processing technologies such as extrusion, gel, or tape casting," García-Tuñon points out. "Our goal was to print graphene structures (not composites) using small amounts of additives and water-based systems. In this way it could be easily scaled up in a manufacturing process."

Graphene is very hydrophobic so it is not possible to formulate a water-based ink directly. The researchers therefore used chemically modified graphene—also known as graphene oxide—instead. Graphene oxide can be processed in water to build the desired architectures.

Once the structure is made, it is thermally treated in a special atmosphere to recover the properties of graphene.

The mechanical stability of the printed parts may allow the use of additional treatments (*e.g.*, chemical or electrochemical reduction) to further manipulate the properties without compromising the structure.

The team is working on developing new formulations as well as the development of specific applications for example in flexible electronics and oil adsorption.

"I think there are still many challenges to overcome in both additive manufacturing and graphene technologies," notes García-Tuñon. "The buzzword 3D printing is now everywhere; we can find many examples of commercially available 3D printers to make your own Hello Kitty iPhone cases, and all sort of plastic models. But there is still a long way to go from here to the use of 3D printing for a wide variety of materials in multicomponent and practical devices."

Featured scientist: Dr Esther García-Tuñon (http://www.imperial.ac.uk/people/esther.garcia-tunon)
Organization: Centre for Advanced Structural Ceramics, Imperial College London (UK)
Relevant publication: E. García-Tuñon, S. Barg, J. Franco, R. Bell, S. Eslava and E. D'Elia, *et al.*, Printing in Three Dimensions with Graphene, *Adv. Mater.*, 2015, 27(10), 1688, DOI: 10.1002/adma.201405046

3.2.4 Fully 3D-Printed Quantum Dot LEDs to Fit a Contact Lens

To date, the 3D printing of electronic components has been limited to the printing of batteries, strain sensors, interdigitated-electrode capacitors and passive metallic structures such as interconnects and antennas on surfaces.

The ability to directly and seamlessly incorporate materials with a range of diverse functionalities with 3D printing is particularly attractive as it could allow the simultaneous, comprehensive, and direct printing of structural, biological, and electronic materials that capture the complete spectra of material properties.

The free-form generation of active electronics in unique architectures, which transcend the planarity inherent to conventional microfabrication techniques, has been an area of increasing scientific interest. Yet, attaining seamless interweaving of electronics is challenging due to the inherent material incompatibilities and geometrical constraints of traditional micro-fabrication processing techniques.

At the fundamental level, 3D printing should be entirely capable of creating spatially heterogeneous multi-material structures by dispensing a wide range of material classes with disparate viscosities and functionalities, including semiconducting colloidal nanomaterials, elastomeric matrices, organic polymers, and liquid and solid metals.

"The big push in 3D printing these days is to try to print two or more polymers at once," says Michael McAlpine, at the time of this work an assistant professor of mechanical and aerospace engineering at Princeton University, now a professor in the Mechanical Engineering Department at the University of Minnesota. "In our research, we go way beyond that. We show that we can print interwoven structures of quantum dots, polymers, metal nanoparticles, *etc.*, to create the first fully 3D printed LEDs, in which every component is 3D printed."

This demonstration represents a proof of concept in combining active nanoelectronic components with the versatility of 3D printing, which enables the three-dimensional free-form fabrication of active electronics.

"Using this approach, we can create unique structures, such as 2 × 2 × 2 arrays of LEDs, in which the electrical wiring runs horizontal and vertical, to create a multi-color 3D stack of LEDs," notes Yong Lin Kong, a graduate student in McAlpine's group who led this project. "We also use 3D scanning to carefully scan a contact lens and store the specific topology of that lens, and then alter our 3D printing to adjust to that topology, allowing us to conformally 3D print LEDs on a contact lens. This may find future uses in electronic contact lenses or bionic eye applications in the future."

"This work outlines an exciting breakthrough that enables the direct printing of functional, embedded, active 3D nanoelectronics using only a 3D printer," he adds. "Indeed, this is the first time to our knowledge that semiconducting nanoparticles have been 3D printed, and the first time that such a broad array of diverse functional materials has been fully interwoven entirely using a 3D printer."

The team's approach consists of three key steps. First, it identifies electrodes, semiconductors, and polymers that possess desired functionalities and exist in printable formats. Next, great care is taken to ensure that these materials are dissolved in orthogonal solvents so as not to compromise the integrity of underlying layers during the layer-by-layer printing process. Finally, the interwoven patterning of these materials is achieved *via* direct dispensing in a CAD-designed construct.

As a proof of concept of this approach, the researchers demonstrate the 3D printing of quantum dot light-emitting diodes (QLEDs), which involves the design, integration and printing of five classes of materials with distinct material properties.

"Specifically, we demonstrate the seamless interweaving of (1) emissive semiconducting inorganic nanoparticles; (2) an elastomeric matrix; (3) organic polymers as charge transport layers; (4) solid and liquid metal leads; and (5) an UV-adhesive transparent substrate layer," explains Kong. "The printed QLEDs exhibit excellent performance characteristics. The combination of 3D scanning and 3D printing allows for the direct printing of active functional electronics onto the precise topology of a non-flat object (Figure 3.3)."

He points out that this approach allows for the free-form fabrication of multi-dimensional nanoelectronics within a complex, interwoven architecture such as a 3D array of embedded QLEDs.

The QLEDs printed by McAlpine's team capture the unique properties of quantum dots: tunable and pure color emission. Further, combining a complementary 3D light-scanning technique with this approach allows for the fabrication of electronics topographically tailored to curvilinear surfaces.

"We anticipate that this general strategy can be expanded to 3D printing other classes of active devices, such as MEMS devices, transistors, solar cells, and photodiodes," says McAlpine. "Our results suggest a number of exciting applications, including the generation of geometrically tailored devices containing LEDs and multimodal sensors to provide a new tool for optogenetics for studying neural circuitry."

Co-printing of active electronics with biological constructs could also lead to new bionic devices, such as prosthetic implants that optically stimulate nerve cells.

According to the team, future work will address a number of key challenges. These include: (1) increasing the resolution of the 3D printer so that smaller devices can be printed; (2) improving the performance and yield of the printed devices; and (3) incorporating other classes of nanoscale functional building blocks and devices, including semiconductor, plasmonic, and ferroelectric.

Featured scientists: Prof. Michael McAlpine Research Group (http://www.mcalpineresearchgroup.com/)

Organization: Department of Mechanical and Aerospace Engineering, Princeton University (USA)

Relevant publication: Y. Kong, I. Tamargo, H. Kim, B. Johnson, M. Gupta and T. Koh, *et al.*, 3D Printed Quantum Dot Light-Emitting Diodes, *Nano Lett.*, 2014, **14**(12), 7017, DOI: 10.1021/nl5033292

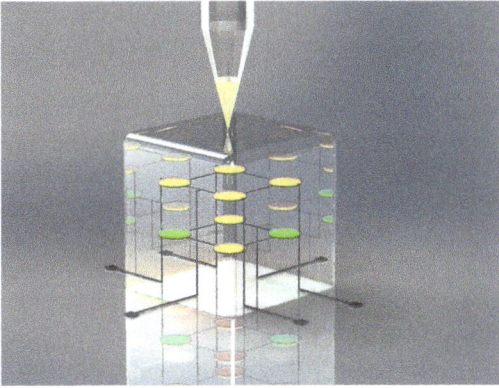

Figure 3.3   3D printed 2 × 2 × 2 multidimensional array of embedded QLEDs. Layout of the multi-color 3D QD-LED array design. (Image: McAlpine Group.)

3.2.5   3D-Printed Programmable Release Capsules

McAlpine's group has also developed a novel 3D-printing-based method to produce highly monodisperse core/shell capsules that can be loaded with biomolecules such as therapeutic drugs.

"Our method provides us with robust control over particle properties, passive release kinetics, and particle distributions throughout a 3D matrix," says McAlpine. "Furthermore, we render these capsules stimuli-responsive by incorporating gold nanorods into the polymer shell, allowing for highly selective photothermal rupture and triggered temporal release of the biomolecular payload."

McAlpine and his group have worked with researchers from Washington University in St. Louis to develop the use of 3D printing to hierarchically order polymers, biomolecules, and nanomaterials into hybrid functional materials in a scalable manner.

"Our technique is enabled *via* the 3D printing of multiplexed arrays of biomolecule-loaded capsules, along with tunable and orthogonal laser-triggered rupture and release of active biomolecules," Maneesh K. Gupta, a former member of McAlpine's group and now a research scientist at Air Force Research Laboratories, explains. "One can imagine filling the capsules with molecules such as drugs, nucleic acids, enzymes, growth factors, cell markers and other functional proteins, into a hydrogel ambient."

Fanben Meng, a postdoc researcher in McAlpine's lab, adds: "The advantages of our 3D printing-based method include: (1) highly monodisperse capsules; (2) efficient encapsulation of biomolecular payloads; (3) precise spatial patterning of capsule arrays: (4) 'on the fly' programmable reconfiguration of gradients; and (5) versatility for incorporation in hierarchical architectures."

An interesting feature of these capsules is the control of the spatial and temporal release of payloads. This is achieved by incorporating gold nanorods in the shell. Using specific wavelengths determined by the length of the nanorods, the rods can be heated, which subsequently ruptures the shell.

The 3D printing technique itself also provides precise control over the capsule volumes and architectures.

"This work provides a promising solution to generating multiplexed spatio-temporal molecular gradients in 3D architectures, which is significant to mimic the dynamic microenvironment surrounding cells in natural tissues, as living organisms guide tissue development through highly orchestrated gradients of biomolecules that direct cell growth, migration, and differentiation in 3D matrices," notes Meng (Figure 3.4).

The researchers expect that this platform of 3D printed programmable-release capsules will be useful in applications such as dynamic tissue

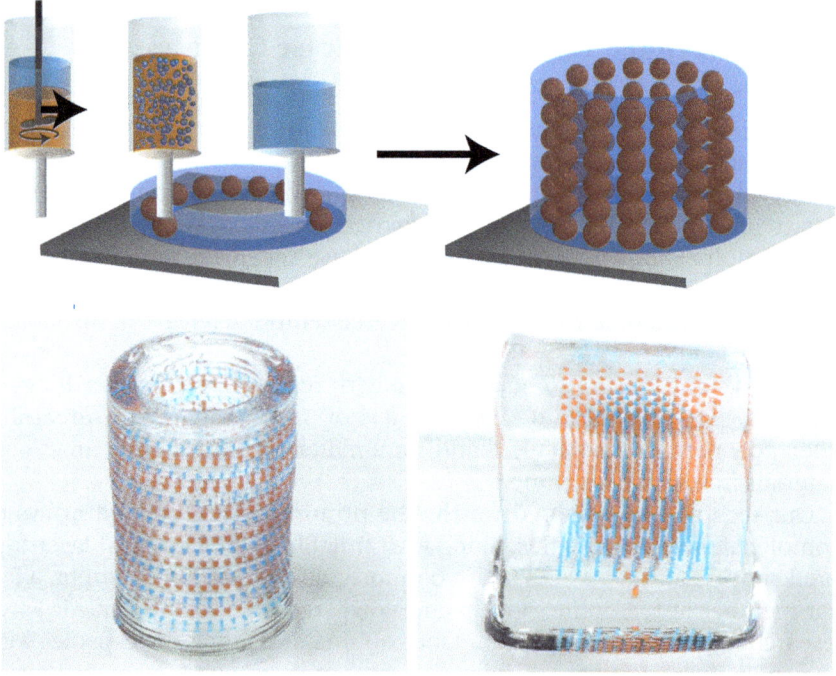

Figure 3.4   3D printing of hierarchically multiplexed capsule arrays. Top row: schematic illustrating an emulsion ink-based method to 3D print complex capsule arrays. The emulsion ink is prepared by directly dispersing the aqueous core in the PLGA solution. The hydrogel and emulsion inks are sequentially printed in a layer-by-layer manner to form a 3D structure. Bottom row: optical images of 3D multiplexed capsule arrays directly printed in cylindrical and square hydrogel matrices, respectively (colors of the capsules are from food dyes in the dispersed cores). (Image: McAlpine Research Group, University of Minnesota.)

engineering, 3D printed drug delivery systems, synthetic/artificial tissues, programmable matter, and bionic nanosystems.

Beyond this, another important application area could be combinatorial screening of biomolecular gradients—drugs, toxins, pollutants, *etc.*—against cell types.

A particularly far-reaching example would be to imagine having a collection of stem cells, which could be triggered using a red laser to develop into a heart, or with a green laser into a liver.

"Our work was motivated by the fact that living systems utilize exquisite control of biomolecular gradients to control cell fate and ultimately enable complex functional tissues," says McAlpine. "We believe that replicating such control is a key to many future advances in bioengineering."

"There has been tremendous prior work on utilizing microfluidics, particle encapsulation, and stimuli-responsive materials to address some of these challenges," he adds. "Our findings offer a novel perspective by offering a 3D printing based approach to solve these challenges, which has advantages in precision control over volumes, spatial distributions, and diversity of materials including nanomaterials and biomolecules."

One of the next stages of the investigation requires studies regarding the biocompatibility and feasibility of using 3D printed capsules to control cell fate, from the individual cell level to the level of tissue engineering.

As the team points out, beyond biological studies, there is a tremendous amount of work to be done to enhance the material and functional properties of the capsules. For example, improving the resolution and spatial alignment of these printed capsules; developing stimuli-responsive shells with reversible payload release properties; and developing other approaches to stimulate the capsules, such as using electrical signals rather than light triggering.

Featured scientists: Prof. Michael McAlpine Research Group (http://www.mcalpineresearchgroup.com/)
Organization: Mechanical Engineering Department, University of Minnesota, Minneapolis, MN (USA)
Relevant publication: M. Gupta, F. Meng, B. Johnson, Y. Kong, L. Tian and Y. Yeh, *et al.*, 3D Printed Programmable Release Capsules, *Nano Lett.*, 2015, **15**(8), 5321, DOI: 10.1021/acs.nanolett.5b01688

3.2.6 Embedded 3D-Printing for Soft Robotics Fabrication

The field of robotics has come a long way from its early days of bulky mechanical machines used on industrial assembly lines. As robots penetrate the physical world more and more, engineers have looked for ways to make soft robotic systems safer for human interaction. Without bulky, breakable hardware, soft robots also might be able to explore hard-to-reach spaces and traverse bumpy terrain.

A significant challenge in soft robotics involves fabricating soft sensors and actuators, which, so far, have been very tedious to produce. Building soft sensors used by roboticists usually requires a multi-step, manual molding/lamination/sealing/infilling process.

As a result, the design and fabrication process is cumbersome; the sensor form factors are unnecessarily limited; and there are issues with mechanical robustness.

A research team led by professors Jennifer A. Lewis and Robert Wood from the School of Engineering and Applied Sciences and the Wyss Institute for Biologically Inspired Engineering at Harvard University, has demonstrated a new method for creating highly stretchable sensors based on embedded 3D printing of a carbon-based resistive ink within an elastomeric matrix (for which they coined the term *e-3DP*).

"By implementing e-3DP, we can vastly accelerate the design and implementation of soft sensors," notes Lewis. "We believe that this manufacturing technique has potential application in many fields. Ideally, with appropriate material development, e-3DP could be implemented to create embedded electronic devices that are currently made *via* layer-by-layer processing and where material interfaces limit their performance. Specifically, soft robots, wearable electronics, and soft actuators are the most immediate applications. In the future, we plan to extend e-3DP to create novel 3D microfluidic devices and well beyond."

To enable e-3DP, the team developed a multicomponent materials system composed of an ink, reservoir, and filler fluid. These constituents are tailored to exhibit the desired rheological properties required to maintain high-fidelity geometries throughout the embedded printing and curing process.

To pattern the sensing elements of the printed 3D devices, the researchers used carbon conductive grease, an off-the-shelf, inexpensive, environmentally benign suspension of carbon black particles in silicone oil as the functional ink.

"Several criteria must be fulfilled when designing an appropriate supporting reservoir and filler fluid for e-3DP," explains Lewis. "First, the reservoir must facilitate patterning the desired ink filaments without breakup. Second, any defects that arise during printing as the nozzle translates within the reservoir must be rapidly healed by incorporation of the filler fluid. Finally, after printing the embedded sensing elements, the infilled reservoir must be transformed by curing into a monolithic, highly extensible, conformal elastomer matrix."

To meet these requirements, the team modified a commercially available silicone elastomer to create both the reservoir and filler fluid.

One of the key results of this work is the demonstrated ability to pattern chemically dissimilar materials within each other. Specifically, the deposition of a conductive ink into an elastomeric reservoir that serves as the sensing element and matrix, respectively, for the printed soft sensors.

While attention has focused on 3D printing of rigid materials such as plastics and metals, this work demonstrates the printing of soft materials, which opens up myriad applications—including soft robots, actuators, and wearable sensors.

The researchers note that their e-3DP method significantly simplifies the fabrication of soft sensors. Soft sensors have been the subject of much research for their potential to interface with the human body. However, problems such as limited extensibility, high cost, poor durability, or lack of manufacturing scalability have prevented their wide-spread adoption.

"e-3DP allows sensors to be created in a highly programmable and seamless manner," Lewis points out. "Print path control coupled with the supporting reservoir enables arbitrary geometries in planar and 3D motifs to be created, while interfaces between layers are eliminated—significantly improving device reliability."

Going forward, the team will address two of the currently most pressing challenges of the e-3DP process: further improving the conductive ink performance, and improving interconnects to rigid components.

Featured scientists: Prof. Jennifer A. Lewis (http://lewisgroup.seas.harvard.edu/), Prof. Robert Wood's Harvard Microrobotics Lab (http://micro.seas.harvard.edu/)
Organizations: School of Engineering and Applied Sciences and the Wyss Institute for Biologically Inspired Engineering at Harvard University, Cambridge, MA (USA)
Relevant publication: J. Muth, D. Vogt, R. Truby, Y. Mengüç, D. Kolesky and R. Wood, *et al.*, Embedded 3D Printing of Strain Sensors within Highly Stretchable Elastomers, *Adv. Mater.*, 2014, **26**(36), 6307, DOI: 10.1002/adma.201400334

References

1. L. Wang, F. Li, M. Kuang, M. Gao, J. Wang and Y. Huang, *et al.*, Interface Manipulation for Printing Three-Dimensional Microstructures Under Magnetic Guiding, *Small*, 2015, **11**(16), 1900, DOI: 10.1002/smll.201403355.
2. S. van Reenen, P. Matyba, A. Dzwilewski, R. Janssen, L. Edman and M. Kemerink, A Unifying Model for the Operation of Light-Emitting Electrochemical Cells, *J. Am. Chem. Soc.*, 2010, **132**(39), 13776, DOI: 10.1021/ja1045555.
3. H. Keum, A. Carlson, H. Ning, A. Mihi, J. Eisenhaure and P. Braun, *et al.*, Silicon micro-masonry using elastomeric stamps for three-dimensional microfabrication, *J Micromech. Microeng.*, 2012, **22**(5), 055018, DOI: 10.1088/0960-1317/22/5/055018.
4. J. Lin, S. Wang, P. Huang, Z. Wang, S. Chen and G. Niu, *et al.*, Photosensitizer-Loaded Gold Vesicles with Strong Plasmonic Coupling Effect for Imaging-Guided Photothermal/Photodynamic Therapy, *ACS Nano*, 2013, **27**(6), 5320, DOI: 10.1021/nn4011686.

5. See the original 1970 paper: A. Ashkin, Acceleration and Trapping of Particles by Radiation Pressure, *Phys. Rev. Lett.*, 1970, **24**(4), 156, DOI: 10.1103/PhysRevLett.24.156.
6. L. Bosanac, T. Aabo, P. Bendix and L. Oddershede, Efficient Optical Trapping and Visualization of Silver Nanoparticles, *Nano Lett.*, 2008, **8**(5), 1486, DOI: 10.1021/nl080490+.
7. S. Reihani and L. Oddershede, Optimizing immersion media refractive index improves optical trapping by compensating spherical aberrations, *Opt. Lett.*, 2007, **32**(14), 1998, DOI: 10.1364/OL.32.001998.

CHAPTER 4

The Future is Flat—Two-Dimensional Nanomaterials

The bulk properties of materials often change dramatically with nanoscale ingredients. Composites made from particles of nano-size ceramics or metals smaller than 100 nanometers can suddenly become much stronger than predicted by existing materials-science models. For example, metals with a so-called grain size of around 10 nanometers are as much as seven times harder and tougher than their ordinary counterparts with grain sizes in the hundreds of nanometers. The causes of these drastic changes stem from the weird world of quantum physics. The bulk properties of any material are merely the average of all the quantum forces affecting all the atoms. As you make things smaller and smaller, you eventually reach a point where the averaging no longer works.

The properties of materials can be different at the nanoscale for two main reasons:

First, nanomaterials have a relatively larger surface area when compared to the same mass of material produced in a larger form. This can make materials more chemically reactive (in some cases materials that are inert in their larger form are reactive when produced in their nanoscale form), and affect their strength or electrical properties.

Second, quantum effects can begin to dominate the behavior of matter at the nanoscale—particularly at the lower end—affecting the optical, electrical and magnetic behavior of materials. Materials can be produced that are nanoscale in one dimension (for example, very thin surface coatings), in two dimensions (for example, nanowires and nanotubes) or in all three dimensions (for example, nanoparticles such as quantum dots).

Nanotechnology: The Future is Tiny
By Michael Berger
© Michael Berger 2016
Published by the Royal Society of Chemistry, www.rsc.org

Nanomaterials fall into two categories: the nanoscale version of bulk materials (titanium, gold, ceramics...) that exhibit new properties and allow new functionalities. And secondly, entirely new material structures that were discovered or synthesized over the past 30 years. These new materials include carbon nanotubes, fullerenes, quantum dots, and a new class of single-atomic-layer materials such as graphene.

In this section we will not look at "conventional" nanomaterials but at two new classes of novel materials with exciting properties and promising applications such as invisibility cloaks: two-dimensional materials and metamaterials.

Today, all these materials are being synthesized in laboratories and researchers are beginning to find ways to integrate them into devices and make them practical enough to find their way into industrial manufacturing processes.

4.1   Graphene

Carbon comes in many different forms, from the graphite found in pencils to the world's most expensive diamonds. In 1980, we knew of only three basic forms of carbon, namely diamond, graphite, and amorphous carbon. Then, fullerenes and carbon nanotubes were discovered and, in 2004, graphene joined the club. Graphene is an atomic-scale honeycomb lattice made of carbon atoms.

Existing forms of carbon basically consist of sheets of graphene, either bonded on top of each other to form a solid material like the graphite in your pencil, or rolled up into carbon nanotubes (think of a single-walled carbon nanotube as a graphene cylinder) or folded into fullerenes.

Graphene is undoubtedly emerging as one of the most promising nano-materials—often hyped as a "wonder material"—because of its unique combination of superb properties, which opens the way for its exploitation in a wide spectrum of applications ranging from electronics to optics, sensors, and biodevices.

The reason nanotechnology researchers are so excited is that graphene and other two-dimensional crystals (it's called 2D because it extends in only two dimensions: length and width; as the material is only one atom thick, the third dimension, height, is considered to be zero) open up a whole new class of materials with novel electronic, optical and mechanical properties.

Early experiments with graphene have revealed some fascinating phenomena that are highly relevant for researchers working towards molecular electronics. For instance, it was found that graphene remains capable of conducting electricity even at the limit of nominally zero carrier concentration because the electrons don't seem to slow down or localize. The electrons moving around carbon atoms interact with the periodic potential of graphene's honeycomb lattice, which gives rise to new quasiparticles that have lost their mass, or "rest mass" (so-called *mass-less Dirac fermions*). That means that graphene never stops conducting.

4.1.1 New Synthesis Method for Graphene Using Agricultural Waste

The four major synthesis methods for producing graphene are: chemical vapor deposition (CVD) on metallic films; epitaxial growth on silicon carbide; liquid exfoliation of graphite crystals; and the chemical reduction of graphene oxide.

Using these processes, it is possible to produce high-quality and large-area graphene in large quantities. However, they all come with drawbacks as well: the complexity of the CVD process—including high temperatures and expensive substrates—makes it not very suitable for bulk production; epitaxial growth provides wafer-scale graphene, but silicon carbide is expensive and this process requires a high temperature above 1500 °C; and exfoliation or chemical reduction of graphite oxides can produce graphene in a scalable manner although the use of toxic chemical agents as well as complex processing requirements tends to complicate the scaling up of such processes.

Researchers have also proposed an alternative way of making graphene from rice husk.

This research, using an ordinary synthetic apparatus and abundant agricultural waste, suggests that low-cost graphene materials could be easily and cheaply synthesized on an industrial scale.

The annual global rice production reaches almost 700 hundred million tons, and a part of the waste from rice husks—the outer, protective covering of a rice kernel which is about 20 wt% of the entire kernel, or roughly 120 million tons a year—can be a massive resource as feed material for graphene production.

Due to its abundance, risk husk has already received much attention as a starting material in generating high-value-added materials such as silica and porous carbon.

Hiroyuki Muramatsu, an assistant professor in the Faculty of Engineering at Shinshu University in Japan, and Yoong Ahm Kim, an associate professor in the School of Polymer Science and Engineering at Chonnam National University in Korea, have demonstrated the ability of synthesizing bulk amounts of crystalline graphene with nanosized domains in a rapid, reliable, scalable, and cost-effective manner, by chemically activating (with potassium hydroxide) agricultural waste such as rice husk ash.

Although the production of activated carbon from rice husk ash has a long tradition, this is the first time that graphene structures have been observed in rice husk-derived activated carbon.

"We have also noted for the first time that highly crystalline and atomically clean edges are present in the synthesized materials, even though our graphene sample is prepared at a relatively low temperature (850 °C)," Muramatsu and Kim point out.

The nano-sized crystalline graphene exhibited a monolayer or multilayer structure with clean edges, whereas the corrugated graphene consisted of domains a few nanometers in size (200–300 carbon atoms), showing clear grain boundaries.

The researchers note that their findings clearly demonstrate that rice husk ash could be converted to high-value-added graphene in a rapid, reliable, scalable, and cost-effective manner.

They add: "The presence of clean and stable edges might possess novel physicochemical properties that make our material very attractive when fabricating high-performance carbon-based energy storage and conversion devices—*e.g.*, supercapacitor and hydrogen storage—next-generation water filter and various nanocomposites."

According to the researchers, there are still some issues to resolve. First, the process used (calcination and KOH treatment) makes it difficult to assess the detailed growth mechanism of graphene with its unique structures. Secondly, the production process needs potassium hydroxide treatment to synthesize graphene. The team is keen to find an alternative activation process without using strong alkali compounds.

Featured scientists: Prof. Hiroyuki Muramatsu (http://soar-rd.shinshu-u. ac.jp/profile/en.yafVbpkh.html); Prof. Yoong Ahm Kim (https://scholar. google.com/citations?user=91PIrw0AAAAJ&hl=ja)
Organizations: Faculty of Engineering, Shinshu University (Japan); Chonnam National University, Gwangju (Republic of Korea)
Relevant publication: H. Muramatsu, Y. Kim, K. Yang, R. Cruz-Silva, I. Toda and T. Yamada, *et al.*, Rice Husk-Derived Graphene with Nano-Sized Domains and Clean Edges, *Small*, 2014, **10**(14), 2766, DOI: 10.1002/ smll.201400017

4.1.2   Inkjet Printing of Graphene

Graphene has a unique combination of properties that is ideal for next-generation electronics, including mechanical flexibility, high electrical conductivity, and chemical stability. Combine this with inkjet printing, already extensively demonstrated with conductive metal nanoparticle ink, and you get an inexpensive and scalable path for exploiting these properties in real-world technologies.

Although liquid-phase graphene dispersions have been demonstrated, researchers are still struggling with sophisticated inkjet printing technologies that allow efficient and reliable mass production of high-quality graphene patterns for practical applications. There are several challenges that need to be overcome:

- good ink should possess proper fluidic properties, in particular the right viscosity and surface tension;
- the graphene concentration in these solvents is often quite low so that several tens of print passes are required to obtain functional films, reducing the efficiency of the technique;
- graphene flakes aggregate easily in inks or during solvent evaporation, which decreases the ink stability and/or degrades the film/device performance;
- ideal solvents for graphene dispersions are toxic so that their corresponding inks cannot be used in an open environment;
- most studies published so far on inkjet printing of graphene have actually been based on graphene oxide inks, not graphene inks.

Work by researchers at the KTH Royal Institute of Technology in Sweden has addressed these issues and they propose an approach to overcome these problems. The team, led by Max Lemme and Mikael Östling, professors at the School of Information and Communication Technology at KTH, demonstrates a mature but simple technology for inkjet printing of high-quality few-layer graphene.

The approach is based on the team's previously published distillation-assisted solvent exchange technique to prepare high-concentration graphene dispersions.[1] They first exfoliate graphene from graphite flakes in dimethylformamide (DMF), and then DMF is exchanged by terpineol through distillation by virtue of the large difference between their boiling points. Therefore, graphene can be significantly concentrated if terpineol is of much lower volume than DMF. More importantly, the solvent is changed from low-viscosity and toxic DMF to high-viscosity (about 40 cP at 20 °C) and environmentally-friendly terpineol.

They note, though, that the disadvantages of the technique in the previous work—a short stable period (the dispersion can only be stable for about 10 hours) and severe flake aggregation during solvent evaporation—prevent the dispersions from being practical inks. "In our follow-up work, we have improved the ink formulation mainly through polymer stabilization. Before distillation, a small amount of polymer (ethyl cellulose) is added into the harvested graphene/DMF dispersion to protect the graphene flakes from agglomeration. After printing, the stabilizing polymers can be effectively removed through a simple annealing process."

The resulting graphene dispersion had a stable period of at least several weeks. The researchers point out that the inks provide well-directed and constant jetting out of all nozzles at an even velocity, which is comparable to the performance of commercially available inks.

To investigate the quality of the printed graphene, the team fabricated large-area centimeter-scale graphene thin films with between 1 and 6 printing layers on glass slides.

"Printed transparent conductive films attain a sheet resistance around 200 kΩ sq^{-1} at a transmittance of about 90%," the team summarizes its results. "Printed narrow-line resistors exhibit a resistance range from a few kΩ to several MΩ. Printed few-layer graphene thin film transistors can be modulated by the electric field effect. Printed micro-supercapacitors achieve a high specific capacitance of 0.59 mF cm^{-2} and a rapid frequency response time around 13 ms."

The scientists conclude that the present technology provides an efficient and low-cost method to fabricate a variety of graphene electronic devices with good performance and is a promising alternative for future commercial applications in printed and flexible electronics.

Featured scientists: Prof. Max Lemme (http://www.kth.se/en/ict/forskning/professorer/max-christian-lemme-1.280545); Prof. Mikael Östling (http://www.kth.se/en/ict/forskning/professorer/mikael-ostling-1.281414)
Organization: School of Information and Communication Technology, KTH Royal Institute of Technology, Stockholm (Sweden)
Relevant publication: J. Li, F. Ye, S. Vaziri, M. Muhammed, M. Lemme and M. Östling, Efficient Inkjet Printing of Graphene, *Adv. Mater.*, 2013, 25(29), 3985, DOI: 10.1002/adma.201300361

4.1.3 Graphene from Fingerprints

Over the past few years, researchers have developed numerous methods for synthesizing graphene. The synthesis of high-quality graphene is commonly prepared by a complex and costly process—epitaxial growth on transition metal surfaces *via* chemical vapor deposition (CVD) using high-purity hydrocarbons as precursors.

"However, this kind of synthesis is in conflict with one of the desired applications—transistor arrays—since a metal support short-cuts the graphene layer," says Dr Frank Müller, a researcher in the Condensed Matter Physics group of the Department of Experimental Physics at the University of Saarland. "To date, however, the presence of a metal surface is a basic requirement for the CVD synthesis of graphene because its catalytic effect is necessary for the process."

Müller explains that one way to circumvent the problem is to firstly grow graphene on a thin metal film lying on an insulating surface. Secondly, the metal film is removed by chemical etching. That way, the graphene layer gets in direct contact with the insulator. Etching, however, introduces a new challenge, since it can induce defects in the graphene layer and can spoil its unique electronic properties.

"The challenge to be overcome can be reduced by use of metals with very small chemical interaction with the graphene layer," notes Müller. "Unfortunately, on some of these metals—such as silver—the performance of carbon uptake during a CVD process decreases with decreasing interaction. Therefore, we have used liquid precursor deposition (LPD) synthesis, which is an adequate alternative to CVD."

In LPD, the precursor molecules are provided *via* the liquid phase, *e.g.* by rinsing the metal surface with an organic solvent. In this process, weakly interacting metals take enough carbon from the precursor molecules to ensure graphene growth.

In their study, Müller and his fellow researchers demonstrate the reliability of graphene growth by LPD using an unconventional liquid—a human fingerprint.

Müller concedes that, on first sight, graphene formation by a fingerprint may seem like a useless experiment. "Nobody would seriously consider the application of this synthesis route in a technological production step. However, our experiments prove the reliability of LPD synthesis because graphene growth from fingerprints, although starting with a highly uncontrolled way to deposit carbon, achieves the same results as the LPD synthesis route using pure synthetic precursors."

In all experiments by the University of Saarland team, LPD synthesis by using synthetic precursors and by using fingerprints provided the same results in terms of graphene monolayer formation. The precursors just affect the initial amount of material that is provided after deposition.

"Since the fingerprint contains a broad mixture of different chemicals, its excellent performance for graphene formation is an interesting result in terms of suitable or unsuitable precursor," Müller points out. "Our experiments show that graphene does not only grow from organic 'waste' but also self-assembles directly out of the 'waste', showing that the synthesis of monolayer graphene on transition metals is an extremely robust process."

LPD may give access to a wider range of substrate materials for graphene growth. The results from this study may contribute to a possible future synthesis route for graphene on insulators, thereby disposing all the problems in handling the current transfer techniques.

Future research on graphene will focus on sandwich structures that combine several layers of 2D materials, *i.e.* the so-called van der Waals heterostructures. They form a new class of materials with tunable electronic and mechanical properties.

Featured scientist: Dr Frank Müller
Organization: Condensed Matter Physics group, Department of Experimental Physics, University of Saarland, Saarbrücken (Germany)
Relevant publication: F. Müller, S. Grandthyll, S. Gsell, M. Weinl, M. Schreck and K. Jacobs, Graphene from Fingerprints: Exhausting the Performance of Liquid Precursor Deposition, *Langmuir*, 2014, **30**(21), 6114, DOI: 10.1021/la500633n

4.1.4 Graphene Laminate Drastically Changes Heat Conduction of Plastic Materials

Graphene laminate—multilayer stacks of graphene layers piled on top of each other—is a promising material for thermal coating applications.

"Graphene by itself has very high intrinsic thermal conductivity, exceeding that of diamond, which is around 2000 W mK^{-1} at room temperature," explains Alexander A. Balandin, Professor of Electrical Engineering and Founding Chair of Materials Science and Engineering at University of California – Riverside (UCR). "Depending on the size of the flake and quality, suspended graphene samples revealed the thermal conductivity values in the range from 2000 W mK^{-1} to 5000 W mK^{-1} near room temperature. When placed on a substrate the thermal conductivity of graphene reduces due to phonon scattering from the rough interfaces but remains relatively high."

Balandin notes that the overall values of the thermal conductivity of graphene laminate at room temperature are substantially lower (~90 W mK^{-1}) than those measured for large suspended graphene samples. This is explained by the fact that the thermal conductivity of the laminates is limited not by the lattice dynamics of the graphene flakes but by their size, attachment to each other, and orientation with respect to the heat flux. However, the heat conduction ability of graphene laminate is still substantially better than that of plastics, which explains the practical relevance of the research.

Balandin's group, together with scientists from the University of Manchester, investigated thermal conductivity of graphene laminate films deposited on polyethylene terephthalate (commonly known as PET) substrates. They found that the compressed laminates have higher thermal conductivity for the same average flake size owing to better flake alignment.

The researchers prepared graphene laminate by deposition of an aqueous dispersion of graphene flakes on PET substrates. Graphene laminate films can be prepared in large quantities in industrial environments.

They studied two types of graphene laminate—as deposited and compressed—in order to determine the physical parameters affecting the heat conduction the most (Figure 4.1).

"We measured the thermal conductivity of graphene laminate in the range from 40 W mK^{-1} to 90 W mK^{-1} at room temperature," says Balandin. "We

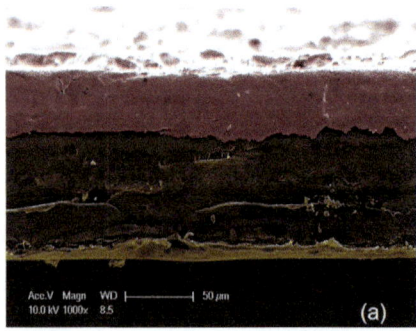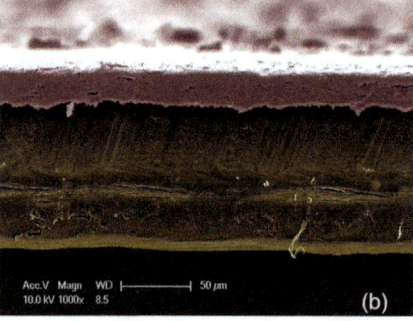

Figure 4.1   Cross-sectional SEM images of the (a) uncompressed and (b) compressed graphene laminate on PET samples. The pseudo colors are used to indicate the graphene laminate (burgundy) and PET (yellow) layers. (Image: Nano-Device Laboratory, UC Riverside.)

found—quite unexpectedly—that the average size and the alignment of graphene flakes are more important parameters defining the heat conduction than the mass density of the graphene laminate."

Interestingly, the team found that the thermal conductivity scales up linearly with the average graphene flake size in both uncompressed and compressed laminates.

This means that the compressed laminates have higher thermal conductivity for the same average flake size owing to better flake alignment.

Balandin points out that the team's results are important because they show the possibility of enhancing the thermal conductivity of plastic materials up to 600-times by coating them with the thin graphene laminate films.

The thermal conductivity of PET is extremely low, in the range from 0.15 to 0.24 W mK^{-1}, at room temperature. Other plastic materials also suffer from low thermal conductivity. This prevents many possible applications, which otherwise would benefit from the low cost, durability and light weight of plastics. Present work demonstrates that a few micrometer-thick graphene laminate films deposited on plastic films can drastically change their heat conduction properties.

This means that graphene laminate coatings could substantially increase the range of practical applications for plastic materials. Examples of such applications include electronic component packaging and solid-state lighting.

The team used a non-contact optothermal Raman method for the thermal studies. This is a direct steady-state measurement technique, which determines thermal conductivity directly without the need to calculate it from the thermal diffusivity data. In this technique, originally used for measuring the thermal properties of graphene, the micro-Raman spectrometer is used as a thermometer to determine the local temperature rise. The Raman excitation laser is also used as a heater.

"Our study utilized a relatively simple theoretical model to explain how thermal conductivity of graphene laminate depends on the parameters of laminate such as size of the flakes and concentration of impurities," says Balandin. "We will now develop a more detailed theoretical model, *e.g.* based on multi-scale simulation of heat transport, to assist in optimization of graphene laminate synthesis for thermal management applications."

Featured scientist: Prof. Alexander A. Balandin's Nano-Device Laboratory (http://ndl.ee.ucr.edu/)
Organization: Bourns College of Engineering, University of California – Riverside (USA)
Relevant publication: H. Malekpour, K. Chang, J. Chen, C. Lu, D. Nika and K. Novoselov, *et al.*, Thermal Conductivity of Graphene Laminate, *Nano Lett.*, 2014, **14**(9), 5155, DOI: 10.1021/nl501996v

4.1.5 Graphene Quantum Dot Band-Aids Disinfect Wounds

A growing body of medical nanotechnology research deals with the development of antibacterial applications, ranging from nanotechnology-based approaches for diagnosing superbugs to antimicrobial surface coatings and wound treatment with antibacterial nanomaterials.

Especially silver nanomaterials have been used effectively against different bacteria, fungi and viruses but also carbon nanomaterials such as nanotubes and graphene.

By combining graphene quantum dots (GQDs) with a low dose of a common medical reagent, hydrogen peroxide H_2O_2, researchers have designed a novel antibacterial system.

By using GQDs, a high concentration of H_2O_2—which is harmful to healthy tissue and even delays wound healing—can be avoided for wound disinfection. With the assistance of GQDs, H_2O_2 with a concentration that is 2–3 orders lower than that commonly used, can kill bacterial effectively.

"We find that the peroxidase-like activity of GQDs originates from their ability to catalyze the decomposition of H_2O_2, generating hydroxyl radicals," says Xiaogang Qu, a professor of chemistry at Changchun Institute of Applied Chemistry. "Since the hydroxyl radicals have a higher antibacterial activity, the conversion of H_2O_2 into hydroxyl radicals improves the antibacterial performance of H_2O_2 at a low level, which makes it possible to avoid the toxicity of H_2O_2 at high levels in wound disinfection."

Researchers have already demonstrated that functional inorganic analogues, such as V_2O_5 and magnetic iron oxide, can be used to assist H_2O_2 for antibacterial application, *i.e.* killing bacteria and destroying bacterial biofilms.

Qu points out that the V_2O_5 and magnetic iron oxide were considered to be toxic and inappropriate for the application *in vivo* unless *via* elaborate surface functionalization.

"Considering the excellent biocompatibility and high peroxidase-like activity of GQDs, we conducted our work using GQDs to improve the antibacterial activity of H_2O_2 for the application of wound disinfection."

Qu's team showed that their system provides antibacterial properties against both Gram-negative (*E. coli*) and Gram-positive (*S. aureus*) bacteria *in vitro*.

The researchers also designed Band-Aids containing graphene quantum dots and found in their experiments that these GQD-Band-Aids showed excellent antibacterial property *in vivo* with the assistance of low concentrations of H_2O_2.

"Our results indicate that GQD-Band-Aids have potential use for wound disinfection," says Qu.

A future extension of this work could be the design of biocompatible functional nanoparticles with high peroxidase-like activity to further improve the antibacterial activity of H_2O_2 at low concentrations.

As Qu notes, the synergy of peroxidase-like activity and other antibacterial mechanisms to achieve better effects may be a particular, but promising, challenge in this field.

Featured scientist: Prof. Xiaogang Qu
Organization: Changchun Institute of Applied Chemistry, Changchun (PR China) (http://english.ciac.cas.cn/)
Relevant publication: H. Sun, N. Gao, K. Dong, J. Ren and X. Qu, Graphene Quantum Dots-Band-Aids Used for Wound Disinfection, *ACS Nano*, 2014, 8(6), 6202, DOI: 10.1021/nn501640q

4.1.6　A Nanomotor that Mimics an Internal Combustion Engine

Efficient and perhaps autonomous molecular motors play an essential part in many visionary goals of nanotechnology, just like electric motors can be found in many appliances today. These nanomachines could perform functions similar to the biological molecular motors found in living cells, activities such as transporting and assembling molecules, or facilitating chemical reactions by pumping protons through membranes.

Although most nanomotor applications are still part of futuristic scenarios, there already is a fast-growing body of nanotechnology research that is dedicated to molecular machinery; and the results of this very early-day nanomachine research are spectacular: well-designed molecules or supramolecules show different kinds of motion—fueled by different driving forces such as light, heat, or chemical reactions—resulting in molecular shuttles, molecular elevators and rotating nanoscale motors.

While most of these nanomotors are powered by quantum or, in most cases, catalytic chemical processes, the nanoscale equivalent of conventional internal heat engines that are so prevalent in our daily life has been missing. The exception is the prototype of such a nanoscale single-ion heat engine that has been constructed by a team in Germany.[2]

In work performed at the National University of Singapore (NUS), researchers suggest a new type of ultra-thin graphene engine, which mimics an internal combustion engine system.

"Our graphene engine consists of only a few parts—functionalized graphene for high-elastic piston, chemisorbed chlorine trifluoride (CIF_3) molecules for recyclable high-power volume changeable actuator, and laser light as the 'ignition plug'—which would make it simple to work with," says Barbaros Özyilmaz, a professor in the Department of Physics at NUS. "This is the first time anyone has used graphene for this type of motor."

This is how the engine works: several layers of chemisorbed CIF_3 films that are sandwiched between single layers of graphene, supported by a silicon substrate, are exposed to a 532 nm laser beam. This leads to an explosion-like increase in pressure due to a rapid volume expansion of the CIF_3 molecules and delaminates graphene from the substrate in an upward blister-like bulge. After the laser is turned off, graphene goes back to its original flat state because the high reactivity of CIF_3 molecules causes them to chemisorb back onto the graphene.

"The rapid increase in pressure followed by the volume expansion and contraction under the graphene sheet are equivalent to the explosion and motion of the piston of an internal combustion engine," explains Özyilmaz. "Key to the performance of this graphene-based engine is that the high pressure between the substrate and graphene can be sustained due to its high Young's modulus, gas impermeability, and high adhesion energy."

The internal pressure of the blister is estimated to be about 10 MPa per expansion cycle of the engine. The working speed is extremely fast—less than 0.001 seconds per cycle. Even after 10 000 cycles, the graphene engine does not degrade, proving exceptional reliability.

The team also notes that the blister size is easily controllable by modifying the ignition parameters.

And, compared to our life-size combustion engines, this graphene engine does not generate any by-products.

"Another compelling factor is that we do not need too specific a working condition, so for varied applications, this is a viable engine for nanomachines," says Özyilmaz.

He notes that the bulging motion of this graphene nanoengine can be used as a pump or valve for nanofluidic applications. The device could be easily integrated into various applications by combining MEMS or NEMS techniques to transfer the generated force to each of the components. "So, consequently, the next stage would be to make and study this simple application of the nanoengine."

Featured scientist: Prof. Barbaros Özyilmaz (http://graphene.nus.edu.sg/barbaros/)
Organization: Department of Physics, National University of Singapore
Relevant publication: J. Lee, J. Tan, C. Toh, S. Koenig, V. Fedorov and A. Castro Neto, *et al.*, Nanometer Thick Elastic Graphene Engine, *Nano Lett.*, 2014, **14**(5), 2677, DOI: 10.1021/nl500568d

4.1.7 The Most Effective Material for EMI Shielding

Sensitive electronic devices such as cell phones and computers require shielding from electromagnetic interference (EMI). Such shielding—which must be electrically conductive—has traditionally been made of metal, which poses a weight problem in the push to miniaturize and lighten electronics.

Previous research has already demonstrated that ultra-lightweight carbon nanostructure-based nanocomposite materials outperform conventional metal shielding due to their light weight, resistance to corrosion, flexibility, and processing advantages.[3]

In new work, scientists in Korea have demonstrated that single-layer graphene is an excellent choice of material for high-performance EMI shielding. They found that CVD-synthesized graphene shows an EMI shielding effectiveness (in terms of decibels) more than seven times greater than gold film of the same thickness.

A team led by Byung Jin Cho, a professor in the Department of Electrical Engineering at the Korea Advanced Institute of Science and Technology (KAIST), suggests the feasibility of manufacturing an ultra-thin, transparent, weightless, and flexible EMI shield by a single or a few atomic layers of graphene.

"Besides the first experimental reports on EMI shielding effectiveness of monolayer graphene, we also studied the dominant mechanism of the shielding through modeling," says Cho. "According to our model, the ideal monolayer graphene can surprisingly shield as much as 97.8% of incident waves; our actual CVD-synthesized graphene can shield about 40%. This finding is significant because it means that graphene is the most effective material for EMI shielding in terms of shielding effectiveness per mass."

There have been experimental reports and theoretical analyses about the shielding effectiveness of graphene. But in these previous works, the researchers used graphene flakes obtained by the reduction of graphene oxide, not a single layer graphene as in this work.

For accurate assessment of the EMI shielding capability of a graphene layer, the researchers needed a true monolayer graphene over the entire area of the shield. In their experiments, they fabricated a waveguide with a graphene shield measuring 8.6 cm × 4.3 cm.

"To have such a high quality and large size graphene sheet with an excellent thickness uniformity is not an easy job," Cho points out. "Another difficulty is to ensure accuracy of the EMI measurement, eliminating interference from all other sources."

So far there have been no theoretical studies on the mechanism of EMI shielding of graphene. With the new model developed by the KAIST team, researchers can now predict the EMI shielding effectiveness of graphene for a specific application.

Since graphene has negligible mass and is ultra-thin, transparent, and flexible, it will be an excellent choice of EMI shielding material for portable electronic devices, transparent electronics, automobiles, and EMI isolation of 3D integrated circuits.

Featured scientists: Prof. Byung Jin Cho's Nano Electronics & Energy Device Lab (https://need.kaist.ac.kr/)
Organization: Division of Electrical Engineering, Korea Advanced Institute of Science and Technology, Yuseong-gu, Daejeon (Republic of Korea)
Relevant publication: S. Hong, K. Kim, T. Kim, J. Kim, S. Park and J. Kim, *et al.*, Electromagnetic interference shielding effectiveness of monolayer graphene, *Nanotechnology*, 2012, **23**(45), 455704, DOI: 10.1088/0957-4484/23/45/455704

4.1.8   Eavesdropping on Cells with Graphene Transistors

The integration of biological components with electronics, and more specifically, the interfacing of complex biological systems, is one of the current challenges on the path towards bioelectronics (or bionics for short).

Basically, there are three levels where electronics and biology could interface: molecular, cellular and skeletal. For any implanted bionic material, it is the initial interactions at the biomolecular level that will determine longer-term performance. While bionics is often associated with skeletal level enhancements (*e.g.* artificial muscles), electronic communication with living cells is of interest with a view to improving the results of tissue engineering or the performance of implants such as bionic ears or eyes.

Pioneering researchers, such as Peter Fromherz[4] from the Max-Plank-Institute of Biochemistry in Germany, have worked for more than 20 years on interfacing neurons and silicon devices. They are experimenting with individual neurons from different parts of the brain by cultivating them and trying to establish *ex vivo* neural networks. The goal is to stimulate neurons with electric signals and observe how the live network reacts and modifies itself.

One goal of such research could be neural prostheses that augment or restore damaged or lost functions of the nervous system. To do that, they need to be able to perform two main functions: stimulate the nervous system and record its activity. To that end, researchers are working on the development of brain implants on flexible substrates, which can record with high sensitivity the electrical and chemical activity of neurons.

Currently, most of the implants are based on metal electrodes. However, in addition to some issues with biocompatibility and stability under the harsh environment conditions of *in vivo* implants, metal electrodes have a limited electronic functionality. In that respect, the use of field-effect transistors would enable additional electronic functionality, due to the inherent amplification function of these devices. The sensing mechanism of these devices is rather simple: variations of the electrical and chemical environment in the vicinity of the field effect transistor (FET) gate region will be converted into a variation of the transistor current. The amplification factor is known as the device transconductance.

"Up to know, and due to its technology maturity, most of the work (pioneering work of P. Fromherz and A. Offenhäusser) has been done based on Si-FET technology," says Jose Antonio Garrido, a researcher at the Walter Schottky Institut at Technical University of Munich. "However, there have been some issues related to this technology which prevented a more successful implementation into real applications. One of them has been the relatively high electric noise of silicon devices—resulting from traps and defects at the silicon/oxide interface. The other is the difficulty of integrating silicon technology with flexible substrates. Further, silicon material is not stable under physiological conditions."

Therefore, there is interest in using other materials that can overcome the limitations of silicon technology. Carbon nanotubes and graphene are obvious candidates. Garrido and his team have demonstrated, for the first time, that CVD-grown graphene can be employed to fabricate arrays of transistors that are able to detect the electrical activity of electrogenic cells (*i.e.* electrically active cells).

Not only that; they also show that these devices exhibit a noise level which is as low as that of ultra-low-noise silicon devices. "And it is worth noting that the silicon technology has been improved over decades, whereas our graphene devices will surely improve heavily over the next years," Garrido points out.

This work is part of the team's overall goal of developing a graphene-based electronic platform for a future generation of bioelectronic systems, with special focus on the electronic interface to living cells and nerve tissue.

In previous work, they already demonstrated that graphene solution-gated field-effect transistors, with their facile technology, high transconductance, and low noise promise to far outperform state-of-the-art silicon-based devices for biosensor and bioelectronic applications.[5]

"Very recently, graphene has been proposed as the new frontier material which could lead to a new generation of bioelectronic systems," says Garrido. "Among others, the properties that make graphene so attractive are its high chemical stability, high mobility of charge carriers, low-noise, sensitivity to surface charge and molecular adsorption, and its ability to be integrated onto flexible substrates. The high performance of graphene in terms of carrier mobility, results in FET devices with very high transconductance, *i.e.* gate sensitivity. Further, high quality graphene is a 2D material which shows low reactivity, and thus is expected to be rather stable in a physiological environment."

In their recent work, the team reports on the extracellular detection of action potentials from electrogenic cells employing arrays of graphene-based solution-gated field-effect transistors (G-SGFETs). The arrays were fabricated using large-area graphene films grown by CVD on copper foil. Cardiomyocyte-like HL-1 cells were cultured on these arrays and exhibited a healthy growth. After characterizing the transistors in an electrolyte, the action potentials of these cells could be detected and resolved by the G-SGFETs under the cells.

The two main issues that the team has to address concern improving the noise performance, and optimization of the transfer of technology to flexible substrates. These two aspects involve a heavy focus on graphene technology, both in the preparation of films by CVD as well as on the fabrication of devices and transfer to flexible substrates.

"Given that parylene and kapton are employed for *in vivo* implants, we are focusing on these two substrates," says Garrido. "Surely, improving the growth conditions of CVD graphene will result in less defective films, which will have a positive impact on the device performance; thus, the expected lower noise and higher carrier motilities of the improved devices will lead to an enhanced signal-to-noise ratio."

Another important aspect, as he points out, is the biocompatibility of graphene. There have been some concerns on the biocompatibility of this nanocarbon material. However, in contrast to nanotubes and carbon nanoparticles such as fullerenes, graphene is a 2D material which in principle should not have the same problems: *i.e.*, due to the large extension in two dimensions, cells are not expected to incorporate graphene.

In fact, Garrido's team, in cooperation with Amel Bendali and Serge Picaud from the Vision Institute in Paris, have been studying the biocompatibility of graphene layers, by using cultures of retinal neuron cells. The results, which are being prepared for publication, demonstrate that CVD graphene exhibits excellent biocompatibility; retinal cells show a healthy growth on graphene, as good as that shown on their standard culture substrates.

"Of course," says Garrido, "much more work is required along this direction; and in particular, investigating the biocompatibility of graphene *in vivo*".

Currently, the researchers have just demonstrated a proof-of-concept platform. The exciting research comes now:

"We are part of a recently granted European project, NEUROCARE, aiming at developing brain-implants based on flexible nanocarbon devices," says Garrido. "Our role in this project is to develop ultra-low-noise graphene FET sensors which can record the electrical activity of nerve tissue. The electrical activity will be recorded to feedback and control the stimulation of nerve tissue, for retinal, cortical, and cochlear implants. Thus, arrays of graphene FETs will be fabricated on kapton and parylene substrates, and later on implanted in lab animals."

In terms of fundamental aspects, the researchers are particularly interested in understanding the cell/transistor interface since the signal recorded with the transistor strongly depends on such an interface.

Garrido notes that several concepts will be used to improve the cell–transistor contact, for instance by modifying the graphene surface with biomolecules—such as extracellular matrix proteins—which can provide a more suitable chemical environment for the cell.

Further, the use of biomolecules tethered to graphene can provide a unique opportunity to measure the chemical activity of cells. Here, the scientists are particularly interested in the selective chemical detection of neurotransmitters using graphene devices. In fact, the combination of biomolecules and graphene FETs can enable new applications of graphene for biosensing.

"In addition," says Garrido, "we are aiming at improving the space resolution of our recording devices, which would require the preparation of graphene transistors with an active area of 100×100 nm^2. For graphene, it is a great challenge to fabricate devices in the nanometer scale without degrading electronic properties such as the carrier mobility. In these small devices, defects in the graphene edges have been observed to strongly decrease the carrier mobility. We will have to develop passivation procedures to deactivate these edge defects."

The bottom line is that a successful resolution of the various challenges of using graphene devices could lead to the cure of illnesses with electronic prostheses for organs such as the eye or ear. Additionally, it could also enable novel solutions to substitute a motor, sensory or cognitive modality that might have been damaged as a result of an injury.

"There are many challenges ahead before this type of application is readily available," Garrido concludes. "And there are also other competing materials. However, graphene appears as a very promising material which could enable important breakthroughs in the field of bioelectronics and neural prosthesis in a not too far future."

Featured scientist: Jose Antonio Garrido (http://www.wsi.tum.de/People/Profile/tabid/287/Default.aspx?id=64eda115-8a72-4062-9963-5de0e26a37d4)
Organization: Walter Schottky Institut, Technische Universität München (Germany)
Relevant publication: L. Hess, M. Jansen, V. Maybeck, M. Hauf, M. Seifert and M. Stutzmann, *et al.*, Graphene Transistor Arrays for Recording Action Potentials from Electrogenic Cells, *Adv. Mater.*, 2011, **23**(43), 5045, DOI: 10.1002/adma.201102990

4.1.9 Graphene Beats Polymer Coatings in Preventing Microbially-Induced Corrosion

The huge economic impact of the corrosion of metallic structures is a very important issue for all modern societies. Estimates for the cost of corrosion degradation run to about €200 billion a year in Europe and over $270 billion a year in the U.S.[6] The annual cost of corrosion consists of both direct costs and indirect costs. Worldwide, it is estimated that these costs approach $1 trillion annually. The direct costs are related to the costs of design, manufacturing, and construction in order to provide corrosion protection, and the indirect costs are concerned with corrosion-related inspection, maintenance and repairs.

Given the huge economic incentives, corrosion prevention and protection is a major business. The advanced materials that are being developed and used in modern industries require increasingly sophisticated coatings for improved performance and durability.

Take, for example, the case of microbially-induced corrosion (MIC)—one of the lesser-understood forms of corrosion where micro-organisms manifest metallic surfaces and induce substantial damage that often goes unnoticed until there is a loss in the component functionality.

New research features graphene as a promising novel surface coating that can be used to minimize metallic corrosion under harsh microbial conditions.

"The most significant finding of our work is that graphene coating offers 100-fold improvement in corrosion resistance compared to commercial polymer coatings available in the market," says Dr Ajay Krishnamurthy, at the time of this work a guest researcher at the engineering lab (EL) at National

Institute of Standards and Technology (NIST) and now a research scientists at Swinburne University of Technology in Australia. "This finding is remarkable considering that graphene is nearly 4000 times thinner than several commercial coatings but offers more than an order of magnitude higher resistance to microbial attack."

Ultra-thin graphene coatings have already been demonstrated as corrosion-resistant coatings for metals. However, as Krishnamurthy notes, studies have provided some very interesting observations on the failure of graphene coatings on copper substrates.

"CVD graphene coatings have come under a lot of scrutiny recently. This is because the CVD process of producing graphene often introduces surface defects such as dangling bonds, Stone–Wales defects, point defects *etc.* Coupled with its superior electrical conductivity, these sites act as means of charge transport and lead to localized (pitting) corrosion of the metallic surfaces over long periods of time."

With their study, the international research team, which included scientists from Rensselaer Polytechnic Institute, South Dakota School of Mines and Technology, Oklahoma State University, and Shenyang National Lab for Materials Science, hypothesized that the defective sites on the graphene sheet are clogged due to the presence of polysaccharides and other microbial debris which lead to overall reduction in metallic corrosion rates over extended periods of time.

In a publication two years prior,[7] the team had compared a bare nickel surface to a few-layer CVD graphene coated nickel surface and showed lower charge transfer and lower Ni ion dissolution due to the presence of graphene coating.

They then took this study one step further and tried to understand how the microbial communities that colonize the protection system can affect the corrosion rates.

"We have used the graphene coating as a benchmark to compare to other popular polymer coatings such as parylene and polyurethane," Krishnamurthy points out. "Though initially promising, these coatings failed due to various reasons including microbial attack—fermentation, acid production *etc.*—and the non-conformity of hand-applied coatings."

Despite complete surface colonization by microbes, the graphene-coated Ni maintains its integrity without noticeable surface corrosion, while the parylene- and polyurethane-coated surfaces both rapidly corroded with visible (green and gooey) by-products on the surface.

"Having established that CVD-grown few-layer graphene offers superior resistance to MIC, we sought to characterize the defectiveness of the as-grown graphene coating when compared to graphene films that are transferred (wet-transfer) onto surfaces that are incompatible with CVD growth," says Krishnamurthy.

For this purpose, to cater to practical applications of using graphene coatings on surfaces where it can't be CVD grown, the researchers transferred 1–3 layers of graphene onto SiO_2 surfaces and examined the effects of the transfer process on introducing surface defects in the graphene sheets.

"We mapped the defect peak intensity ratios using confocal Raman spectroscopy," Krishnamurthy describes the investigation. "We concluded that the wet etch chemical methods introduce significant amount of defects, while the CVD growth process in itself is less deterrent."

This work shows that microbial conditions provide a unique dimension to corrosion research. The microbes are capable of forming insulating polysaccharide films on the graphene surface that can further mitigate metallic corrosion. The scientists hypothesize that the graphene–microbe interaction is providing a unique benefit to corrosion applications.

The next stages of the team's investigations are to test these coatings for other technologically relevant metals including mild steel.

"Our past research work evaluated graphene coatings for porous metallic electrodes under immersed conditions," concludes Venkata Gadhamshetty, an associate professor at South Dakota School of Mines & Technology, who is continuing this work on MIC in collaboration with Professor Nikhil Koratkar at Rensselaer Polytechnic Institute. "We will evaluate graphene coatings for flat metal sheets typical to the construction industry, and also aggressive atmospheric conditions including C5-I category as defined by ISO."

Featured scientist: Dr Ajay Krishnamurthy (http://www.swinburne.edu.au/science-engineering-technology/staff-profiles/view.php?who=akrishnamurthy)
Organization: Engineering lab, National Institute of Standards and Technology, Boulder, CO (USA) (http://www.nist.gov/el/)
Relevant publication: A. Krishnamurthy, V. Gadhamshetty, R. Mukherjee, B. Natarajan, O. Eksik and S. Ali Shojaee, *et al.*, Superiority of Graphene over Polymer Coatings for Prevention of Microbially Induced Corrosion, *Sci. Rep.*, 2015, **5**, 13858, DOI: 10.1038/srep13858

4.1.10 Janus Separator: A New Opportunity to Improve Lithium–Sulfur Batteries

Lithium–sulfur (Li–S) batteries, which employ sulfur as a cathode and metallic lithium as an anode, are emerging as promising alternatives to the widely used lithium-ion batteries because theoretically they can render 3–6 times higher energy density (2600 Wh kg^{-1}). In practice, though, it has proven challenging to approach this theoretical value. In addition, most lithium–sulfur batteries suffer from capacity loss under moderate current densities and self-discharge.

"The two major problems with Li–S batteries stem from the intrinsic inert reaction kinetics of sulfur redox and the unique 'shuttle' mechanism described as that soluble intermediates—polysulfides consisting of lithium-terminated sulfur-chains — diffuse between the cathode and anode, thus being consumed within the battery instead of being utilized," explains

Qiang Zhang, an associate professor at the Department of Chemical Engineering at Tsinghua University. "To solve these issues and to improve battery performance relies on not only the electrode materials but also other cell components such as the separator."

The separator—which in secondary batteries and supercapacitors separates the cathode and anode to prevent shorting; while in flow batteries and fuel cells, it should selectively let specific ions pass through while rejecting others—is a key component in energy storage devices and its unique functionalities are indispensable.

"In a lithium–sulfur battery, the separator should take more responsibility," says Zhang. "In fact, as the result of polysulfide shuttle, the discharge products are easily deposited between the separator and the cathode instead of within the cathode. An insulating film is formed on the electrode surface and thereby prevents the further reaction. This might be a negligent reason for low capacity and its fading."

"Previous works tried to retard the diffusion of polysulfides usually by modifying the sulfur cathode," he adds. "But we think the separator is the key to control the shuttle effect and engineered a special separator to improve our lithium–sulfur batteries—and we called it *Janus Separator*."

Janus is a two-faced god in Roman mythologies and has been used in science to describe asymmetric structures. A "Janus structure" can often realize new properties that are inconceivable for homogeneous or symmetric structures.

Accordingly, as the scientists at Tsinghua University report, their Janus Separator is made up of two different layers. One is a polymeric porous membrane and the other is a thin film of nanostructured carbon.

"The carbon layer we designed for our Janus Separators is composed of a kind of mesoporous but graphitic carbon," says Hong-Jie Peng, a member of the team. "It looks like a micrometer-sized honeycomb where many graphene cages are compactly packed together and interconnected. The size of each cage is around 7–8 nm. Compared to previously reported microporous carbon or carbon blacks, this carbon material exhibits much larger pore volume so that it can accommodate more solid products while remains the ion channel."

He points out that the graphitic nature also leads to superior electrical conductivity for fast charge transport.

"Actually the Janus Separator outperformed separators made of other carbon materials tested in our experiments," he says.

Dai-Wei Wang, a team member, tested the lithium–sulfur battery prototype with a Janus Separator and routine sulfur cathode blended with carbon blacks. Both the capacity and cyclic stability is improved by 40% when compared with a routine separator.

"The most exciting result for us is that, accompanied with improvements in capacity and stability, the passivated film induced by shuttle phenomenon completely disappear by employing our Janus Separator," he notes. "We are happy that our hypothesis on capacity decay has been unambiguously supported."

In most reported lithium–sulfur cells, the sulfur content in the cathode and the areal loading amount are both very low as a result of adding high amounts of carbon in order to gain higher capacity than based on sulfur only. However, since only sulfur stores the energy, its low content strongly impedes the overall energy density.

"People might argue that by employing our Janus Separator we actually add more carbon to the whole system so that the battery shows a better performance," says Hong-Jie. "But in the lithium–sulfur batteries we fabricated, the areal loading of sulfur is 5.3 mg cm^{-2}, which is 2–10 times higher than in most reports. And the loading of carbon in Janus Separators is only 0.3 mg cm^{-2}. By employing the Janus Separator with such little carbon addition, the areal capacity is improved from 2.6 mA h cm^{-2} to 5.5 mA h cm^{-2}."

"In the materials and chemistry world, sometimes not too much change can make a really big difference," comments Qiang.

The concept of a Janus Separator by engineering a carbon layer between the polymer membrane and the sulfur cathode might only be the beginning. According to the team, the next steps could lead to further device designs towards anode protection or smart management of mass transportation.

"I believe the Janus Separator will open new opportunities for improving the electrochemical energy storage devices (EESDs)," says Qiang. "Not only for lithium–sulfur batteries but also for other EESDs where the redox materials are mobile and under complicated phase change, the Janus Separator will play a significant role. I am looking forward to more and more progress on novel separators and applications."

Featured scientist: Prof. Qiang Zhang (http://www.chemeng.tsinghua.edu.cn/scholars/zhangqiang/Qiang-English.htm)
Organization: Beijing Key Laboratory of Green Reaction Engineering and Technology (FLOTU), Department of Chemical Engineering, Tsinghua University, Beijing (PR China)
Relevant publication: H. Peng, D. Wang, J. Huang, X. Cheng, Z. Yuan and F. Wei, *et al.*, Janus Separator of Polypropylene-Supported Cellular Graphene Framework for Sulfur Cathodes with High Utilization in Lithium-Sulfur Batteries, *Adv. Sci.*, 2016, 3(1), 1500268, DOI: 10.1002/advs.201500268

4.2 Beyond Graphene

The fascination with two-dimensional materials that started with graphene has spurred researchers to look for other 2D structures, for instance metal carbides and nitrides. The unique qualities of 2D materials, such as their reduced dimensionality (a 2D material is entirely made up of its surface) and symmetry, lead to the appearance of phenomena that are very different from those of their bulk material counterparts.

One particularly interesting analogue to graphene would be 2D silicon—silicene—because it could be synthesized and processed using mature semiconductor techniques, and more easily integrated into existing electronics than graphene is currently. Another material of interest is 2D boron, an element with worlds of unexplored potential. And yet another new 2D material—made up of layers of crystal known as molybdenum oxides—has unique properties that encourage the free flow of electrons at ultra-high speeds.

Having a new family of 2D structures with a wide range of chemistries can open the door to a better understanding of the differences between the properties of 2D and 3D materials, leading to identification of useful properties of 2D carbides, nitrides, oxycarbides and other related structures, and finally resulting in new applications.

4.2.1   MAX Phases Get Two-Dimensional as Well

Clean and affordable energy generation and storage is one of the most significant challenges that our world is facing in the 21st century. Materials are going to play a crucial role in the generation and storage of renewable energy. While searching for new materials for electrical energy storage, materials scientists have discovered a new family of two-dimensional compounds proposed to have unique properties that may lead to ground-breaking advances in energy storage technology.

"Our research team transformed three-dimensional titanium-aluminium carbide—a typical representative of a large family of layered ternary carbides called MAX phases—into a two-dimensional structure with greatly different properties," says Yury Gogotsi, a professor in the Department of Materials Science and Engineering at Drexel University and Director of the A. J. Drexel Nanotechnology Institute. "This work opens the door for a wide range of metal carbide and/or nitride compositions in the form of 2D sheets."

A team led by Gogotsi and Michel Barsoum, A. W. Grosvenor Professor of Materials Science and Engineering, recounts their ability to transform 3D titanium-aluminium carbide, a typical representative of a family of layered ternary carbides called MAX phases, into a two-dimensional structure with greatly different properties.

MAX phases, known as ductile and machineable ceramics, have been researched by Barsoum's lab for more than a decade and dozens of layered carbides, nitrides and carbonitrides with a variety of properties have been synthesized. As he explains on his website: "As a consequence of their layered structure, these materials kink and delaminate during deformation and also exhibit an unusual, and sometimes unique, combination of properties; they are not sure whether they want to be metals or ceramics. While they conduct heat and electricity like metals, they are elastically stiff, strong, brittle, and heat-tolerant like ceramics. They are resistant to chemical attack, readily machineable and thermal shock, are damage tolerant, and sometimes fatigue, creep, and oxidation resistant."

However, these ceramics have always been produced as 3D materials, until Drexel PhD student Michael Naguib placed titanium-aluminium carbide (Ti_3AlC_2) powders in hydrofluoric acid at room temperature to selectively

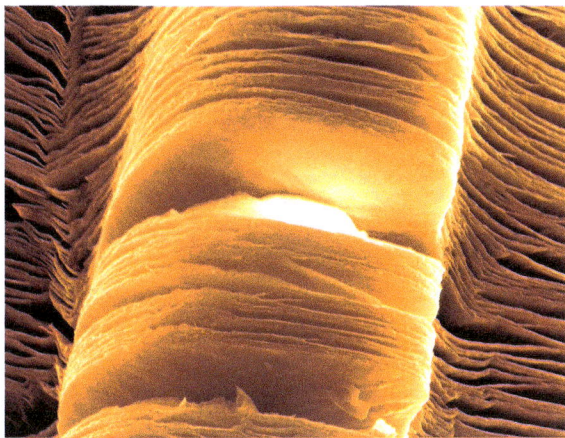

Figure 4.2   Scanning electron micrograph of exfoliated MXene nanosheets. (Image: Babak Anasori, Drexel University.)

remove the aluminium. The result of this chemical process—referred to as exfoliation—essentially spreads out the layered carbide material and yields two-dimensional Ti_3C_2 nanosheets, which have since been coined MXene, as a kin to graphene (Figure 4.2).

"To the best of our knowledge, metal carbides and/or nitrides have never been exfoliated," says Gogotsi. "This is the first time for a metal carbide to be produced in 2D sheets. We believe that this is a major breakthrough in materials science. Discovery of a new family of materials has always been an important event. If MXene attracts interest of ceramic scientists, chemists, physicists, and computational material scientists, we may expect new exciting discoveries in the near future."

Most notable is that the exfoliated material shows many features of graphene. For example, it can roll into nanotube-like scrolls, some with diameters of less than 40 nanometers, having the potential for a broad range of applications, ranging from energy storage devices to biomedical applications and composites.

For instance, MXene could be used in energy storage devices such as electrodes of Li-ion batteries, pseudo capacitors, *etc.* The researchers also envision its use as reinforcement in composites, similar to clays or graphene, which increase the mechanical properties and decrease the gas permeability of polymers. A variety of surface chemistries, the presence of transition metal oxides and a high surface area make MXene potentially attractive for catalytic applications.

Gogotsi points out that the implications and importance of this work extend far beyond the results shown in this paper.

"We are talking about a large family of 2D metal carbides and nitrides, so exploring different structures to find the optimum chemistry for each application is the next step in our work," he says. "Our students have exfoliated several other MAX phases and are working on their property characterization. The challenge now is controlling the surface chemistry of those 2D sheets; because we found that changing the surface chemistry will change the electrical properties of MXene."

Featured scientists: Prof. Yury Gogotsi's Nanomaterials Group (http://nano.materials.drexel.edu/); Prof. Michel Barsoum's MAX/MXene Research Group (http://max.materials.drexel.edu/)
Organization: Department of Materials Science and Engineering, Drexel University; A. J. Drexel Nanotechnology Institute (http://nano.drexel.edu/)
Relevant publication: M. Naguib, M. Kurtoglu, V. Presser, J. Lu, J. Niu and M. Heon, *et al.*, Two-Dimensional Nanocrystals Produced by Exfoliation of Ti_3AlC_2, *Adv. Mater.*, 2011, 23(37), 4248, DOI: 10.1002/adma.201102306

4.2.2   Transistor Made from All-2D Materials

Much hope (and hype) rides on graphene as a "post-silicon" material for fabricating next-generation nanoelectronic devices. However, graphene's Achilles heel is its lack of an energy band gap—a specific property of semiconductor materials that separates electrons from holes and allows a transistor implemented with a given material to be completely switched off. The result of this lack is leaky transistors not suitable for digital electronics.

Therefore, graphene must be modified to produce a band gap if it is to be used in electronic devices. Various methods of making graphene-based field effect transistors (FETs) have been exploited, including doping graphene, tailoring graphene like a nanoribbon, and using boron nitride as a support.

Using a new approach, researchers from the University of California, Berkeley, and Lawrence Berkeley National Laboratory have demonstrated the operation of an all-2D transistor, using a transition metal dichalcogenide (TMDC) channel material, hexagonal boron nitride (h-BN) gate dielectric, and graphene source/drain and gate contacts.

This is one of the first demonstrations of a transistor built only with 2D materials.

Featuring the same flat hexagonal, honeycombed structure as graphene and many of the same electrical advantages, these TMDCs—unlike graphene—have direct energy band gaps. This facilitates their application in transistors and other electronic devices.

"Our devices show no degradation of mobilities at high electric fields," says Tania Roy, a postdoc researcher in Prof. Ali Javey's research group at Berkeley. "Transistors with conventional three-dimensional materials show degradation of mobility at high electric fields due to surface roughness scattering. Our observations demonstrate that all-2D transistors do not suffer from this problem."

She points out that the interfaces are clean and devoid of trap states that kill the mobility. "Thus, all-2D transistors can offer a solution to the problems faced by transistors built with conventional materials systems."

The researchers present a unique platform for utilizing heterostructures of user-defined layered materials with atomically uniform and digitally controlled thicknesses for functional devices.

An important result from this work is the fact that the mobility of these all-2D transistors does not degrade at high electric fields, showing little surface roughness scattering in these devices, which would translate to faster and more power-efficient devices.

The research team also shows that they can digitally control the thickness of their transistors and the materials that compose the transistors. For example, they can very accurately control the number of layers of 2D material—in this case, transition metal dichalcogenide MoS_2—that will make up the channel.

Similarly, the thickness of the gate and all the electrodes can also be controlled as well.

The problems of interface and surface with conventional materials can be solved using only 2D materials. Roy notes that, since a dry transfer technology was used to make the transistors, there is no lattice mismatch of the channel with the substrate or with the gate dielectric. Also, the channel can be made as thin as a few layers of molecules, without encountering mobility degradation.

For this work, the team used the scotch-tape exfoliation method for building their transistors. "Future work should be towards obtaining large area TMDC materials for the channel and hexagonal boron nitride for the gate," says Roy. "We would like to obtain single crystalline materials, for improved device properties. This is still a major challenge."

She adds that her team used a homemade dry-transfer method for making their transistors, which is ingenious but strenuous. An automated method would be more useful.

Also, TMDC materials are difficult to contact with elemental metals. The scientists are working on obtaining good ohmic contacts to these materials.

Featured scientists: Prof. Ali Javey's Research Group (http://nano.eecs. berkeley.edu/)
Organizations: University of California, Berkeley; Lawrence Berkeley National Laboratory, CA (USA)
Relevant publication: T. Roy, M. Tosun, J. Kang, A. Sachid, S. Desai and M. Hettick, *et al.*, Field-Effect Transistors Built from All Two-Dimensional Material Components, *ACS Nano*, 2014, 8(6), 6259, DOI: 10.1021/nn501723y

4.2.3 Novel Mono-Elemental Semiconductors: Arsenene and Antimonene Join 2D Family

The zero band gap of the electronic structure of graphene, and other elements of the carbon group, seriously reduces these materials' ability to switch current on and off in transistors — even though with surface functionalization and external electric or strain fields, very small band gaps can be achieved.

Researchers have also fabricated and explored monolayers of typical semiconductors such as MoS_2, $MoSe_2$, W_2 and WSe_2, as well as few-layered black phosphorus. Such extreme thinning resulted in substantial changes in these materials' electronic structure and hence in new optoelectronic properties and potential device applications. For example, unlike bulk MoS_2 with its indirect band gap, monolayer MoS_2 was found to be a direct band gap semiconductor and thus displays much stronger luminescence.

They found, however, that the band gaps of most explored 2D systems are below 2.0 eV.

"This situation has greatly impeded the development of 2D semiconductor-based optoelectronic devices with response to the photons with wavelengths less than 620 nm, such as blue and UV light-emitting diodes (LEDs) and photodetectors," Professor Haibo Zeng, Director of the Institute of Optoelectronics & Nanomaterials at Nanjing University of Science and Technology, points out.

By means of density functional theory computations, Zeng and his group identified novel 2D wide-band-gap semiconductors with high stabilities, namely monolayer arsenene and antimonene.

Unlike black phosphorus, both arsenic and antimony are typical semimetals in their natural, layered bulk state. However, monolayered arsenene and antimonene are indirect wide-band-gap semiconductors, and under strain, they become direct band-gap semiconductors.

For arsenene and antimonene, such dramatic transitions of electronic properties could open a new door for nanoscale transistors with high on/off ratio, blue/UV optoelectronic devices, and nanomechanical sensors based on new ultra-thin semiconductors.

"Interestingly, although both bulk arsenene and antimonene are semimetals, when thinning them into a single atomic layer, our calculations indicate that their electronic structures will be transformed into indirect semiconductors with band gaps of 2.49 and 2.28 eV, respectively," says Shengli Zhang, a researcher at the Institute of Optoelectronics & Nanomaterials and first author of a paper on this work. "Further loading of tiny biaxial strain to these monolayers can transfer them from indirect into direct band-gap semiconductors."

He points out that, currently, there are two very important issues beyond the published results. The first one is the relative position of arsenene and antimonene in the family of 2D crystals, *i.e.* how are they special and what are their potentials. The second is the experimental reality, *i.e.* can these 2D materials be fabricated?

"In our lab, we already have fabricated these materials with a thickness of below 10 nm and are now exploring further thinning down to a monolayer," notes Zhang. "We are also exploring the fabrication of transistor devices. Most importantly, though, this work is to induce more researchers to explore these two 2D materials, thus accelerating their development."

After the first experimental demonstration of graphene back in 2004, researchers have mainly focused on 2D materials of the carbon group while ignoring honeycomb sheets of the nitrogen group. Phosphorene, arsenene and antimonene are exciting rediscoveries by the nanomaterials science community and they have quickly become the subject of significant theoretical and experiment investigations.

"We expect that novel 2D mono-elemental semiconductors arsenene and antimonene with wide band gaps can extend the family of semiconductor materials into the nitrogen group," concludes Zeng.

Featured scientist: Prof. Haibo Zeng
Organization: Institute of Optoelectronics & Nanomaterials, Nanjing University of Science and Technology, (PR China) (http://ion.njust-smse.com/en/)
Relevant publication: S. Zhang, Z. Yan, Y. Li, Z. Chen and H. Zeng, Atomically Thin Arsenene and Antimonene: Semimetal-Semiconductor and Indirect-Direct Band-Gap Transitions, *Angew. Chem., Int. Ed.*, 2015, 54(10), 3112, DOI: 10.1002/anie.201411246

4.2.4 Vanadium Disulfide—A Monolayer Material for Li-Ion Batteries

Until early 2014, the investigations of 2D transition metal disulfides (TMDs) have focused mostly on MoS_2, WS_2 and SnS_2. However, these 2D TMDs are all semiconductors and their poor conductivity has limited their electrochemical performances to some extent. Their applications in lithium-ion (Li-ion) batteries are usually combined with the additive of conductive carbon materials such as graphene, which, however, will decrease their gravimetric capacities.

Therefore, metallic 2D TMDs are highly desirable. Since vanadium disulfide (VS_2) few-layered nanosheets have been achieved experimentally,[8] researchers have been motivated to examine the feasibility of using a VS_2 monolayer anode for effective lithium storage. Vanadium dioxide is one of the few known materials that act like an insulator at low temperatures but like a metal at warmer temperatures starting around 67 degrees Celsius.

Excellent electrode materials should have fast lithium diffusion efficiency, high theoretical capacity, and good conductivity. However, it has proven difficult to find a material that meets all these qualifications. The emergence of 2D VS_2 might make it possible to fulfill these requirements in one material.

Using density functional theory computations, researchers have systematically investigated the adsorption and diffusion of lithium on the recently synthesized VS_2 monolayer in comparison with a MoS_2 monolayer and graphite.

"Intrinsically metallic, VS_2 monolayer has a higher theoretical capacity (466 mA h g^{-1}), a lower or similar lithium diffusion barrier compared to MoS_2 and graphite, and has a low average open circuit voltage of 0.93 V," explains Zhongfang Chen, an associate professor in the Department of Chemistry at the University of Puerto Rico. "Our results suggest that VS_2 monolayer can be utilized as a promising anode material for Li-ion batteries with high power density and fast charge/discharge rate."

This work is a joint effort by researchers from the Center for Advanced Nanoscale Materials and the Department of Chemistry at the University of Puerto Rico, and the Key Laboratory of Advanced Energy Materials Chemistry (Ministry of Education) at Nankai University.

"As a new layered inorganic material, VS_2 provides us some basic principles for the enhancement in lithium intercalation and diffusion, as well as development guidance for new Li-ion battery materials," he adds. "Encouragingly, VS_2 few-layers have been realized experimentally, and we strongly believe that 2D VS_2 anodes with excellent electrochemical performances can be achieved in the near future."

Since VS_2 monolayers exhibit many properties similar to MoS_2 monolayers, and its metallic property makes it superior to MoS_2, VS_2 monolayer materials are expected to be rather promising electrode materials in Li-ion batteries, supercapacitors and solar cells.

Although the VS_2 monolayer is predicted to be a highly promising anode material for Li-ion batteries, the large-scale synthesis of high quality VS_2 monolayers is still challenging. Chen says that stronger efforts should be devoted to developing more effective methods for the synthesis of the VS_2 monolayer.

Featured scientist: Prof. Zhongfang Chen (http://chemistry.uprrp.edu/chen/)

Organizations: Department of Chemistry, University of Puerto Rico; Key Laboratory of Advanced Energy Materials Chemistry (Ministry of Education), Nankai University (PR China)

Relevant publication: Y. Jing, Z. Zhou, C. Cabrera and Z. Chen, Metallic VS_2 Monolayer: A Promising 2D Anode Material for Lithium Ion Batteries, *J. Phys. Chem. C.*, 2013, **117**(48), 25409, DOI: 10.1021/jp410969u

4.2.5 Chemically Enhanced 2D Material Makes Excellent Tunable Nanoscale Light Source

Molybdenum disulfide's semiconducting ability, strong light–matter interaction and similarity to the carbon-based graphene make it of interest to scientists as a viable alternative to graphene in the manufacture of electronics, particularly photoelectronics. In particular, MoS_2 has excellent optical properties when deposited as a single, atom-thick layer—unlike graphene it emits light when excited; albeit relatively poorly.

"While a suspended monolayer MoS_2 flake shows photoluminescence efficiencies 100 times higher compared to a suspended bilayer flake, its overall quantum yield is still below 1%, and excitonic emission is spectrally broad at room temperature," says Jason C. Reed, a PhD student in Professor Ertugrul

Cubukcu's Nanoengineered Photonics Group at the University of Pennsylvania. "In contrast, when transferred onto substrates for device integration, quantum yield for MoS_2 monolayers is even lower and only 3.5 times higher than that of a bilayer."

Therefore, in order to realize the potential of atomically thin MoS_2 as a nanoscale active material in a light source, a considerable enhancement of its emission efficiency is necessary.

And that's exactly what the Penn researchers have done. They showed both experimentally and theoretically that two-dimensional MoS_2 can be chemically enhanced and used to make a tunable light source with high spatial and temporal coherence when integrated with a notched microdisk optical cavity.

"What we have done is found a way to chemically enhance the emission intensity of the MoS_2 and incorporate it into a nanoscale structure," Reed, who is first author of a paper on this research, explains. "By enhancing the MoS_2, we made it much brighter. By incorporating it into our nanostructure, we made it more functional."

He notes that the added functionality comes from the nanostructure: the light from the MoS_2 can resonate with the nanostructure, lingering around for much longer than if it was on a flat surface.

The result of this is that when this light is extracted from the defect incorporated into the nanostructure, it can be tuned from light that is "spectrally broad" to light that is very "spectrally narrow".

While it might seem counter-intuitive to modify a light source from having "more colors" to "fewer colors", this light is now higher quality and is also very useful for detecting very small changes in the wavelength of the emission.

"We can also tune the wavelength of light by adding as little as 100 femtograms of material—roughly equal to one seventh the mass of an *E. coli* bacterium—to our nanostructure," says Reed. "This, on the other hand, means that we can also detect equally small additions of material, opening up our work to the great potential for very sensitive detection applications."

The researchers point out that no other work has been done to modify the light emission from MoS_2 to this degree. Not only were they able to chemically enhance the emission of MoS_2, they also achieved a very high level of contrast between the emission peaks and the background.

In terms of pure science, this microdisk cavity light source allows researchers to study the optical properties of a wide range of different 2D materials.

Potential practical applications of this work appear to be promising in the realm of biological sensors, *e.g.* for early detection of certain diseases that may show low concentrations of biomarkers in the early stages.

"Because we can detect very small changes on the surface of our nanostructure, we can also detect small biomolecules interacting with the surface as well," says Reed. "By attaching the correct proteins or antibodies to the surface, we can tailor specific biomolecule attachment at potentially very low detection limits."

Featured scientist: Prof. Ertugrul Cubukcu's Nanoengineered Photonics Group (http://www.seas.upenn.edu/~cubukcu/Cubukcu_Lab/Home.html)
Organization: Department of Material Science and Engineering, University of Pennsylvania, Philadelphia, PA (USA)
Relevant publication: J. Reed, A. Zhu, H. Zhu, F. Yi and E. Cubukcu, Wavelength Tunable Microdisk Cavity Light Source with a Chemically Enhanced MoS$_2$ Emitter, *Nano Lett.*, 2015, **15**(3), 1967, DOI: 10.1021/nl5048303

References

1. J. Li, F. Ye, S. Vaziri, M. Muhammed, M. Lemme and M. Östling, A simple route towards high-concentration surfactant-free graphene dispersions, *Carbon*, 2012, **50**(8), 3113, DOI: 10.1016/j.carbon.2012.03.011.
2. J. Roßnagel, O. Abah, F. Schmidt-Kaler, K. Singer and E. Lutz, Nanoscale Heat Engine Beyond the Carnot Limit, *Phys. Rev. Lett.*, 2014, **112**(3), 030602, DOI: 10.1103/PhysRevLett.112.030602.
3. Y. Yang, M. Gupta and K. Dudley, Towards cost-efficient EMI shielding materials using carbon nanostructure-based nanocomposites, *Nanotechnology*, 2007, **18**(34), 345701, DOI: 10.1088/0957-4484/18/34/345701.
4. http://www.biochem.mpg.de/en/eg/fromherz/.
5. M. Dankerl, M. Hauf, A. Lippert, L. Hess, S. Birner and I. Sharp, *et al.*, Graphene Solution-Gated Field-Effect Transistor Array for Sensing Applications, *Adv. Funct. Mater.*, 2010, **20**(18), 3117, DOI: 10.1002/adfm.201000724.
6. https://www.nace.org/Publications/Cost-of-Corrosion-Study/.
7. A. Krishnamurthy, V. Gadhamshetty, R. Mukherjee, Z. Chen, W. Ren and H. Cheng, *et al.*, Passivation of microbial corrosion using a graphene coating, *Carbon*, 2013, **56**, 45, DOI: 10.1016/j.carbon.2012.12.060.
8. J. Feng, X. Sun, C. Wu, L. Peng, C. Lin and S. Hu, *et al.*, Metallic Few-Layered VS 2 Ultrathin Nanosheets: High Two-Dimensional Conductivity for In-Plane Supercapacitors, *J. Am. Chem. Soc.*, 2011, **133**(44), 17832, DOI: 10.1021/ja207176c.

CHAPTER 5

The Medicine Man of the Future is Tiny

Nanotechnology promises us a radically different medicine compared to the cut, poke and carpet bomb (think chemotherapy) medicine of today. In the future, extremely precise nanomedicine procedures will allow medical professionals to diagnose, intervene, and monitor on the level of individual cells.

A generally accepted definition of *nanomedicine* refers to highly specific medical intervention at the molecular scale for curing disease or repairing damaged tissues, such as bone, muscle, or nerve.

The two major differentiators of nanomedicine will be the tools it uses—the main workhorse will be multifunctional nanoparticles—and the fact that it will enable perfectly targeted and individual treatments: organs and bones, really any body tissue, will one day be diagnosed and treated on a cell by cell basis with precise dosing and monitoring through the use of biomolecular sensors.

Even more advanced, nanotechnology-enabled tissue engineering will become a medical treatment with the goal to replace or restore the anatomical structure and function of damaged, injured, or missing tissue. At the core of tissue engineering is the construction of three-dimensional scaffolds out of biomaterials to provide mechanical support and guide cell growth into new tissues or organs. Experimental efforts are already underway for tissue engineering involving virtually every type of tissue and every organ of the human body.

Although products classified as nanomedicine products have indeed appeared over the past decade, such products have not exactly revolutionized treatment paradigms as envisaged earlier.

In particular, no molecular machine or nanorobot has yet entered clinical trials, although research in these areas is picking up pace. Nevertheless,

Nanotechnology: The Future is Tiny
By Michael Berger
© Michael Berger 2016
Published by the Royal Society of Chemistry, www.rsc.org

advances in micro- and nanoscale engineering in the medical field have led to the development of various robotic designs that one day will allow a new level of minimally-invasive medicine. These micro- and nanorobots will be able to reach a targeted area, provide treatments and therapies for a desired duration, measure the effects and, at the conclusion of the treatment, be removed or degrade without causing adverse effects. Ideally, all these tasks would be automated but they could also be performed under the direct supervision and control of an external user.

Since it looks like the early nanotechnology breakthroughs will occur in the medical field, this section deserves a large coverage in this book.

5.1   Honey, I Swallowed the Doctor

The ideal drug carrier may be something out of science fiction. In principle, it is injected into the body and transports itself to the correct target, such as a tumor, and autonomously delivers the required dose, repeatedly if necessary, at this target. This idealized concept was first proposed by Paul Ehrlich at the beginning of the 20th century and was nicknamed the *magic bullet* concept. With the advent of nanotechnology and nanomedicine, this dream is rapidly becoming a reality.

5.1.1   Magnetic Nanovoyagers in Human Blood

The use of nanomotors to power nanomachines and nanofactories is one of the most exciting challenges facing nanotechnology. The highly successful design shop of Mother Nature has created efficient biomotors through millions of years of evolution and uses them in numerous biological processes and cellular activities.

While nanotechnology researchers have made great progress over the past few years in developing self-propelled nano objects, these tiny devices still fall far short of their natural counterparts' performance. Today, artificial nanomotors lack the sophisticated functionality of biomotors and are limited to a very narrow range of environments and fuels.

In another step towards realizing Richard Feynman's vision of tiny vessels roaming around in human blood vessels working as surgical nanorobots, researchers at the Indian Institute of Science (IISc) in Bangalore have demonstrated, for the first time, externally-driven nanomotors that move in undiluted human blood.

"Most externally—magnetically or acoustically—driven nanomotors realized to date have been actuated in de-ionized water and, in a few cases, in media of biological relevance such as serum," says Pooyath Lekshmy Venugopalan, a PhD student at IISc Bangalore's Centre for Nano Science and Engineering. "The only reported attempt to maneuver a nano-voyager in human blood has been with catalytic microjets, which were moved in human blood diluted 10 times with (toxic) hydrogen peroxide."

The IISc team describes a system of cytocompatible nanopropellers that can be maneuvered in various biological fluids with a small and homogeneous rotating magnetic field. The method of actuation is non-invasive, does not require any chemical fuel, and is therefore ideally suited to *in vivo* applications.

"For artificial nanomotors to be successfully maneuvered in undiluted human blood, two important experimental challenges need to be met," explains Lekshmy. "(1) The thrust generated by the propeller needs to be large enough to overcome the large drag due to the presence of blood cells; and (2) since the large concentration of ions—chlorides, phosphates, *etc.*— in blood can etch most magnetic materials easily, this necessitates a conformal protective coating around the nanomotors, many of which, including the chemically and acoustically powered ones, contain a magnetic material which can be used for controlling their direction of motion."

The researchers overcame these experimental hurdles by using a conformal ferrite coating in conjunction with helical propulsion powered by magnetic fields.

The developed system was also found to be biocompatible, thereby opening up new possibilities in the *in vivo* applicability of artificial nanomotors, which was the team's main goal when they started this project.

Having controlled motion in important biological environments automatically suggests a general platform towards diagnostic and therapeutic applications. Since it is possible to functionalize the nanomotors with appropriate biomolecules, such a system could be used to detect and treat diseases.

"One could also envision bringing the nanomotors in close proximity to a cancerous tissue," notes Lekshmy. "This could have tremendous therapeutic implications, as ferrites—which coat our nanomotors—are commonly used for magnetic hyperthermia. Alternately, by loading the nanomotors with cancer specific drugs, one could localize the treatment significantly."

The team's further research in this area will be directed towards adding functionality to their ferrite-coated nanopropellers, such as using them as sensors for detecting various disease conditions in blood, and to attempt therapeutic applications under *in vivo* conditions.

"For *in vivo* experiments, it may be necessary to image these small objects from a distance," Lekshmy points out. "This is not a trivial task and may require novel imaging methods."

Featured scientist: Pooyath Lekshmy Venugopalan
Organization: Centre for Nano Science and Engineering, Indian Institute of Science, Bangalore (India) (http://www.cense.iisc.ernet.in/nws/)
Relevant publication: P. Venugopalan, R. Sai, Y. Chandorkar, B. Basu, S. Shivashankar and A. Ghosh, Conformal Cytocompatible Ferrite Coatings Facilitate the Realization of a Nanovoyager in Human Blood, *Nano Lett.*, 2014, **14**(4), 1968, DOI: 10.1021/nl404815q

5.1.2 Microrobots to Deliver Drugs on Demand

Several approaches have been explored for the wireless actuation of micro-robots. Among these, magnetic fields have been the most widely employed strategy for propulsion because they do not require special environmental properties such as conductivity or transparency. This approach allows for the precise manipulation of magnetic objects toward specific locations, and magnetic fields are biocompatible even at relatively high field strengths (MRI).

A team of researchers from ETH Zurich and Harvard University has demonstrated that additional intelligence—including sensing and actuation—can be instantiated in these microrobots by selecting appropriate materials and methods for the fabrication process.

"Our work combines the design and fabrication of near infrared light (NIR) responsive hydrogel capsules and biocompatible magnetic microgels with a magnetic manipulation system to perform targeted drug and cell delivery tasks," explains Dr Mahmut Selman Sakar, a research scientist in Bradley Nelson's Institute of Robotics and Intelligent Systems at ETH Zurich.

The team fabricated an untethered, self-folding, soft microrobotic platform, in which different functionalities are integrated to achieve targeted, on-demand delivery of biological agents.

As Sakar explains, the smart carriers, which are designed to be able to deliver biological materials, consist of two parts:

(1) Thermo responsive hydrogel–graphene oxide nanocomposite bilayers, resembling the shape of jellyfish and a Venus flytrap

Stimuli-responsive hydrogels are a class of materials closely resembling biological tissues in their physical and chemical properties. These swollen polymer networks are of interest in research and industry for many biomedical applications due to their unique capability of a reversible volume change in response to different stimuli (temperature, pH, ionic strength, *etc.*). They are used in tissue engineering, drug and cell delivery and wound healing.

Bilayered hydrogel structures fold when two coupled layers swell or contract differently in response to a specific chemical or physical change. The fabricated microstructures can switch from a closed to an open configuration when their body temperature exceeds 40 °C.

"More importantly," says Sakar, "we engineered these microstructures so that a short exposure to a NIR laser source generates the required heat within the target structure. In this way, an external operator can stimulate the microstructures from a certain distance."

This source of actuation was chosen as a controllable trigger mechanism, because it can penetrate body tissue without causing damage even at repeated doses. This part is intended to provide an on-demand drug delivery platform and to protect the internal environment from foreign body reactions.

(2) Magnetic alginate microspheres

These microparticles can be loaded with biological materials and can easily be encapsulated inside the hydrogel bilayers. Alginate, a natural biodegradable polysaccharide, is often chosen as a gelable polymer for long-term and sustained delivery of both drugs and cells. This part of the device allows external magnetic manipulation and sustained delivery.

The number of beads trapped and their position and orientation vary depending on the shape of the structures and the size of the microbeads. Structures resembling Venus flytraps were able to encapsulate more beads along the dominant axis, while jellyfish-like structures folded into a spherical ball filled with uniformly distributed beads. The distribution influenced the magnetic properties of the final platform. In both cases, microbeads remained inside after complete closure of the hydrogel films and throughout the magnetic manipulation experiments.

"These complex structures are manipulated in 3D complex spaces by means of a five-degree-of-freedom (5-DOF) electromagnetic manipulation system, which could potentially be integrated to commercial MRI setups," Sakar describes the automated locomotion of the microrobots. "The system provides precise positioning under closed-loop control with computer vision but can also be used with no visual tracking, relying only on visual feedback to the human operator during direct teleoperation."

He points out that the team's magnetic manipulation system is capable of generating multiple types of state-of-the-art magnetic control techniques including field and gradient propulsion, rotating magnetic fields, and their combinations.

Despite being developed as a prototype, the proposed microrobotic platform possesses all the required features that the researchers envision for biomedical applications including 3D automated magnetic steering; the capacity to carry drugs and cells; and spatiotemporally controlled delivery.

The team employed a flexible and versatile fabrication technique to produce hydrogel-based soft microstructures. The fabrication methods of the self-folding devices are compatible with the production of smaller microcarriers made out of a variety of different materials.

"With further developments, the released magnetic beads could be navigated to very small conduits in the body, allowing a modular approach to the problem of cell and drug delivery," says Sakar.

He adds that, "although some of the individual pieces might have appeared in previous reports, for the first time, our study introduces a complete platform based on self-folding soft polymeric structures that can be used as wirelessly controllable microcarriers. We successfully combined magnetic navigation with light activated actuation on the same body."

The researchers are confident that their unique synthesis and characterization of these biocompatible microcarriers will open up opportunities to perform on-demand, targeted therapeutic interventions.

Featured scientist: Institute of Robotics and Intelligent Systems (http://www.iris.ethz.ch/); Mooney Lab (http://mooneylab.seas.harvard.edu/)
Organization: Department of Mechanical and Process Engineering, ETH Zurich, (Switzerland); John A. Paulson School of Engineering and Applied Sciences, Harvard University, Cambridge, MA (USA)
Relevant publication: S. Fusco, M. Sakar, S. Kennedy, C. Peters, R. Bottani and F. Starsich, *et al.*, An Integrated Microrobotic Platform for On-Demand, Targeted Therapeutic Interventions, *Adv. Mater.*, 2014, **26**(6), 952, DOI: 10.1002/adma.201304098

5.1.3   First Demonstration of Micromotor Operation in a Living Organism

Synthetic micro- and even nanoscale motors—inspired by the 1966 *Fantastic Voyage* movie—were introduced nearly a decade ago. Although major progress has been made toward biomedical applications of these tiny vehicles, there are no reports so far illustrating and examining the operation and behavior of nano/microscale motors in living organisms.

A joint collaborative effort between the teams of professors Joseph Wang and Liangfang Zhang at the Nanoengineering Department at the University of California San Diego, addresses the challenges of operating synthetic motors in living organisms through the use of biocompatible motors that are powered by body fluid (acidic stomach environment). As the zinc body of the motor is dissolved by the acid fuel, the motors self-destroy, leaving no harmful chemicals behind.

This work is the first *in vivo* study of artificial micromotors using a live mouse model.

"Our research demonstrates that the self-propulsion of these microrockets leads a dramatically improved retention of their payloads in the stomach lining compared to the common passive diffusion and dispersion of orally administrated payloads," says Zhang.

"Among the variety of recently developed artificial micromotors, our recently reported zinc-based motors hold great promise for *in vivo* use, particularly for gastric drug delivery, because of their unique features, including acid-powered propulsion, high loading capacity, autonomous release of payloads, and nontoxic self-destruction," explains Wang.

Using established membrane templating processes, the team's zinc-based micromotors can display efficient propulsion in a harsh acidic environment without additional fuel and transport fully loaded cargoes at high speeds.

In their experiments, the researchers inserted their micromotors, loaded with gold nanoparticles as a payload, into the stomachs of living mice and then evaluated their autonomous movement in gastric acid and the motion-induced biodistribution and retention of the micromotors on the stomach wall.

"Our results demonstrate that the self-propulsion of the micromotors leads to a dramatically improved retention of their payloads in the stomach lining compared to the common passive diffusion and dispersion of orally administrated payloads," the scientists note.

The team points out that these findings, along with the absence of toxic effects in the stomach, indicate that the movement of micromotors in the stomach fluid offers potentially distinct advantages for *in vivo* biomedical applications and paves the way for their future clinical studies.

"Our new findings and insights are expected to advance the field of synthetic nano/micromotors and to promote interdisciplinary collaborations towards expanding the horizon of man-made nanomachines in medicine towards the realization of the 'Fantastic Voyage' vision," Wang concludes.

Featured scientists: Prof. Joseph Wang's Laboratory for Nanobioelectronics (http://joewang.ucsd.edu/); Prof. Liangfang Zhang's Nanomaterials & Nanomedicine Laboratory (http://ne.ucsd.edu/faculty/l7zhang/)
Organization: Jacobs School of Engineering, University California San Diego, CA (USA)
Relevant publication: W. Gao, R. Dong, S. Thamphiwatana, J. Li, W. Gao and L. Zhang, *et al.*, Artificial Micromotors in the Mouse's Stomach: A Step toward in Vivo Use of Synthetic Motors, *ACS Nano*, 2015, **9**(1), 117, DOI: 10.1021/nn507097k

5.1.4 Multiplexed Planar Array Analysis from Within a Living Cell

Multiplex biological assays allow simultaneous detection of numerous analytes such as proteins and other molecular components in a single sample. Planar array chips consist of a collection of multiple independent and ordered sensing features immobilized on a solid support, allowing parallel assays. These devices have been a revolution in the field of molecular analysis.

In order to reduce the size of the required samples and speed up detection times, suspended arrays of particles were developed. Here, capture molecules are immobilized onto a microsphere and captured analytes are detected mostly using flow cytometry. However, the advantage of parallel assays in a single device is lost as each microsphere bears only a single molecular probe.

Consequently, there has been a great effort by the scientific community to develop a technique that allows several probes on the same particle.

"Considering that the size of molecular probes in these arrays remains a critical property for many biological applications, there is a continuous need to miniaturize monodisperse particles for further sample reduction and higher throughput," says Dr José Antonio Plaza, who leads the Micro- and NanoTools Group at Instituto de Microelectrónica de Barcelona IMB-CNM (CSIC). "To address this need, we introduced suspended planar-array (SPA) chips whose *in situ* capabilities with a spatial molecular-probe arrangement combine the advantages of both suspended arrays and planar arrays."

This novel technology opens the way towards the multiplexed detection of intracellular biological parameters using a single device in dramatically reduced volumes, such as a living HeLa cell. This work has been done in collaboration with scientists from the Universitat de Barcelona (UB) and the Institut de Bioenginyeria de Catalunya (IBEC) led by Lluïsa Pérez-García, and the Centro de Investigaciones Biológicas CIB(CSIC) led by Teresa Suárez.

The team miniaturized the volume of a regular silicon planar array by a factor of more than 1 billion. This allowed them to demonstrate multiplexed analysis on a 9 ($3 \times 3 \times 1$) μm^3 planar array chip. The chip's volume represents only about 0.35% of the total volume of a typical HeLa cell.

In previous work,[1] the group demonstrated that chips with similar dimensions could be easily internalized inside living HeLa cells and don't affect either the cell viability or the cell division. In follow-up work,[2] the researchers demonstrated a nanomechanical chip that can be internalized to detect intracellular pressure changes within living cells, enabling an interrogation method based on confocal laser scanning microscopy.

"The potential applications of these devices are the molecular multiplexing of extraordinary small volumes in a single chip, to perform multiple molecular analysis in really small volumes simultaneously," says Plaza. "Among them, one of the most fascinating applications is in cell biology. The volume of these chips is so small that they can be internalized inside living cells. Cells are usually analyzed in planar arrays chips. We have succeeded in achieving the opposite: this is the first time a multiplex planar array goes inside a living cell."

The key to fabricating these devices is a combination of a manufacturing technology commonly used in microchip production with a molecular printing technique—polymer pen lithography (PPL). The first one offers a high-throughput of millions of micron-sized chips while the latter offers higher molecular-pattern flexibility in small dimensions.

In their work, the team used PPL to pattern molecules directly on the area-restricted surfaces of the anchored chips.

Some of the potential applications of this chip are the determination of physicochemical intracellular parameters related to genetic determinants of diseases; cellular-function modulators; and dynamic responses of the cell *in vivo*.

In their experiments, the scientists demonstrated multiplexed detection of the intracellular pH changes in HeLa cells. In addition, their results showed that cell viability remains unaffected by the chip.

Plaza and his collaborators are planning to develop chips with new molecular probes that can detect intracellular molecules or even physicochemical parameters relevant for cell functioning, with the added value that this will be done inside intact living cells and by a single device.

"Cells are complex systems," says Plaza. "Traditionally, cell biology has been mainly based on biochemical studies. However, recently, clear evidence emerged that cells' biochemistry is coupled to physical cues, which are obvious in, for instance, mechanotransduction. Thus, the possibility to use intracellular chips to perform biochemical and biophysical studies simultaneously opens a new venue of possibilities in fundamental studies of cell biology."

Featured scientists: Dr José Antonio Plaza's Micro- and NanoTools Group (http://mnt.imb-cnm.csic.es/)
Organization: Micro- and NanoSystems Department at the National Microelectronic Institute of Barcelona IMB-CNM, Barcelona (Spain)
Relevant publication: N. Torras, J. Agusil, P. Vázquez, M. Duch, A. Hernández-Pinto and J. Samitier, *et al.*, Suspended Planar-Array Chips for Molecular Multiplexing at the Microscale, *Adv. Mater.*, 2016, **28**(7), 1449, DOI: 10.1002/adma.201504164

5.1.5 Self-Powered Micropumps Respond to Glucose Levels

In diabetes treatment, drug delivery systems are often based on self-assembled micelles that disintegrate when they get in touch with glucose, thereby releasing the insulin molecules that have been trapped inside. This is a passive mechanism by which the drug release is based solely on diffusion.

In contrast, an active mechanism pumps out the drugs in response to a specific stimulus, *e.g.* an insulin pump responding to blood sugar fluctuations.

Researchers at Penn State, led by Professor Ayusman Sen, have demonstrated an active glucose-responsive self-powered fluidic pump based on a transesterification reaction—the conversion of a carboxylic acid ester into a different carboxylic acid ester—of acyclic diol boronate with glucose.

"The scientific principle of our project is to use well-known glucose/boronate chemistry to design a self-powered micropump device," says Dr Hua Zhang, a member of Sen's group and first author of a paper on this research. "Instead of synthesizing some new molecules with a glucose/boronate reaction, we fabricated a miniature pump that utilizes the energy of this chemical reaction and pumps drugs when glucose levels are high."

Apart from demonstrating a drug delivery mechanism, which actively pumps out small molecules, the team also provided a solution to one of the challenging aspects of current self-powered nanomotor/micropump research—the issue of the power source.

Most micropump devices are simply too small to carry their own fuels. A practical solution is to design these devices so that they can consume fuels from their environment.

"In the past, many of these devices could only use toxic chemical fuels," notes Zhang. "We prove that biofriendly fuel—glucose—can be used in these systems. Our micropump is capable of converting chemical energy into mechanical motion directly."

To fabricate their glucose-responsive micropump, the researchers used a hydrogel material. "This makes the micropump highly hydrophilic so that the glucose molecules in aqueous solution have sufficient access to the hydrophobic acyclic diol boronate moieties in the pump matrix," explains Zhang.

This kind of device might become useful for advanced drug delivery systems, better wound care formulation, and power-free lab-on-chip applications.

Sen, whose group has been investigating self-powered nanomotors and pumps for several years, hopes to compile a library of devices that can perform active transportation and pumping tasks of various complexities.

"We'd like to drive our research in two directions," he says. "One is to investigate the detailed mechanisms of such systems and come up with better and more accurate modeling. The other direction is to bring in other chemistry to build new devices with different capability. In addition, the Defense Threat Reduction Agency (DTRA) is interested in developing these pumps to release antidotes in response to the presence of nerve agents in the environment."

Featured scientists: Prof. Ayusman Sen's research group (http://research. chem.psu.edu/axsgroup/index.html)
Organization: Department of Chemistry, Pennsylvania State University, University Park, PA (USA)
Relevant publication: H. Zhang, W. Duan, M. Lu, X. Zhao, S. Shklyaev and L. Liu, *et al.*, Self-Powered Glucose-Responsive Micropumps, *ACS Nano*, 2014, 8(8), 8537, DOI: 10.1021/nn503170c

5.1.6 Sneaking Drugs into Cancer Cells

A group of researchers, led by professors John Marshall and Jimmy Xu at Brown University, has encapsulated drugs inside carbon nanotubes (CNTs) for drug delivery and shown that these drugs can be released "on command" by inductive heating with an external alternating current or pulsed magnetic field.

"The important advance is that we were also able to encapsulate drugs that are fairly insoluble," Marshall points out. "For example, we encapsulated the C6-ceramide, which can dramatically increase the efficacy of many chemotherapeutic agents including Taxol, doxorubicin and histone deacetylase inhibitors (HDACi). Unfortunately, the systematic use of C6-ceramide is limited because of its hydrophobicity and precipitation as fine lipid suspensions when administered in aqueous solutions. Other approaches, including nanoliposomes have so far been unsuccessful, as a high concentration of nanoliposome ceramide was necessary to achieve anti-tumor effects."

For their work, the team synthesized CNTs with a relatively large inner diameter (~40 nm) in an array with a perfectly aligned vertical orientation. They then opened both ends of the nanotubes with either a mechanical or chemical treatment. By depositing droplets of the drug solution on top of the vertically aligned nanotube array and applying vacuum suction at the bottom, the CNTs were then loaded with the desired drug.

The drug solution was mixed with a temperature-sensitive hydrogel, which, due to the surface tension and viscosity, keeps the gel-drug payload inside the CNTs and also prevents external water from entering the nanotube and thereby displacing the drug payload.

"Our template-synthesized carbon nanotubes are intrinsically conductive but possess higher electrical resistivity than those synthesized by arc-discharging," explains Marshall. "This property is beneficial for our purpose as the electrical resistivity will generate eddy currents and consequential resistive heat *via* magnetic field induction."

The unloading of the molecular drug cargo can be controlled by external stimulation of the CNTs with a small a.c. magnetic field pulse (which has no detectable adverse effect on healthy cells), which can reach deep into any part of the body to generate inductive heat (which also is a potential new strategy for triggering on-command drug release from heat-sensitive nanomaterials).

Marshall notes that for most nanoparticles, the drugs are attached to the outside of the structure or blended into the biodegradable polymer matrix, which can result in drug degradation over the delivery process and produce off-target toxicity.

"A major advantage of carbon nanotubes is that they provide a protective environment for the drug—toxic chemotherapeutic drugs can now be used at 100-fold lower dose, which will greatly reduce drug side-effects," he says.

During *in vitro* experiments, the team tested the ability of C6-ceramide and Taxol to kill three different pancreatic cancer cell lines and found that combining these two agents at a relatively low dose potently produced cell death. They then incubated the same lines of pancreatic cancer cells with CNTs loaded with this drug combination.

The loaded CNTs, which were taken up by the cancer cells, could be opened with a 30-minute application of a 25 kHz a.c. magnetic field. As a result, the encapsulated drugs were released inside the cells, which, in over 70% of the cells, led to cell death.

Marshall points out that, without magnetic field induction, the cells had a viability of almost 98%, indicating the exceptional stability of encapsulated drugs within the CNT throughout the 48-hour-long incubation.

People have been working with similar approaches to sneak drugs into tumors. Previously, metal halides were encapsulated into CNTs, but this process involves a high-temperature (900 °C) molten-phase loading, which is unsuitable for biologics and most drugs.

"We needed a new approach to deliver very toxic drugs to tumor cells and avoid harming healthy cells," says Marshall. It was Jimmy Xu's lab here at Brown who engineered these nanotubes and when he contacted my lab I immediately realized the potential in cancer treatment."

Having demonstrated that their approach works *in vitro*, the next challenge for the scientists is to show it also works in whole animal studies and, of course, eventually people.

Featured scientists: Prof. John Marshall (https://vivo.brown.edu/display/ jomarsha); Prof. Jimmy Xu's Laboratory for Emerging Technologies (http://opto.brown.edu/)
Organizations: Department of Molecular Pharmacology, Physiology and Biotechnology, Brown University; School of Engineering, Brown University; Providence, RI (USA)
Relevant publication: C. Wu, C. Cao, J. Kim, C. Hsu, H. Wanebo and W. Bowen, *et al.*, Trojan-Horse Nanotube On-Command Intracellular Drug Delivery, *Nano Lett.*, 2012, **12**(11), 5475, DOI: 10.1021/nl301865c

5.1.7 Nanoparticle-Corked Nanotubes as Drug Delivery Vehicles

Nitrogen-doped carbon nanotubes (CNTs) have been extensively investigated for fuel cell applications due to their excellent electrocatalytic properties. However, their biomedical applications were comparatively less investigated despite reports of their better biocompatibility.

When considering carbon nanotubes for drug delivery applications, it is desirable to develop strategies that allow utilization of their hollow inner cavities for maximum loading capacity. A small size and facile surface modification are also preferable with regard to their biomedical compatibility.

Nitrogen-doped CNTs already have been demonstrated to have better biocompatibility and mitigated cytotoxicity as compared to traditional undoped pristine CNTs.

Taking advantage of this, Alexander Star, a professor in the Department of Chemistry at the University of Pittsburgh, and his team, used nitrogen doping of CNTs, which resulted in the formation of cup-shaped compartments in CNTs uniquely suitable for encapsulation. The resulting nitrogen-doped carbon nanotube cups (NCNCs or nanocups)—a new cup-shaped carbon nanomaterial with diverse reactivity—can be corked by gold nanoparticles to form enclosed nanocapsules. The nanocups were obtained from fibrous nitrogen-doped CNTs by ultrasonication.

"With their cup shapes, controlled sizes, and versatile chemical properties, the nanocups can serve as ideal drug delivery vehicles with multifunctionality for targeted drug delivery and better biocompatibility," says Star.

This work was motivated by the unique morphology of nitrogen-doped CNTs. "Due to the nitrogen-doping, the tubular structures of CNTs are changed to compartmented structures resembling many individual nanocups stacked up," explains Star. "These hollow nanocups, possessing nitrogen functionalities, can serve as potential nano-containers if successfully isolated out. Using high-resolution transmission electron microscopy (TEM) analysis, we found that the adjacent nanocups have no covalent interaction with each other but are merely physically stacked, which allowed the mechanical separation of the nanocups by ultrasonication ('shaking')."

To be utilized, the stacked cup-shaped nitrogen-doped CNTs need to be separated in order to obtain individual nanocups. Previous attempts at nanocup separation by Star's team involved grinding with a mortar and pestle but resulted only in a limited yield.[3] In this follow-up work, the researchers used high-intensity ultrasonication, which resulted in a much higher yield.

"By adopting the quantitative Kaiser test, we for the first time determined and quantified the amine groups as the major functionalities on the separated nanocups, and we subsequently explored the reactivity and distribution of the amine groups by functionalization with gold nanoparticles and found that the amine groups were preferentially located at the open rims of the nanocups," says Pitt graduate chemistry student Yong Zhao, lead author of a paper on this work. "Taking advantage of this fact, we managed to effectively cork the nanocups by gold nanoparticles with commensurate sizes to create a new type of cup-shaped nanocontainer with a corked opening."

The potential application of this work is to use the nanocups as nanoscale containers, especially as drug delivery vehicles. Star points out that the gold nanoparticle corked nanocups have all the desired properties for biomedical applications, including their hollow, self-confined cup structures, diverse reactivity, and biocompatibility. To make an effective drug delivery vehicle, the surface of nanocups can be functionalized with different moieties such as bio-recognition groups or fluorescent labels. The gold nanoparticle corks can be designed to be open under certain stimuli, to achieve a controlled release of drugs.

There still are some issues the team has to overcome. Although they managed to obtain the individual nanocups at a higher yield compared to previous attempts, incompletely separated nanocups still frequently exist. They appear as short stacks of cups of several units. Star says that size-exclusion separation, such as nanopore-sized filtration, can further increase the concentration of the individual nanocups.

Another problem is that it is hard to cork every separated nanocup with gold nanoparticles due to the scarcity of intrinsic amine groups and the heterogeneous morphology of nanocups. Although gold nanoparticles have a preferential interaction with the opening of the nanocups, the best percentage of corked nanocups was about 23%. Star's team is working on improving the efficiency of their method.

Going forward, the scientists are planning to functionalize the corked nanocups with diverse functional groups for *in vivo* experiments, for example, fluorescent groups that allow the imaging of the nanocups in the tissue.

"In addition, we are investigating the biodegradation behavior of the nanocups using peroxidase enzymes which, as we reported previously, can enzymatically degrade both single-walled and multiwalled carbon nanotubes," says Star. "The challenges facing future research in this area lie in the design of specific cell-targeting and controlled release of drugs in the target cells."

Featured scientists: Prof. Alexander Star's research group (http://www.pitt. edu/~astar/index.htm)
Organizations: Department of Chemistry, University of Pittsburg, PA (USA)
Relevant publication: Y. Zhao, Y. Tang, Y. Chen and A. Star, Corking Carbon Nanotube Cups with Gold Nanoparticles, *ACS Nano*, 2012, **6**(8), 6912, DOI: 10.1021/nn3018443

5.1.8 Plasmonic Nanocrystals for Combined Photothermal and Photodynamic Cancer Therapies

Photothermal therapy (PTT) is a form of cancer treatment where a therapeutic agent absorbs energy from photons and dissipates it partially in the form of heat. When the therapeutic agents, for instance nanoparticles, are located in close vicinity to the tumor site, the temperature increase can lead to cell damage, *i.e.* it kills the cancer cell. Research on PTT has made huge progress thanks to various near-infrared light (NIR) absorbing—*i.e.* plasmonic— nanomaterials that have been developed in the past years.

A similar approach uses light instead of heat and is called photodynamic therapy (PDT). This technique requires the use of a chemical compound— also known as a photosensitizer—with a particular type of light to kill cancer cells. The photosensitizer in the tumor absorbs the light and generates reactive oxygen species (ROS)—such as a hydroxyl radical, singlet oxygen, as well as peroxides—that destroy nearby cancer cells.

While these techniques have been around for years, more recently the use of nanomaterials such as various forms of gold nanoparticles (rods, cages, spheres), quantum dots and iron-oxide nanoparticles has allowed researchers to refine their therapeutic methods with a view to also exploring the mechanisms behind the efficacy.

Scientists also developed combinations of nanomaterial-mediated PTT and organic photosensitizer-mediated PDT to achieve synergistic therapeutic effects. However, most of these approaches achieved only tumor suppression rather than complete destruction, especially under low laser dose conditions.

Among the nanomaterials used, copper sulfide nanocrystals stand out because they can efficiently absorb near-infrared light at the 700–1100 nm range, which is considered as "transparent" to human tissue at this energy level. Another reason that these plasmonic nanocrystals have attracted much attention as materials for PTT is their small size, which leads to the possibility of deeper tumor permeation.

Previous reports have correlated photo induced cell death to the photothermal heat mechanism of copper sulfide nanocrystals, but no evidence of their photodynamic properties had been reported yet.

An international team of researchers led by Drs Teresa Pellegrino, Huan Meng, and Huiyu Liu, used abiotic assays, cultured cancer cells, and a melanoma animal model to demonstrate the PTT activity of copper sulfide nanocrystals. The paper on this research lays out the working principle of colloidal, NIR plasmonic copper sulfide nanocrystals exploitable for both PDT and PTT therapy with NIR activation.

This is the first report that under NIR light radiation, copper sulfide nano-crystals ($Cu_{2-x}S$) achieve efficient cancer destroying efficacy *via* PTT and PDT mechanisms both *in vitro* and *in vivo*.

"Our findings demonstrated the dual functionalities of copper sulfide nanocrystals, which are capable of melanoma cancer inhibition under NIR irradiation *via* photothermal therapy and photodynamic therapy mediated mechanisms," says Huiyu Liu, an associate professor at the Technical Institute of Physics and Chemistry, Chinese Academy of Sciences. "This is the first report demonstrating that the leakage of copper ions from copper sulfide nanocrystals could enhance the ROS generation under NIR light irradiation, which serves as a new mechanism in addition to the sole PTT mechanism."

Investigating the ROS generation mechanism, the researchers were able to show that the reduction of dissolved Cu^{2+} ions leads to Cu^+ ions, which further interact with the biological redox molecules, *i.e.* ascorbic acid and glutathione, and thus trigger the ROS generation.

"Interestingly," Liu notes, "while we worked with a scenario involving nanoparticles, our theory behind the ROS generation is supported by a classic chemistry study, also known as the Haber–Weiss cycle, proposed by Kadiiska *et al.* more than 20 years ago."[4]

Based on the promising effect of photothermal therapies, the research team is confident that their dual functional $Cu_{2-x}S$ nanocrystals could lead to an even more potent platform for cancer treatment.

"We are also considering to perform additional acute and chronic tests for our platform including the use of multiple melanoma animal models to confirm our findings in a B16 murine model," says Liu. "Based on our proof-of-principle results, further studies are required to evaluate and optimize the PTT, or PDT, or dual functional platforms as we demonstrated in various cancer models, *i.e.* melanoma and head and neck carcinoma, with a view to also look at nanosafety in an acute and chronic phase."

She adds that, from a manufacturing perspective, it is necessary to consider the scalability of nanomaterial production as well as quality control.

Featured scientist: Prof. Huiyu Liu (http://sourcedb.ipc.cas.cn/cn/lhs-rck/201209/t20120915_3644823.html)
Organization: Technical Institute of Physics and Chemistry, Chinese Academy of Sciences, Beijing (PR China)
Relevant publication: S. Wang, A. Riedinger, H. Li, C. Fu, H. Liu and L. Li, *et al.*, Plasmonic Copper Sulfide Nanocrystals Exhibiting Near-Infrared Photothermal and Photodynamic Therapeutic Effects, *ACS Nano*, 2015, 9(2), 1788, DOI: 10.1021/nn506687t

5.1.9 Remotely Activating Biological Materials with Nanocomposites

The heating properties of iron oxide nanoparticles have been exploited through the years for use in cancer therapy, gene regulation, and temperature responsive valves. These applications have demonstrated the versatility of iron oxide nanoparticles, but they had rarely, if ever, been used to enhance the activity of thermophilic enzymes.

Thermophilic enzymes are highly stable biomolecular systems that are excellent tools due to their thermostability and long-term activity for extended lifetime uses.

Traditionally, these enzymes are heated in an oven or water bath at a specified temperature. These heating methods are not efficient and much of the heat can be lost due to diffusion limitations of the heat to the sample.

Research work by researchers in the U.S. has addressed the problem of remotely activating biological materials with a higher efficiency than conventional methods such as water baths or convection ovens.

"In our work, we fabricated a hydrogel in which we chemically immobilized a thermophilic dehalogenase and encapsulated iron oxide (Fe_3O_4) nanoparticles," explains Sylvia Daunert, Professor and Lucille P. Markey Chair, Department of Biochemistry and Molecular Biology at the University of Miami. "Previous studies have shown that these nanoparticles heat up when in the presence of an alternating magnetic field (AMF). We used the hydrogel nanocomposites in the AMF to find the field strength for optimum enzymatic activity. Additionally, we heated the hydrogel nanocomposites in a water bath at the optimal temperature for enzymatic activity."

Comparing these results, it was found that the enzyme was two times more efficient when heating in the AMF as compared to heating in a water bath in one-third of the time.

Daunert points out that this heating method is localized as compared to other heating methods. "Furthermore, by using an AMF, which is not limited by penetration depth for uses *in vivo* like other nanomaterials, we envision that this method could be expanded for personalized therapeutics, bioremediation, catalysis, filtering devices, separations, *etc.* In short, this application could be potentially applied to any system where you need to turn on and off bioactivity."

This work was the result of collaboration between Daunert's lab and J. Zach Hilt's group in the Department of Chemical and Materials Engineering at the University of Kentucky.

"Dr Hilt's group used an AMF to heat the Fe_3O_4 nanoparticles for hyperthermia treatment in cancer patients," says Daunert. "My lab focused on recombinant proteins for biomedical and environmental analysis. The enzyme had recently been characterized in my lab and we felt it was an ideal enzyme to show the proof of concept that we could efficiently heat and activate an enzyme remotely using Fe_3O_4 nanoparticles."

Previous research had already shown that by combining the remote heating capabilities of magnetic nanoparticles with the stimuli responsive properties of hydrogels, multifunctional materials can be designed for targeted applications such as drug delivery or hyperthermia cancer.

What is new in this work is that the responsive hydrogel nanocomposite system includes a thermophilic enzyme that can be remotely switched on or off. The team focuses on using a thermophilic dehalogenase for remote controlled dehalogenation of samples. By encapsulating the enzyme into the hydrogel, it allows for the use of the biocatalyst through many enzymatic cycles with different substrates without the need for complicated separation and/or regeneration steps.

"Although this current work focuses on bioremediation, it can be expanded to other applications where thermophilic proteins are employed," says Daunert. "Additionally, the enzyme and nanoparticles in our work were immobilized in a hydrogel network but for future applications, various matrices can be used such as weaving the protein into textiles or incorporation into thin films."

This means that, due to the numerous available methods and chemistries used to functionalize proteins, there are many materials that can be used as platforms to incorporate both the nanoparticles and protein of interest for a variety of applications.

For instance, the scientists envision that their technique could be applied to molecular biology techniques such as using thermostable DNA polymerase for polymerase chain reactions. Remote heating may be plausible in industrial settings where thermophilic enzymes are used for processes such as hydrolysis of starches, cleavage of proteins, or use of thermophilic lipases. The benefit of using a localized heating method allows for less energy input for a greater catalytic output, which could provide cost savings for some of these applications.

Daunert notes that the library of thermophilic organisms that are being discovered is growing. "As this number grows, the breadth of capabilities of this system can be extensive, limited only by the number and type of thermophilic proteins available. Incorporation of thermophilic enzymes into additional platforms beyond hydrogels represents an important advance for designer applications where additional experimental challenges will be met and further research will be required."

Featured scientists: Prof. Sylvia Daunert (http://bm.med.miami.edu/faculty/primary-faculty/sylvia-daunert); Prof. J. Zach Hilt (https://www.engr.uky.edu/research/researchers/james-hilt/)
Organizations: Department of Biochemistry and Molecular Biology, University of Miami, FL (USA); College of Engineering, University of Kentucky, Lexington, KY (USA)
Relevant publication: L. Knecht, N. Ali, Y. Wei, J. Hilt and S. Daunert, Nanoparticle-Mediated Remote Control of Enzymatic Activity, *ACS Nano*, 2012, **6**(10), 9079, DOI: 10.1021/nn303308v

5.1.10   Pre-Coating Nanoparticles to Better Deal with Protein Coronas

A major goal in nanomedicine is to make nanoparticles with the capability of *in vivo* targeting for both imaging and delivery of therapeutic biomolecules in the human body. One strategy to obtain a high targeting yield is to functionalize the surface of nanoparticles with targeting ligands—such as antibodies or aptamers—that enhance the binding of nanoparticles to receptors on the target cells and facilitate nanoparticle uptake by receptor-mediated endocytosis.

Previous reports have shown that when nanoparticles enter the blood stream, the proteins that adsorb onto nanoparticles when they enter the body form a protein corona that hinders interactions between the targeting ligands on the nanoparticles and their binding partners on the cells' surface.

To address this issue, scientists have developed a strategy that enables directing the formation of protein coronas on nanoparticles that are enriched in plasma proteins with natural targeting capabilities. This involves pre-coating nanoparticles with selected proteins prior to their exposure to the desired biological environment.

"One strategy that has been proposed to address this challenge is to exploit the protein corona itself for targeted nanoparticle delivery," says Morteza Mahmoudi, an assistant professor at Tehran University of Medical Sciences, who heads the Laboratory of Nano-Bio Interactions, and a visiting professor at Stanford University's School of Medicine. "We were motivated to investigate the feasibility and limitations of this emerging strategy because few studies have probed whether plasma proteins with natural targeting capabilities could be selectively recruited to the nanoparticles' surfaces and used to direct nanoparticle uptake by desired cell types."

The team's findings show that the surface of a nanoparticle can be engineered to promote the formation of a hard protein corona that is enriched with desired proteins, but a lack of control over the conformations and accessibility of the recruited proteins is currently a limitation of this approach (Figure 5.1).

The team, led by Mary L. Kraft, an associate professor at the University of Illinois at Urbana-Champaign, and Mahmoudi, demonstrated that by pre-coating nanoparticles with proteins that promote the adsorption of desired plasma proteins, they can direct the formation of a protein corona on the surface of a nanoparticle that is enriched in plasma proteins with natural targeting capabilities.

"However, the corona proteins with targeting capabilities did not significantly promote nanoparticle uptake by the desired cells because they were screened by other proteins in the hard corona," says Kraft. "Therefore, to form a functional corona with targeting capabilities, the desired plasma proteins must not only be enriched in the corona, but they must also be

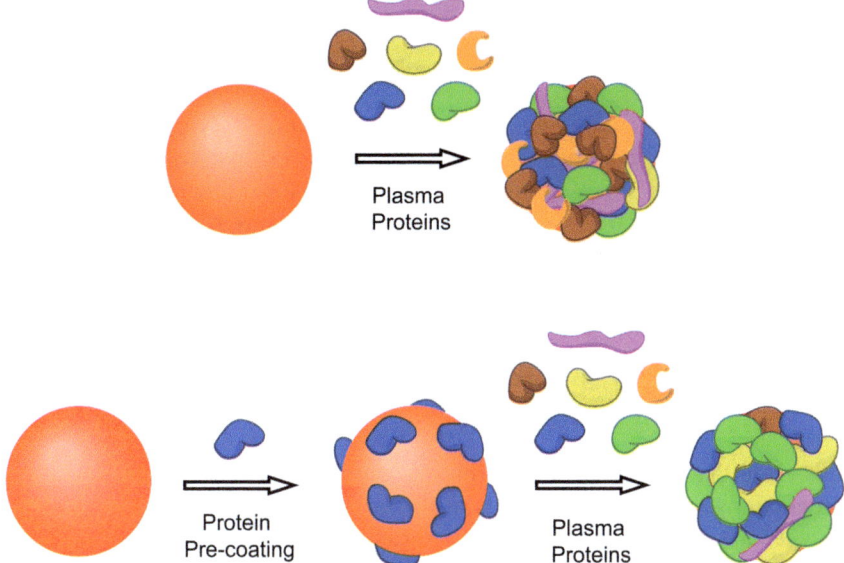

Figure 5.1 Schematic of directing protein corona formation by engineering the surfaces of nanoparticles. Pre-coating nanoparticles with select proteins prior to their exposure to plasma can promote the binding of desired plasma proteins that have targeting capabilities (yellow species). However, the recruitment of plasma proteins with targeting capabilities may not promote nanoparticle uptake because these desired proteins are covered by other plasma proteins (green species). (Image: Laboratory of Nano-Bio Interactions, Tehran University of Medical Sciences.)

positioned such that their binding motifs are accessible to the receptors on the surfaces of the target cells."

These results indicate that strategies for producing protein coronas with targeting capabilities on nanoparticles must not only promote the recruitment of the desired plasma proteins to the nanoparticles' surfaces, but also control their orientations so that they can access binding partners on the target cells.

With further development, nanoparticles functionalized with protein coronas that have targeting capabilities might be used for diagnostic applications—by sensitive targeted imaging—and therapeutics by selectively transporting cytotoxic, *i.e.*, anticancer, drugs to the diseased cells in the body, which would reduce the undesired side effects of these therapies.

"Future directions of our research would be to develop strategies that enable controlling the conformations and accessibility of the desired proteins that are recruited to the protein corona so that they can interact with their binding partners on the target cells," conclude Kraft and Mahmoudi. "We expect that preventing undesired plasma proteins from adsorbing on top of the desired plasma proteins is going to be a major challenge."

Featured scientists: Prof. Morteza Mahmoudi's research group (http://www.biospion.com/); Prof. Mary L. Kraft (http://www.chemistry.illinois.edu/faculty/mary_kraft.html)
Organizations: Tehran University of Medical Sciences (Iran); Department of Chemistry, University of Illinois at Urbana-Champaign, IL (USA)
Relevant publication: V. Mirshafiee, R. Kim, S. Park, M. Mahmoudi and M. Kraft, Impact of protein pre-coating on the protein corona composition and nanoparticle cellular uptake, *Biomaterials*, 2016, 75, 295, DOI: 10.1016/j.biomaterials.2015.10.019

5.2   Sensors and Nanoprobes for Everything—Down to Single Molecules

Future medical diagnostics will rely on nanotechnology-enabled sensors to detect changes in individual cells, for instance cell surface charge, to detect diseases at their earliest stage. Since diseased cells, such as cancer cells, frequently carry information that distinguishes them from normal cells, accurate probing of these cells is critical for early detection of a disease.

Early and accurate detection of cancer is critical for successful cancer therapies. In most cases, a tissue biopsy is the initial means of making a diagnosis after some form of cancer screening has come up with a positive result. However, tumor antigen-specific autoantibodies—antibodies produced by the immune system that are directed against one or more of the individual's own proteins—are known to appear months, even years, before clinical diagnosis of cancer, and autoantibodies have been found in many types of cancer.

With increasing accuracy, nanosensor platforms for detecting cancer biomarkers are becoming viable complements to invasive biopsies of metastatic tumors; although, for the time being, it still appears likely that biopsies will have to be used for a final diagnosis.

5.2.1   A Quick and Simple Blood Test to Detect Early-Stage Cancer

Unfortunately, despite recent advances in molecular diagnostics, non-invasive screening tests for early-stage detection of most cancer types are almost non-existent.

"Most studies are attempting to identify tumor-specific antigens and detect antibodies that are specific to individual tumor-associated antigens," says Qun Huo, an associate professor in the NanoScience Technology Center at the University of Central Florida. "Different from these approaches, the novel nanoparticle test that we developed detects an overall increase of human immunoglobulin G (IgG), including the tumor-specific autoantibodies, adsorbed to a gold nanoparticle surface."

"On one hand, this test may not be able to identify the specific type of cancer; on the other hand, this test may potentially be able to detect early stage tumor-induced immune responses associated with a broad spectrum of cancer types, making this test potentially a universal screening test for cancer risk assessment," she points out.

Huo worked as part of an interdisciplinary team that developed the extremely facile new blood test: it comprises two simple liquid mixing processes. The test results are straightforward, easy to interpret, and do not require any further data processing or statistical analysis. The cost of the test is low: the material cost is less than $1 per test. The test results are obtained within minutes.

There are two innovative elements in this work, and both innovations are based on the unique properties of gold nanoparticles.

The first one is the development of a gold nanoparticle-enabled dynamic light scattering assay (NanoDLSay) technique for molecular detection.

The second one is the discovery of cancer biomarker molecules from the "protein corona" formed on the gold nanoparticles upon adsorption of blood serum proteins to the nanoparticle surface.

"With these two innovations, we developed a simple blood test that detects molecular differences between cancer patients and healthy controls," says Huo. "Early cancer detection faces two-fold challenges: one is to discover the biomolecules that can become cancer biomarkers, and the second one is the development of suitable techniques for the detection of these cancer biomarker molecules. Our development opens the door for potential solutions to both challenges."

The new blood test exploits the unique optical property of gold nanoparticles, well known for their strong light scattering at the visible light region. Researchers have been using this optical property of gold nanoparticles for biomolecular and cell imaging for a while, because gold nanoparticles can easily stand out from the background scattering of biological cells under the dark-field optical imaging mode.

Dynamic light scattering (DLS) is a technique used routinely for nanoparticle size characterization and analysis. DLS relies on the light scattering property of nanoparticles for size measurement. Most researchers working in the nanoparticle field either have DLS instruments in their own laboratories or have ready access to DLS instruments at their central facility. Moreover, DLS is a relatively inexpensive and easy-to-use instrument.

"Our unique contribution is to combine the strong light scattering property of gold nanoparticle probes with the size measurement capability of DLS, and to create a quantitative analytical technique, nanoparticle-enabled dynamic light scattering (NanoDLSay) for chemical and biomolecular detection and quantitative analysis," notes Huo.

NanoDLSay detects chemical and biological target molecules and species by simply measuring the size change of the gold nanoparticle probes upon binding with target analytes (Figure 5.2).

The surface chemistry of gold nanoparticles can be designed to contain specific target-recognizing molecular species such as antibodies, DNA probes,

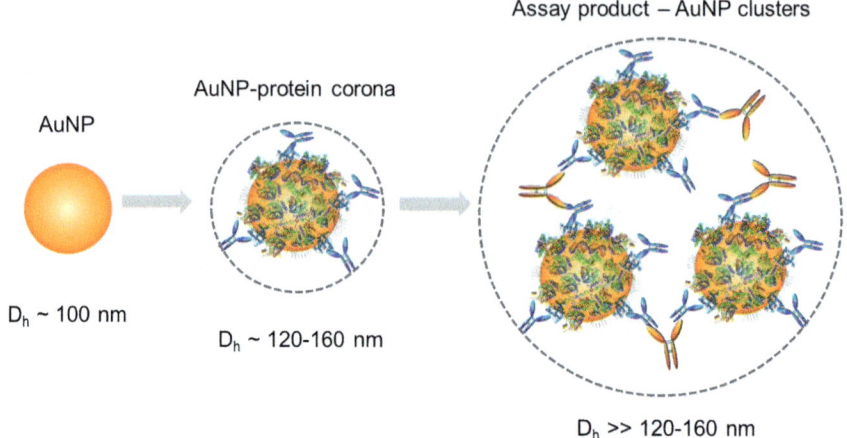

Figure 5.2   Detection of autoantibodies and other target proteins in the gold
                  nanoparticle protein corona by measuring the average particle size
                  (hydrodynamic diameter D_h) change of the nanoparticles upon binding
                  with specific antibodies to the target proteins. (Image: Dr Huo, Univer-
                  sity of Central Florida.)

carbohydrate binding proteins, and metal-chelating chemical ligands. Upon
binding with the target analytes, the analytes cause the gold nanoparticle
probes to cluster together. As a result, the average particle size of the assay
solution is increased, and such size increases can be readily detected by DLS.

"We and other research groups have so far successfully used NanoDLSay
for a wide range of target analyte detection and quantitative analysis, including proteins, DNAs, antibodies and antigens, viruses, carbohydrates, toxic
metal ions, small molecules, food and environmental toxins," says Huo.
"The published studies show that as a quantitative technique, NanoDLSay is
highly sensitive and reproducible."

The above-mentioned unique surface chemistry of gold nanoparticles
also plays another important role in the way this blood test works. It has
been known that when nanoparticles, such as citrate-capped gold nanoparticles, are mixed with blood, the proteins and other biomolecules will adsorb
to the nanoparticle surface to form a layer of "protein corona". Although
the concept of protein corona is most frequently used, other non-protein
biomolecules such as lipids, DNAs, and carbohydrates could also be part of
the corona.

The team's contribution is that they first proposed the hypothesis that the
molecular composition of this protein corona could reveal the cancer status.

"When a tumor starts to develop in the human body, many chemical and
biochemical reactions will occur in the tissue and the blood," explains Huo.
"Our rationale was that the overall composition and the relative quantity of
biomolecules in a cancer patient's blood should differ from a healthy person,
therefore, when a gold nanoparticle is mixed with a blood serum, the type

and amount of proteins and other biomolecules adsorbed to the nanoparticle surface could differ between cancer patients and a healthy person."

With this hypothesis in mind, the researchers applied their new Nano-DLSay technique to study the proteins in the gold nanoparticle protein corona, and indeed, they discovered that a major blood protein, human IgG, is increased in prostate cancer patients. Through proteomic analysis, they also discovered other proteins that may serve as potential cancer biomarkers.

IgG is one of the most abundant proteins in the blood. IgG antibody is a key component of our immune system. Cancer immunology research has revealed that upon tumorigenesis, our immune system will launch defensive actions against the tumor by producing autoantibodies. The team's finding of increased immune activity from prostate cancer patients provided additional evidence of the potential links between cancer and the immune system.

This study addresses a specific diagnostic problem that exists in the prostate cancer field: to distinguish early stage prostate cancer from non-cancer prostate conditions with the smallest possible error rate. One of the major problems in prostate cancer screening is the over-diagnosis of cancer, because the widely used PSA test has poor specificity in distinguishing early stage prostate cancer from BPH (benign prostate hyperplasia). In their study, the scientists found that their new test can distinguish prostate cancer patients from BPH patients with more than 90% specificity.

Whereas the team's analytical technique—NanoDLSay—can be used to address a wide range of biomedical research problems, their current focus is to identify cancer biomarker molecules from gold nanoparticle protein corona, and to develop new blood tests suitable for early stage cancer detection and diagnosis.

"We have so far mainly worked on prostate cancer," says Huo. "We are extending our study to other types of cancer. Our vision is to develop an array of blood tests for early detection and diagnosis of all major cancer types, and these blood tests are all based on the same technique and same procedure."

Going forward, the team's research will focus on two directions: one is continuing new cancer biomarker discovery using the NanoDLSay technique; and the second one is to conduct more extensive clinical studies to validate the new cancer biomarkers discovered in this study.

Featured scientists: Prof. Qun Huo (http://nano.ucf.edu/faculty/huo.php)
Organization: NanoScience Technology Center, University of Central Florida, Orlando, FL (USA)
Relevant publication: T. Zheng, N. Pierre-Pierre, X. Yan, Q. Huo, A. Almodovar and F. Valerio, *et al.*, Gold Nanoparticle-Enabled Blood Test for Early Stage Cancer Detection and Risk Assessment, *ACS Appl. Mater. Interfaces*, 2015, 7(12), 6819, DOI: 10.1021/acsami.5b00371

5.2.2 Nanoparticles Allow Simple Monitoring of Circulating Cancer Cells

With increasing accuracy, "liquid biopsies"—where circulating tumor cells (CTCs) are isolated from blood samples—are becoming a viable complement or even alternative to invasive biopsies of metastatic tumors. CTC is of great interest for evaluating cancer dissemination, predicting patient prognosis, and also for the evaluation of therapeutic treatments, representing a reliable potential alternative to invasive biopsies and subsequent proteomic and functional genetic analysis. Unfortunately, given that human blood is a complex fluid that contains a variety of cells and metabolites, the fast detection of CTCs is quite a difficult task.

"The main techniques reported for CTC detection consist in their labeling with tagged antibodies (immunocytometry) followed by fluorescence analysis or the detection of the expression of tumor markers by reverse-transcriptase polymerase chain reaction," says Dr Alfredo de la Escosura Muñiz, a senior researcher in the Nanobioelectronics & Biosensors Group at the Institut Català de Nanotecnologia. "However, the previously reported isolation of CTCs from human fluids are limited to complex analytic approaches that often result in a low yield and purity."

Escosura and his collaborators, led by Arben Merkoçi, describe a rapid and simple electrochemical biosensing strategy to quantify circulating tumor cells based on the simultaneous use of antibody-coated magnetic beads, which selectively bind to the cancer cells for subsequent magnetic isolation, and antibody-coated gold nanoparticles, to also selectively bind to the cancer cells for final electrochemical detection.

"We combine, for the first time, the use of both micro- and nanoconjugates and the simple electrochemical detection methodology for the sensitive and selective quantification of circulating tumor cells," says Escosura.

EpCAM (epithelial cell adhesion molecule) and CEA (carcinoembryonic antigen) proteins, both expressed by Caco2 cells, were selected as targets for cell detection using biofunctionalized electrochemical labels. For the preparation of these labels, the team synthesized 20 nm gold nanoparticles and then conjugated them with anti-EpCAM or anti-CEA antibodies.

For the magnetic capture of Caco2 cells, EpCAM glycoprotein was used as a target for microbead capture, and mouse monoclonal anti-EpCAM was chosen to create biofunctionalized microbeads (uniform 2.8 µm polystyrene beads with a magnetic core) for the capture of CTCs.

Escosura points out that the main advantages of the presented method, compared with ones already reported for circulating tumor cell detection (*i.e.*, optical methods, ELISA, RT-CPR), rely on the sensitive and quantitative electrochemical detection technique used.

"The sensitivity, simplicity, low cost, easy-to-use mode, and miniaturization/portability of the electrochemical detection system makes it ideal for point-of-care applications," he says. "Furthermore, the use of gold nanoparticle electrocatalytic labels gives rise to very simple electrochemical records that allow the rapid quantification of the cells in the sample."

The team envisions the application of their method to the quantification of circulating tumor cells in real human samples where, besides cells (cancerous and noncancerous ones), proteins and metabolites are also present. This method can also be adapted for other cancer cells by redesigning both micro- and nanoconjugates with the appropriate antibodies.

Future research could take advantage of this technique for isolation, labeling and sensitive electrochemical detection/quantification of circulating tumor cells by incorporating it into lab-on-a-chip systems, which could contribute to the desired standardization of circulating cancer cell detection technologies.

Featured scientists: Dr Alfredo de la Escosura Muñiz, Prof. Arben Merkoçi
Organization: Nanobioelectronics & Biosensors Group at the Institut Català de Nanotecnologia, Barcelona (Spain) (http://www.nanobiosensors.org/)
Relevant publication: M. Maltez-da Costa, A. de la Escosura-Muñiz, C. Nogués, L. Barrios, E. Ibáñez and A. Merkoçi, Simple Monitoring of Cancer Cells Using Nanoparticles, *Nano Lett.*, 2012, **12**(8), 4164, DOI: 10.1021/nl301726g

5.2.3 Multiplexing Biosensors on a Chip for Human Metabolite Detection

The convergence of nanotechnology, biology, and photonics opens the possibility of entirely new classes of biosensing and imaging nanodevices. The application of nanomaterials such as carbon nanotubes, nanowires, and graphene to biosensor design and fabrication promises to revolutionize diagnostics and therapy at the cellular and even molecular level. These nanoprobes and nanosensors have the potential for a wide variety of medical uses down to the cellular level.

So far, there have been very few research reports on single electrode materials that enable the simultaneous detection of different metabolites—such as glucose, urea, cholesterol, and triglycerides—in whole blood. Moreover, it is a considerable challenge to integrate all required materials and devices on a single chip to ultimately produce a multiplexing biosensor array.

In a significant step towards fabricating integrated sensor arrays on a chip to recognize different metabolites with the simplified technology of only one single electrode material system, researchers have demonstrated that biosensors based on conducting polymer hydrogels (CPHs) enable the precise and full-range detection of different metabolites in human blood.

"We wanted to demonstrate that there is a kind of nanomaterial that can meet the requirements of a biosensing platform to detect all metabolites simultaneously with high sensitivity while being compatible with simple patterning technology," says Guihua Yu, an assistant professor of materials science at the University of Texas at Austin. "Synergizing the advantages of both organic conductors and hydrogels, nanostructured CPHs present excellent electrochemical performance and good biocompatibility. They also provide an advantageous interface between the substrates and electrode. All these advantages make CPHs excellent candidates for high-performance biosensors—as we confirmed in our work."

To develop their biosensor platform, Yu's team fabricated a platinum nanoparticle modified CPH electrode. Polyaniline hydrogels provide hierarchically porous, nanostructured matrices, and particularly, solvated surfaces resulted from the hydrophilic nature of the hydrogels. The team homogeneously dispersed enzymes and platinum nanoparticles in the hydrogel matrix to catalyze the electrochemical oxidization of the hydrogen peroxide molecules generated in the enzymatic reaction between the substrates and enzymes and to enhance the current collection in electrochemical processes.

"Owing to the unique features of conducting polymer hydrogels, such as high permeability to biosubstrates and rapid electron transfer, our biosensors demonstrate excellent sensing performance with a wide linear range (uric acid, 0.07–1 mM; cholesterol, 0.3–9 mM, and triglycerides, 0.2–5 mM), high sensitivity, low sensing limit, and rapid response time of about 3 seconds," says Lijia Pan, a professor at the School of Electronic Engineering, Nanjing University, who, together with Yu, led the project.

In their experiments, the researchers demonstrated the detection of different metabolites separately on the same conductive-polymer-based bioelectrode platform.

However, there still are some hurdles to overcome before we will see multiplexing biosensors on a chip that are fabricated at low cost and on a large scale: eliminating the mutual interference between the various metabolites is the next key challenge to solve. The related work is ongoing in Yu's lab.

"We envision that, once it is possible to successfully fabricate these multiplexing biosensors, we will see a very sensitive yet inexpensive 'vital reader' on the market that can tell people accurately the values of 3–4 key human metabolites with only one single drop of blood," says Yu.

His team is already working towards building such multiplexing biosensors on a chip and designing and implementing the seamless connection of the biosensor and a portable terminal (*e.g.* a smartphone) by wireless communication.

Featured scientists: Prof. Guihua Yu's research group (http://yugroup. me.utexas.edu/); Prof. Lijia Pan (http://ese.nju.edu.cn/faculty. php?name=panlijia)
Organizations: University of Texas at Austin (USA); School of Electronic Engineering, Nanjing University (PR China)
Relevant publication: L. Li, Y. Wang, L. Pan, Y. Shi, W. Cheng and Y. Shi, *et al.*, A Nanostructured Conductive Hydrogels-Based Biosensor Platform for Human Metabolite Detection, *Nano Lett.*, 2015, **15**(2), 1146, DOI: 10.1021/nl504217p

5.2.4 Multimodal Biosensor Integrates Optical, Electrical, and Mechanical Signals

A prime example of how the integration of multiple disparate nanotechnology fields allows the realization of novel or expanded functionalities, researchers at the University of Pennsylvania have demonstrated a multimodal sensing device, which integrates the functionalities of three traditional single mode sensors.

Specifically, the team, led by Ertugrul Cubukcu, an assistant professor in the Department of Materials Science and Engineering (he moved to the University of California – San Diego in December 2015), fabricated a graphene-based multimodal biosensing device, capable of transducing protein binding events into optical, electrical, and mechanical signals.

So rather than focusing on improving the performance of single-mode devices by developing ever more sophisticated designs and better binding-receptor molecules, the researchers' tool is a lateral approach that combines several modes of operation on a single device platform.

This means that instead of detecting or analyzing biological samples either *via* optical, electrical or mechanical means, as most sensors do, it is possible to obtain information regarding all three properties from just a single sample.

This work uniquely integrates a nanoelectronic graphene field effect transistor based sensor and a nanoantenna based optical sensor on a nanomechanical resonator based sensor. As the team points out, this multimodal biosensor combines all the advantages of the respective single-mode sensors and achieves a 100 times improvement in the sensing dynamic range.

"Our device consists of a freestanding low stress silicon nitride membrane clamped on all sides to a silicon frame, a configuration capable of supporting high quality (Q) factor mechanical resonance modes which are highly sensitive to adsorbed mass," Fei Yi, a postdoctoral researcher in Cubukcu's group, describes the device structure. "It also serves as a structural support for the subsequent introduction of plasmonically active gold nanodisk antennae and graphene monolayer transferred onto the top surface. The nanoantennae enable surface refractive index sensing *via* their spectrally resonant electromagnetic nearfields, while the graphene acts as a traditional field effect transistor (FET) sensing channel and bioactive interface for protein adsorption. A 100 nm thin metal coating on the underside of the membrane serves as the gate electrode—with the membrane itself as the gate dielectric."

Yi points out that, to do this without perturbing the individual performances of each component is not as straightforward as it seems, particularly in the realm of nanoscale devices.

"When you are talking about co-locating active components that are individually highly sensitive to perturbations, it is very likely that performance degrades as a result," says Yutong Zhu, another member of the team. "However, with the proper choice of materials and design architecture, this can be overcome, and overall performance can even be enhanced by establishing synergies between the different sensing modes."

This is the reason graphene is a key material critical to this device design: not only does it act as a functional element—graphene FET sensor for electrochemical detection of analytes—but its atomic thickness does not significantly perturb the optical near-fields and mass response of the plasmonic and mechanical counterparts.

Furthermore, graphene's affinity to biological molecules compared to inorganic surfaces such as gold or silicon more than compensates for the perturbations associated with integrating these components.

This integrated sensor approach has two distinct benefits: it expands upon current sensor capabilities because for the same amount of analyte, three (or potentially even more) pieces of information can be obtained. It is also very straightforward to have each mode of sensing targeting a different concentration regime, thereby effectively extending the linear dynamic range of the device by orders of magnitude.

Multimodal biosensors promise to create entirely new functionalities. For instance, it becomes feasible to differentiate and, ideally, identify multiple distinct biomolecules from a mixture containing numerous unknowns, as opposed to only trying to detect for the presence of a few types of known molecules.

"The only way to do this is to have real time, simultaneous measurements of a single analyte sample across several parameters—which hopefully are sufficient to uniquely establish the identity of the molecule," says Yi. "In this work, we have shown that the latter is possible, while the former—simultaneity and in real time—present unique technical challenges of their own, and is something we are working towards."

Featured scientists: Prof. Ertugrul Cubukcu's Nanoengineered Photonics Group (http://www.seas.upenn.edu/~cubukcu/Cubukcu_Lab/Home.html)
Organizations: Department of Materials Science and Engineering, University of Pennsylvania, Philadelphia, PA (USA) (Prof Cubukcu moved to UC San Diego in December 2015)
Relevant publication: A. Zhu, F. Yi, J. Reed, H. Zhu and E. Cubukcu, Optoelectromechanical Multimodal Biosensor with Graphene Active Region, *Nano Lett.*, 2014, **14**(10), 5641, DOI: 10.1021/nl502279c

5.2.5   Detecting Damaged DNA with Solid-State Nanopores

DNA is constantly being damaged in our cells by radiation and other random sources. There is a long list of structural and chemical variations to DNA that have an impact on human health. One of the major forms of this damage is called depurination, or the selective loss of A and G bases from the double helix structure.

In our cells, there is a system in place to fix depurination, called the base excision repair (BER) pathway. It usually is quite successful at repairing the damage, but can sometimes make mistakes that result in mutations. As a result, depurination is directly linked to a host of diseases, including anemia and cancer.

Better tools for the study of depurination could therefore be important in studying these diseases or potentially in their early diagnosis.

Scientists from the Hall Lab at Wake Forest University School of Medicine have shown that DNA depurination can be detected electrically using solid-state nanopores.

Depurination has conventionally been studied indirectly—*e.g.* determining the replicative ability of DNA—or with expensive bulk techniques such as liquid chromatography.

"Our technique is fast, requires little starting material, and can offer a rough assessment of depurination at the single-molecule level," says Adam R. Hall, an assistant professor of Biomedical Engineering at Virginia Tech-Wake Forest School of Biomedical Engineering and Sciences. "As with other single-molecule approaches, this allows the measurement of small populations that would be masked or averaged out in bulk techniques (Figure 5.3)."

The team's solid-state nanopore devices are single fabricated, nanometer-scale (5–6 nm) holes in thin SiN membranes through which individual molecules can be threaded electrophoretically. As they pass, their presence and certain aspects of their structure can be determined, much like a film strip being run through a projector.

"What we have found is that the structural changes resulting from depurination tend to act as 'breaks', slowing the threading process down," explains Hall. "The more damage is present in a molecule, the slower it moves through the nanopore, and this can be observed from the electrical signal."

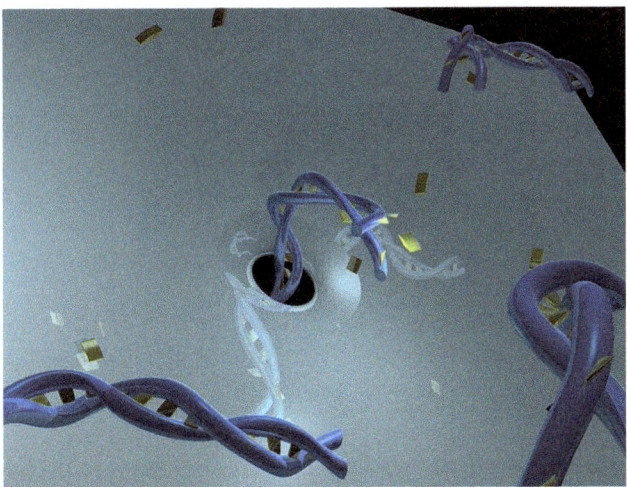

Figure 5.3 Schematic of the system: a short duplex DNA strand is pushed through a nanopore. (Image: The Hall Lab, Wake Forest University School of Medicine.)

In their experiments, the researchers found that depurinated DNA translocates up to an order of magnitude more slowly than undamaged molecules, on average.

These results advance the study of depurination and DNA damage by providing a new diagnostic tool that has extremely high sensitivity, is fast, label-free, and can be used for a coarse determination of either the overall level of depurination within a collection of dsDNA or the degradation of individual translocating molecules.

As depurination is also a natural process, occurring at a certain rate over time, the scientists expect that the technique might be useful in diverse areas such as forensics.

Hall points out that a great deal of focus has been put on solid-state nanopores as a potential way to sequence DNA ultrafast. While biological nanopores have made tremendous strides in this context, artificial nanopores have not been developed as rapidly.

"This is due largely to limitations in fabrication and physical limitations of the system," he says. "However, we feel that there are plenty of areas that solid-state nanopores can access more readily—many with clinical relevance."

The experiments conducted by Hall's group are part of a push in their field to expand and generalize the capabilities of solid-state nanopore systems and pursue applications matched to their strengths, including speed of measurement, small required sample sizes, and flexibility in the types and sizes of the analytes that can be probed.

For example, other groups have been applying solid-state nanopores to DNA epigenetics and drug interactions. These pursuits are more likely to yield useful tools and protocols in the near term.

"Going forward, global assessment of depurination levels is an important capability, even more powerful would be the ability to localize the damage to specific genes," Hall says. "Solid-state nanopores may be able to achieve this as well, with their ability to spatially map structural variations at the single-molecule level. We are looking into ways to explore this."

Featured scientists: Prof. Adam R. Hall's research group (http://www.the-halllab.org/)
Organizations: Wake Forest University School of Medicine, Winston-Salem, NC (USA)
Relevant publication: M. Marshall, J. Ruzicka, E. Taylor and A. Hall, Detecting DNA Depurination with Solid-State Nanopores, *PLoS ONE*, 2014, **9**(7), e101632, DOI: 10.1371/journal.pone.0101632

5.2.6 Wearable Graphene Strain Sensors Monitor Human Vital Signs

Monolayer graphene exhibits exceptional electronic and mechanical properties, making it, among other applications, a very promising material for nanoelectromechanical devices. For instance, researchers demonstrated that the piezoresistive effect in graphene in a nanoelectromechanical membrane configuration can provide direct electrical readout of pressure to strain transduction.

Graphene's piezoresistive effect, combined with its other properties such as ultra-translucency, superior mechanical flexibility and stability, high restorability, and carrier mobility, enables the use of graphene in high-sensitivity strain sensors.

Potential application areas for these sensors could be found in flexible display technology, robotics, smart clothing, electronic skin, *in vitro* diagnostics, implantable devices, and human physiological motion detection—which has been considered as an effective approach to evaluate human health.

To demonstrate this application, researchers at Tsinghua University and Nanchang University in China, have reported on a method to monitor human motions.

"We prepared a simple-structured and low-cost graphene woven fabrics (GWFs) strain sensor, which can readily distinguish various strain levels of human motion signals," Hongwei Zhu, a professor at Tsinghua University, explains. "As we applied stress on the strain sensor, high-density cracks generated in the network, leading to the decrease of current pathways and the increase of resistance."

The team obtained their GWFs by using chemical vapor deposition (CVD) to grow graphene on the surface of crisscross copper meshes—similar to graphene growth on a copper foil substrate. Subsequently, the copper mesh was etched away and the remaining graphene fabric transferred to a pretreated film composited with medical tape and PDMS glue, which is a flexible, biocompatible, shape controllable material. Completing the strain sensor was the connection of silver wires to the graphene mesh.

When used as a human motion sensor, the signals of GWFs resistance change depend on deformation strain, which is formed by the motion. The stronger the motion, the larger the strain, and the easier the motion signals can be recorded.

"Our GWF strain sensor can readily distinguish various strain levels of human motion signals," says Zhu. "Thanks to its special crisscross configuration compared to conventional graphene films, GWF possesses an extremely high gauge factor, which we calculated to be ~103 under 2–6% strains, 106 under higher strains (>7%), and ~35 under small strain of 0.2%. This indicates that the sensor could be stretched by a tiny deformation of 0.2% with obvious resistance change, enough to be recorded."

In other words, the signals of any weak motions, including vital signs such as pulse and breathing, phonation, expression changes, or blink of an eye, can be detected. The sensor could also endure a large deformation of 30% with a completely reversible electrical property.

"Owing to its distinctive features of high sensitivity and reversible extensibility, our graphene-based piezoresistive sensors have wide potential applications in the fields of flexible displays, robotics, fatigue detection, body monitoring, *in vitro* diagnostics, and advanced therapies," Zhu concludes.

Featured scientist: Prof. Hongwei Zhu (http://learn.tsinghua.edu.cn:8080/2008990024/Homepage2/index_e.html)

Organizations: School of Materials Science and Engineering, Tsinghua University, Beijing (PR China)

Relevant publication: Y. Wang, L. Wang, T. Yang, X. Li, X. Zang and M. Zhu, *et al.*, Wearable and Highly Sensitive Graphene Strain Sensors for Human Motion Monitoring, *Adv. Funct. Mater.*, 2014, **24**(29), 4666, DOI: 10.1002/adfm.201400379

5.2.7 Biosensor Detects Biomarkers for Parkinson's Disease

Recently, nanotechnology researchers have begun to work with graphene foams—three-dimensional (3D) structures of interconnected graphene sheets with extremely high conductivity. Applications range from stretchable electronics and highly sensitive gas sensors to extremely water-repellent coatings.

Since graphene foam possesses a high porosity of close to 100%, this offers the opportunity to use it as a scaffold for other nanomaterials to generate synergistic effects. One example is the integration of zinc oxide (ZnO) nanowires to enhance the sensitivity of electrochemical biosensors.

A research team, led by Professor Young Hee Lee, Director of the Center for Integrated Nanostructure Physics, Institute for Basic Science at Sungkyunkwan University in Korea, has fabricated vertically aligned ZnO nanowire arrays on 3D graphene foam and used this electrode to selectively detect uric acid (UA), dopamine (DA), and ascorbic acid (AA) by a differential pulse voltammetry (DPV) method.

The team demonstrated that their uniquely implemented structure is able to detect uric acid in a reliable statistical level from the serum of Parkinson's disease patients.

Explaining the background for this research, Lee notes that abnormal levels of UA are symptomatic of several diseases, including gout, hyperuricemia, and Parkinson's disease (PD). "Dopamine is an important neurotransmitter that is widely distributed within the mammalian central nervous system. Low levels of DA are related to neurological disorders such as PD and schizophrenia. Ascorbic acid is a vital vitamin in the human diet and is well-known for its antioxidant properties. It is also well-known that UA, DA, and AA coexist in the extracellular fluid of the central nervous system and serum. However, it is difficult to simultaneously detect each species in a mixture with high selectivity and sensitivity when using conventional solid electrodes because their oxidation potentials overlap, the surface area is insufficient, and/or the kinetic accessibility of each species is limited."

Currently, diagnosis of PD essentially relies on the assessment of clinical symptoms and a blood test for PD is undoubtedly a major goal for researchers. Being able to accurately test for biomarkers associated with the disease with simple blood tests would be a major breakthrough in diagnosing PD in the early stages when treatments are most likely to be effective.

"Our optimized ZnO nanowire/graphene foam electrode provides high selectivity with a detection limit of 1 nM for UA and DA," says Lee. "The key features of our structural design are a large surface area with mesoporous 3D graphene structures to facilitate ion diffusion easily; high conductivity from 3D graphene foam; and active sites of ZnO surface for high selectivity."

In their experiments, the researchers oxidized UA, DA, and AA biomolecules found in serum extracted from human peripheral blood of healthy individuals as well as PD patients. This process involved proton and electron generation at the surface of the ZnO nanowire arrays whereby electrons are transferred to the electrode. Using DPV measurements with the ZnO nanowire/graphene foam electrode, the samples were analyzed for UA levels.

"The average UA concentrations for the healthy individuals and the Parkinson's disease patients were 355 ± 30 and 265 ± 20 μM, respectively," says Lee. "This clear reduction in UA levels in the serum of PD patients with reliable statistics ($p < 0.001$) strongly implies that our approach is a significant step forward, which we believe will be beneficial for diagnosing PD and monitoring disease progression."

These results with a ZnO/graphene electrode design for high sensitivity and high selectivity biosensors are promising for a future where test results for serious diseases can be achieved simply with a drop of blood. Improving on their design, the team hopes to be able to selectively and simultaneously detect other disease biomarkers and biomolecules accurately with high sensitivity. Since the oxidation potential of some biomolecules may overlap, there is no guarantee that they can always succeed. Therefore, they will proceed by testing individual diseases, including cancers, in order to improve their sensor design and make it as generally applicable as possible.

Lee points out that in addition, due to structural advantages, there is also the possibility that this electrode material may be used for many other applications such as other types of biosensors, gas sensors, methanol oxidation reactions in fuel cells, solar cells, and supercapacitors.

Featured scientists: Prof. Young Hee Lee's Carbon Nanotube Research Laboratory (http://shb.skku.edu/nanotube/)
Organizations: Institute for Basic Science, Sungkyunkwan University (Republic of Korea)
Relevant publication: H. Yue, S. Huang, J. Chang, C. Heo, F. Yao and S. Adhikari, *et al.*, ZnO Nanowire Arrays on 3D Hierachical Graphene Foam: Biomarker Detection of Parkinson's Disease, *ACS Nano*, 2014, 8(2), 1639, DOI: 10.1021/nn405961p

5.2.8 Breath Nanosensors for Diagnosis of Diabetes

Human breath contains a number of volatile organic components (VOCs). An accurate detection of a specific VOC—*i.e.*, a biomarker for a particular disease—in the exhaled breath can provide useful information for diagnosis of various diseases. For example, acetone, H_2S, ammonia, and toluene can be used as biomarkers for evaluating diabetes, halitosis, kidney malfunction, and lung cancer, respectively. Consequently, breath analysis has been recognized as an increasingly accurate diagnostic method to link specific gaseous components in human breath to medical conditions and exposure to chemical compounds. Sampling breath is also much less invasive than testing blood, can be done very quickly, and creates as good as no biohazard waste.

The critical advantage of exhaled breath analysis is that it allows for noninvasive disease diagnosis. For this reason, several techniques such as gas chromatography, mass spectroscopy and optoelectronic analysis have been adopted to detect sub-ppm (parts per million) level VOCs in exhaled breath. However, these techniques are limited for use in real-time diagnosis with portable devices due to bulky device size and complex measuring processes.

"One of the promising detection tools for exhaled breath are metal oxide based chemiresistive gas sensors," explains Il-Doo Kim, an assistant professor in the Department of Materials Science and Engineering at Korea Advanced Institute of Science and Technology (KAIST). "When chemiresistive sensors are exposed to oxidizing or reducing analyte gases, sensor resistivity is changed as a function of temperature as well as of trace gas concentration. The advantage of chemiresistive sensors is that they can offer greater usability for portable real-time breath sensors thanks to their miniaturized size, low cost, easy fabrication, and simplicity of operation."

The most challenging issue in the field of chemiresistive sensors in diagnosis of diseases is selectivity. In general, exhaled breath consists of tens of VOCs. The oxide materials used in chemiresistive sensors can react with several types of gases.

So far, researchers have focused on finding optimum materials and structural modifications to increase sensor responses. For example, one-dimensional (1D) oxides, which exhibit a high surface area, have attracted much attention because the mechanisms of chemiresistive sensors are based on a surface reaction between target gas molecules and adsorbed oxygen on oxide. The utilization of these 1D nanostructure materials, however, has shown little progress in applications to gas sensors for diagnosis of diseases because most disease-related gases found in exhaled breath are in very low concentrations. In particular, gas response speed is rather low (>1 min) and is not suitable for real-time diagnosis.

Kim and his collaborators focused on a method modifying 1D structures to maximize gas sensor responses. They proved that a chemiresistive sensor can work as a VOC sensing device to detect very low concentrations of acetone if the sensing materials have optimized morphology and microstructure (Figure 5.4).

"What we have discovered is the efficacy of porous SnO_2 nanofibers consisting of multiple coaxial thin-tubes," says Kim. "This unique morphological feature, with a high surface-to-volume ratio and porous interior structures, can make multiple sensing layers within a single fiber accessible, quickly and effectively."

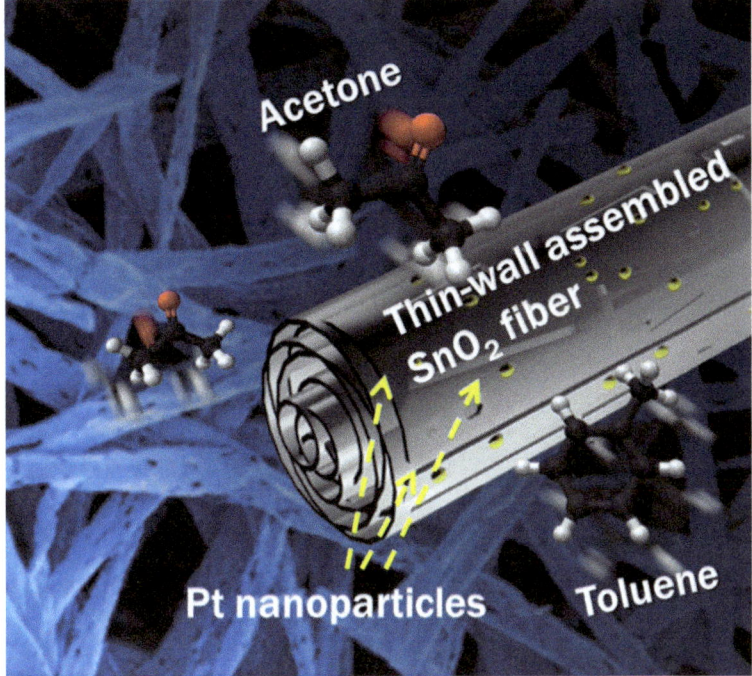

Figure 5.4 Ultrafast acetone sensors using thin-wall assembled SnO_2 nanofibers functionalized by catalytic Pt nanoparticles for diagnosis of diabetes. (Image: Dr Kim, KAIST.)

The team's sensor test results for acetone detectability are around 0.1 ppm, a level, which is eight times lower than the required gas-sensing level for diagnosing diabetes, and is the best result among reported SnO_2 sensors.

"From our result, we can see a positive prospect for the utilization of this material as an exhaled breath sensor for diagnosing diabetes," adds Kim. "In addition, using the catalytic platinum nanoparticle decoration on the surface of multiple coaxial SnO_2 thin-tubes, we can dramatically reduce the gas response time to less than 20 seconds."

The researchers synthesized the thin-walled SnO_2 fibers—which are composed of wrinkled SnO_2 nanotubes—by electrospinning with controlled phase separation between precursor-rich phases and polymer-rich phases. Electrospinning is a versatile way to produce polymeric, metallic, or metal-oxide fibers; it has been widely studied for application in chemical sensors, bio tissue-engineering, energy storage materials, electrochemical capacitors, photocatalysts, *etc.*

When injecting high-dielectric-constant-solutions containing polymers into a sharp needle under a current of tens of thousands of volts, the liquid jet is dragged down to the grounded plate and solidified into polymeric fibers. In particular, researchers have paid attention to an interesting phenomenon called micro phase-separations between polymers and other dissolved solutes.

As Kim explains, the primary feature of micro phase-separation is the possibility to synthesize irregularly shaped fibers such as bundle-type, hollow, and bumpy fibers, which are advantageous for certain applications that require a large surface area; this is because the phase separation behaviors, for example, between the Sn precursors and the polymers, can increase the porosity of the SnO_2 fibers during heat treatment, during which polymers are burned and left with empty voids between oxidized SnO_2 crystals.

"So far, the known factors that induce phase-separation are miscibility differences, solvent properties, humidity, *etc.*," he says. "In addition to these factors, in our work we successfully demonstrate that the variation of flow rate—electrospinning solution feeding rate—can trigger micro phase-separation. The resulting morphology is unprecedented and clearly distinguished from other regular forms of fibers."

Unlike the other forms of phase-separated fibers, this fiber has a unique morphology with a directional property, which the team proves is the result of an increase in the flow-rate. For instance, a high flow-rate intensifies the surface tension and elongation of polymer chains in the longitudinal direction.

Furthermore, the increase in solution supply leads to a higher solvent evaporation rate, which as a whole pushes concentration gradients to the radial direction. Taking two directional phenomena into account, the resulting unique morphology can exhibit elongated open pores along the axis direction and lamellar wall structures along the radial direction.

The researchers believe that these unique characteristics and the mechanisms of the fabrication can be beneficial in various applications that require complex 1D structures. They also expect that other types of oxide materials can provide superior sensing capacity to react with very low concentrations of VOCs when a similar morphology is properly synthesized.

The primary research objective of Kim's group is to give oxides selectivity to a particular target gas. In previous work, they investigated the gas selectivity of WO_3, given by the addition of catalytic nanoparticles on the oxide surface.[5]

Decoration of Pt and IrO_2 nanoparticles gives WO_3 selectivity to acetone and to hydrogen sulfide, which is another type of VOC. Even though challenges remain regarding selectivity to acetone in more complex systems with various VOCs, we can look ahead to a future of utilizing exhaled breath chemiresistive sensors to diagnose diabetes.

"We concede that we still have a long way to go," says Kim. "Through further optimization in the areas of material selection, micro-structuring, and functionalizing with various catalysts, however, we expect in the foreseeable future to make significant progress in the development of exhaled breath sensors with high accuracy and superior selectivity."

Eventually, the final goal is to integrate these novel exhaled breath sensors into smartphones, allowing users to easily monitor their health anytime, anywhere.

Featured scientist: Prof. Il-Doo Kim's Advanced Nanomaterials and Energy Lab (http://advnano.kaist.ac.kr/new/sp_main/main.php)
Organization: Department of Materials Science and Engineering, Korea Advanced Institute of Science and Technology,
Relevant publication: J. Shin, S. Choi, I. Lee, D. Youn, C. Park and J. Lee, *et al.*, Thin-Wall Assembled SnO_2 Fibers Functionalized by Catalytic Pt Nanoparticles and their Superior Exhaled-Breath-Sensing Properties for the Diagnosis of Diabetes, *Adv. Funct. Mater.*, 2013, **23**(19), 2357, DOI: 10.1002/adfm.201202729

5.2.9 Ultrafast Sensor Monitors You While You Speak

Researchers from the Nokia Research Center in Cambridge, UK, have shown that the distinctive 2D structure of graphene oxide (GO), combined with its superpermeability to water molecules, leads to sensing devices with unprecedented speed.

"It has been well known that graphene oxide can be very sensitive to water, and that this material is specifically permeable to water molecules," Stefano Borini, a principal researcher at the Nokia Research Center, points out. "However, the ultimate performance in terms of speed had not been demonstrated yet."

The Nokia team reported the experimental observation of the unparalleled response speed of humidity sensors based on graphene oxide, which are—to the best of the scientists' knowledge—the fastest humidity sensors ever reported.

Such exceptional performance is intimately related to some unique properties of graphene oxide: its two-dimensional nature and its super-permeability to water.

Borini and his colleagues began to study graphene oxide based sensors because they were looking for scalable, low-cost sensor solutions that could also exhibit good performance. In the course of this work, they realized the possibility of making ultra-thin films in a very easy way, such as by spraying, and discovered the exceptional sensing performance of the material.

In their experiments, the researchers demonstrated response and recovery times—defined as the time to go from 10% to 90% of the high humidity value and *vice versa*—of less than 100 ms, with the thinnest films being able to respond up to 40–50 Hz. They found that the response rate is a function of the thickness of the GO films, for instance the 25-nm-thick film is clearly slower than the 15-nm-thick one. This means that the tuning of the film thickness would allow engineering of humidity sensors with optimized speed.

"The ultrafast response speed of these sensors allows us to observe the modulation of moisture in a user's breath, and we demonstrate this feature in a whistled tune recognition application," says Borini. "Moreover, the same concept may be explored in the case of different 2D materials, such as functionalized graphene and 2D transition metal dichalcogenides, in interaction with different vapors and gases for the realization of ultra-thin nanoporous films for sensing applications."

While humidity sensing may be useful for a lot of applications, some of them are of limited use due to the fact that the standard response and recovery times are too long. Now, though, the fast response times of these GO humidity sensors enables entirely new user interfaces to be designed, *e.g.* sensing the moisture from a user's breath. This would make it possible to monitor both breathing and speaking.

The graphene oxide sensing film that the team developed is extremely thin (the thinnest is about 15 nm), optically transparent, and mechanically flexible. This would enable humidity sensors in a wide variety of form factors including small wearable devices or even Band-Aid-type skin stickers.

Borini notes that, before we see commercial applications, there are still a few questions concerning graphene oxide films that need answering, including assessing the material's stability over time.

Nokia has filed several patent applications regarding this work.

Featured scientist: Dr Stefano Borini
Organization: Nokia Research Center, Cambridge (UK)
Relevant publication: S. Borini, R. White, D. Wei, M. Astley, S. Haque and E. Spigone, *et al.*, Ultrafast Graphene Oxide Humidity Sensors, *ACS Nano*, 2013, 7(12), 11166, DOI: 10.1021/nn404889b

5.2.10   Detecting Flu Viruses in Exhaled Breath

Among the nanotechnology-based platforms for label-free and ultrasensitive virus detection are gold nanoparticles, carbon nanotubes, and silicon nanowires. Using the latter, researchers at Peking University, China, have configured a silicon nanowire field effect transistor as an ultra-sensitive influenza virus detection device for exhaled breath samples by chemically linking an antibody to its surface. Upon virus binding to the antibody, the conductance undergoes a discrete change and thereby transforms a biological presence of a certain virus in exhaled breath into electrical signals.

Studies have shown that exhaled breath from a flu patient contains influenza viruses but, although the use of silicon nanowire (SiNW) sensors for virus detection is not new, no studies have been conducted to apply silicon nanowire technology to the diagnosis of flu.

"We have successfully demonstrated the direct and selective detection of influenza viruses (H3N2) in diluted exhaled breath condensate samples collected from the flu patients within minutes using our silicon nanowire sensor," says Maosheng Yao, a professor at the College of Environmental Sciences and Engineering at Peking University. "Our work suggests that the SiNW sensor device, when calibrated by virus standards and exhaled breath condensate controls, can be reliably applied to the diagnosis of flu in a clinical setting with two orders of magnitude less time compared to the gold standard method RT-qPCR."

"Commercialization of the technology described in this work as a hand-held device, which is entirely doable, opens up outstanding opportunities of revolutionizing flu diagnosis in a clinical setting," says Yao.

According to the World Health Organization, annual influenza (mostly H3N2 and H1N1) epidemics worldwide are responsible for three to five million cases of severe illness, eventually leading to 250 000 to 500 000 deaths. For diagnosis of viral infections, health professionals empirically rely on the white blood cell levels from a routine blood test and accompanying clinical symptoms such as headache, cough, and arthralgia.

However, as Yao points out, this practice lacks scientific evidence. "In certain cases, employing RT-qPCR for viral detection in nasal swabs and bronchoalveolar lavages, exhibited higher detection rates than culturing and immuno-detection methods. Apart from other compounds, influenza and papilloma viruses previously have been detected in EBC samples by RT-qPCR and ELISA methods. However, these methods are impacted by the false-negatives for low-level influenza viruses and longer detection time of up to several hours as a result of the labor-extensive procedures such as RNA extraction and amplification for RT-qPCR," he says.

Accordingly, this procedure is not practical in a clinical setting given the large number of hospital visits and the patients' limited wait time, especially during an influenza outbreak or a bio-terror attack.

The team fabricated the silicon nanowires for their sensor device using a chemical vapor deposition method. Subsequently, the nanowires were aligned on silicon substrates with microfluidic channels placed over them. Then, electrical contacts to individual SiNWs were formed by standard photolithographic processes and thermal evaporation. After that, these nanosensor devices were functionalized with influenza A H3N2–H1N1 subtype or a biomarker (control) antibodies.

With their sensors, the researchers found they were able to selectively detect influenza A viruses down to ~30 viruses per µL in clinical exhaled breath condensate (EBC) samples within minutes.

"For 90% of the cases, we have observed that EBC samples tested positive or negative by the RT-qPCR method generated corresponding positive or negative SiNW sensor responses with an unparalleled detection speed," says Yao. "While for those EBC samples with very low quantity of viruses, antibody modified magnetic beads were shown capable of efficiently concentrating the viruses for enhanced direct detection by the SiNW sensor device."

In a previously published study,[6] the team achieved the detection of air-borne flu viruses within minutes by integrating their SiNW sensor with an air sampling device and a microfluidic channel.

Yao notes that one limiting factor with their technology is the cross-reaction between antibodies and non-target antigens, which is a challenging problem common to immuno-based detection methods. "One of our hurdles to move this technology forward is to develop highly specific virus subtype antibodies that could avoid the cross reactions between viruses and non-specific antibodies," he says.

The researchers at Peking University are confident that their SiNW sensor system can be readily extended to other types of viruses or biomarkers present in EBC samples. "Integration of different virus antibody modified SiNW sensor devices in a single chip using a micropipette technology would offer the opportunity to simultaneously detect different viruses in a single EBC sample," says Yao.

Featured scientists: Prof. Maosheng Yao's Bioaerosol Lab (http://openwetware.org/wiki/Yao-Lab)
Organization: College of Environmental Sciences and Engineering, Peking University (PR China)
Relevant publication: F. Shen, J. Wang, Z. Xu, Y. Wu, Q. Chen and X. Li, *et al.*, Rapid Flu Diagnosis Using Silicon Nanowire Sensor, *Nano Lett.*, 2012, **12**(7), 3722, DOI: 10.1021/nl301516z

5.2.11   Nanosensor for Advanced Cancer Biomarker Detection

Epigenetic mechanisms are chemical changes in DNA that do not alter the actual genetic code but can influence the expression of genes, and can be passed on when cells reproduce. One of the most important is DNA methylation, where methyl groups—small structures of carbon and hydrogen—are appended to specific locations on a DNA strand. This plays a critical role in determining which genes are active in a cell at any given time; which in turn plays an important role in embryonic development, cell growth and reproduction, and many diseases.

Numerous methods have been developed for the examination of DNA methylation in the hope of detecting biomarkers as early warning signs in various diseases, including cancer. However, current methods for methylation analysis are costly, complex and less quantitative in determining methylation states at individual CpG sites (most DNA methylation occurs in CpG dinucleotides).

Both biological and synthetic nanopores have been proposed for DNA methylation detection.

Researchers at the University of Missouri employed protein nanopores to investigate a novel metal ion-bridged DNA interstrand lock, and explore its potential in location-specific methylation detection.

This work provides a powerful biophysical tool to explore the interactions of metal ions and nucleic acids in humans and other living organisms.

"The nanopore single-molecule approach presented in our work is label-free, does not require DNA sequencing, and provides a rapid, low cost and accurate tool with multiplex detection capability for epigenetic biomarker discovery and diagnostic set," Liqun Gu, an associate professor in the Department of Bioengineering and Investigator in the Dalton Cardiovascular Research Center, summarizes his team's work. "The focus of our research is single-molecule and single-base investigation of a base-pair specific metal-ion/nucleic-acid interaction. Such an interaction can be used to discriminate methylated and unmethylated cytosine, with a potential in DNA methylation analysis."

What the scientists found was that the uracil–thymine mismatch at a CpG site can be bound with a divalent mercuric ion. Gu explains that the metal binding creates a reversible interstrand lock, called MercuLock, which enhances the hybridization strength by nearly two orders of magnitude.

"In contrast, the 5-methyl cytosine–thymine mismatch cannot form such MercuLock," says Gu. "Thus uracil and methylated cytosine can be discriminated from MercuLock signatures in the nanopore. Because uracil is converted from unmethylated cytosine by the bisulfite treatment, the identity of uracil corresponds to an unmethylated cytosine. Therefore the methylation status in the original gene is determined."

This scientific finding and its potential applications might be useful for the discovery of epigenetic biomarkers and cancer diagnosis.

Any analytical tool that is sensitive to the change in hybridization stability can be used to detect the interstrand lock. For example, according to Professor Gerald A. Meininger, Director of the Dalton Cardiovascular Research Center at the University of Missouri, an atomic force microscope (AFM) could be used to study the dehybridization of DNA carrying multiple interstrand locks. AFM is unique in measuring the distance between interstrand locks. The sensitivity of an AFM is unique in its capability of measuring distance and force and could be capable of detecting interstrand locks, thus determining the position of each methylated and unmethylated cytosine.

In this recent work, though, a nanopore is utilized to study how the binding of a metal ion to a single base-pair stabilizes the nucleic acids' hybridization. According to Gu, this leads to three significant findings:

- The resulting single-nucleotide discrimination capability of the nanopore is particularly significant in site-specific genetic and epigenetic detection;
- Mercury ions not only bind to the thymine–thymine (T–T) mismatch, but to the uracil–thymine (U–T) mismatch as well. This is a significant contribution to the field of the metal/nucleic acids interaction, and expands the nanopore single-base discrimination ability from DNA to RNA;
- Completely different from traditional mercury biosensor applications, the team explored the potential application of this single-base interstrand lock for epigenetic methylation detection.

"We first uncovered the binding of Hg^{2+} ion to the U–T mismatch, which led to the unprecedented application of mismatch-specific interstrand locks in the epigenetic detection," notes Gu. "The demonstrated ability to discriminate a single mismatched base pair bound with the metal ion renders the nanopore a powerful analytical tool to investigate the mechanism governing the metal–nucleic acids interactions."

In past years, various metal–nucleic acids interactions have been extensively studied but their application has been limited to metal ion biosensors.

A well-known example is the mercury sensor that measures Hg^{2+} binding to the T–T mismatch. This work now opens an avenue to the application of metal–nucleic acids interactions in genetic and epigenetic detection.

Prior to this work, researchers reported a different nanopore method for methylation detection. In that report,[7] the targeted DNA was biotinylated and attached to a streptavidin for immobilization in the pore. The methylation site needs to have a fixed distance to the biotin terminal, such that upon immobilization, the methylation site is located within the sensing zone to change the nanopore current.

The core of the method developed by Gu's group is the base-specific interstrand lock. The methylation status can be determined by examining whether an interstrand lock is formed.

"The 'locked' dsDNA shows greatly enhanced hybridization stability, resulting in a nanopore signature that is ~30-fold prolonged compared with dsDNA without the interstrand lock," says Gu. "Such signatures are easy to discriminate, allowing one to easily recognize the methylation status. By using a series of probes, each probe targeting a specific CpG cytosine, the methylation status for multiple CpG sites can be analyzed."

The team points out that the interstrand locks they propose could find useful applications in nucleic acids nanotechnology. They can tune the hybridization strength when constructing nucleic acids nanostructures such as origami. They may also be used to improve the binding of regulators such as microRNAs to the target gene to control gene expression.

Featured scientist: Prof. Liqun Gu (http://bioengineering.missouri.edu/faculty/gu-a.php)
Organization: Department of Bioengineering, University of Missouri, Columbia, MO (USA)
Relevant publication: I. Kang, Y. Wang, C. Reagan, Y. Fu, M. Wang and L. Gu, Designing DNA interstrand lock for locus-specific methylation detection in a nanopore, *Sci. Rep.*, 2013, **3**, 2381, DOI: 10.1038/srep02381

5.2.12   Optical Detection of Epigenetic Marks

Understanding the purpose of epigenetic markers—and how they change over time—will be crucial in understanding biological processes ranging from embryo development to aging and disease. But just how the markers work, and what different markers mean, is painstaking work that still has a long way to go.

Deciphering the epigenetic code is a massive mapping exercise, but will provide important information on how epigenetic markers differ among cell types and between healthy and sick individuals. Actually detecting the markers, however, has proven difficult.

Advancing this research field, scientists have reported the direct visualization of individual epigenetic modifications in the genome. This is a technical and conceptual breakthrough as it allows not only quantification of the amount of modified bases but also pinpoints and maps their position in the genome.

"Being able to count individual modifications results in the most sensitive detection of genomic hydroxymethylcytosine ever reported," says Yuval Ebenstein, who leads the NanoBioPhotonix research group at Tel Aviv University. "Our findings open up research into numerous human cell types, including human blood cells that so far have not been accessible due to low levels of epigenetic modification."

In this work, the scientists demonstrate the utility of optical sensing for the detection and quantification of 5-hydroxymethylcytosine (5hmC) residues in genomic DNA.

Since its discovery in human genomic DNA in 2009, this modification has been the focus of numerous studies and has been established as an important epigenetic mark linked to stem-cell differentiation, aging, and neurodegenerative disease.

"In our work, we utilize enzymatic glucosylation followed by click labeling of a fluorescent reporter to specifically tag individual 5hmC sites along DNA molecules," Ebenstein explains. "The result is a simple yet highly sensitive 5hmC quantification assay that is compatible with bulk DNA analysis as well as with 5hmC detection on individual DNA molecules."

With this report, the group extends their previous work[8,9] on mapping DNA binding proteins to mapping epigenetic DNA modifications.

The immediate impact of these results is the demonstration that a simple and quick UV-vis measurement can replace the tedious radioactive, mass spectrograph and affinity-based methods that exist today—basically enabling any lab equipped with a UV-vis spectrophotometer to quantify hydroxymethylation levels.

"The more exciting prospect is to be able to profile individual genomic molecules and detect distinct hmC patterns that may serve as biomarkers for disease," says Ebenstein.

"Optical mapping of DNA has mainly focused on genetic analysis and tremendous progress has been made in recent years," he adds. "Our work enables the addition of an epigenetic overlay to existing fluorescence barcoding technologies in order to map epigenetic patterns in the context of the underlying sequence. The main challenge now is to make sense of the information and new analysis algorithms must be developed."

Featured scientist: Prof. Yuval Ebenstein's NanoBioPhotonix Lab (http://www.nanobiophotonix.com/)

Organization: Department of Chemical Physics, School of Chemistry, Tel Aviv University (Israel)

Relevant publication: Y. Michaeli, T. Shahal, D. Torchinsky, A. Grunwald, R. Hoch and Y. Ebenstein, Optical detection of epigenetic marks: sensitive quantification and direct imaging of individual hydroxymethylcytosine bases, *Chem. Commun.*, 2013, **49**(77), 8599, DOI: 10.1039/C3CC42543F

5.2.13   Nanosensor Tattoo on Teeth Monitors Bacteria in Your Mouth

Detection of very small amounts of a chemical contaminant, virus or bacteria in food systems is an important potential application of nanotechnology. Bacterial contamination in general is a major problem affecting both developing and developed countries and the rapid and sensitive detection of pathogenic bacteria at the point of care is extremely important; especially given that more and more pathogens are becoming immune to antibiotics.

The limitations of most conventional diagnostic methods are the lack of ultrasensitivity and the delays in getting results. One problem is that many infectious agents have very low minimum infective doses, requiring detection mechanisms that offer extremely high sensitivities. Another problem is the time delay between taking samples and getting the results: current methods for the detection of pathogenic contaminants involve the collection and pre-processing of analyte samples—biological specimens, food samples *etc.*—and then performing laboratory-based assays. It will be extremely advantageous to have sensing systems that can be directly integrated with the sources of contamination or points of infection to provide an *in situ* monitoring/detection of bacterial presence.

A team of scientists, led by Fiorenzo Omenetto at Tufts University and Michael C. McAlpine (at the time of this work at Princeton University, now University of Minnesota), has developed a novel approach to interfacing passive, wireless graphene nanosensors onto biomaterials *via* silk bioresorption.

"Graphene is capable of highly sensitive analyte detection," says Manu S. Mannoor, a graduate student in McAlpine's group. "We demonstrate that graphene can be printed onto water-soluble silk. This in turn permits intimate biotransfer and direct interfacing of graphene nanosensors with a variety of substrates including biological tissues and hospital IV bags to provide *in situ* monitoring and detection of bacterial contamination and infection."

The nanoscale nature of graphene allows for high adhesive conformality after biotransfer and highly sensitive detection. The team demonstrated their nanosensor by attaching it to a tooth for battery-free, remote monitoring of respiration and bacteria detection in saliva.

Such an approach of direct interfacing of nanosensors onto biomaterials could revolutionize areas ranging from health quality monitoring to adaptive threat detection.

Mannoor explains the fabrication technique: "First, we printed graphene nanosensors onto water-soluble silk thin-film substrates. The graphene is then contacted by interdigitated electrodes, which are simultaneously patterned with an inductive coil antenna. Finally, the graphene/electrode/silk hybrid structure is transferred to biomaterials such as tooth enamel or tissue, followed by functionalization with bifunctional graphene–AMP (antimicrobial peptides) biorecognition moieties."

Silk thin-films serve as an ideal "temporary tattoo" platform owing to their optical transparency, mechanical robustness, biotransferability, flexibility and biocompatibility. Omenetto's research group has been conducting research with silk composite materials for biosensing applications for a while now and has already reported[10] a promising path towards the development of a new class of metamaterial-inspired implantable biosensors and biodetectors.

The resulting device architecture is capable of extremely sensitive chemical and biological sensing, with detection limits down to a single bacterium, while also wirelessly achieving remote powering and readout. Potential applications for this kind of "bio-transferrable" sensor could include on-body health quality monitoring, hospital sanitation monitoring, and food safety analysis and may provide a first line of defence against pathogenic threats at the point of contamination.

"What we were able to demonstrate is only a prototype, 'first generation' platform that served as a proof of concept for the *in situ* bacterial contamination monitoring by direct interfacing of graphene nanosensors with a variety of substrates including biological tissues," says Mannoor. "Future challenges mainly include improving the selectivity of the detection system to be able to distinguish between various species of pathogenic bacteria. Reducing the sensor form factor is also another challenge facing the future development of the sensor."

Featured scientists: Prof. Fiorenzo Omenetto's research lab (http://ase.tufts.edu/biomedical/unolab/home.html); Prof. Michael C. McAlpine's research group (http://www.mcalpineresearchgroup.com/)
Organizations: School of Engineering, Tufts University, Medford, MA (USA); Department of Mechanical Engineering, University of Minnesota, Minneapolis, MN (USA)
Relevant publication: M. Mannoor, H. Tao, J. Clayton, A. Sengupta, D. Kaplan and R. Naik, *et al.*, Graphene-based wireless bacteria detection on tooth enamel, *Nat. Commun.*, 2012, **3**, 763, DOI: 10.1038/ncomms1767

5.2.14 Tracking Nanomedicines Inside the Body

One of the key issues in the development of novel nanomedicines is the ability to track nanoscale drug carriers inside the body to evaluate where they go and how they get there. In order to achieve products that get regulatory approval and can be administered to patients, scientists need to come up with non-invasive tests for the biodistribution of nanomedicines.

"Virtually all previous preclinical studies in this area of research rely on 2D fluorescence reflectance imaging (FRI)," says Dr Twan Lammers, a researcher at the Department of Experimental Molecular Imaging, Helmholtz Institute for Biomedical Engineering, RWTH-Aachen University in Germany. "Given the limitation of 2D FRI in not being able to detect the fluorescence in deep-seated organs and tissues, 3D fluorescence molecular tomography (3D FMT) emerged as an alternative. However, the lack of anatomical information—to assign the fluorescence signal to specific organs—was an important barrier hindering the routine use of standard 3D FMT for *in vivo* imaging of nanomedicines."

This situation is very different for activatable probes used as molecular diagnostics, which were already used relatively extensively in conjunction with 3D FMT.

Such a hybrid FMT-based imaging protocol to enable more meaningful and more quantitative *in vivo* analyses has been developed by a team, led by Lammers, that involved scientists from Aachen University, University of Twente, Utrecht University, and the Academy of Sciences of the Czech Republic.

Specifically, the team used computer tomography (CT) to enable a more accurate analysis of probe localization in non-superficial organs—something that is impossible using 2D FRI and standard 3D FMT.

They were able to visualize and quantify the biodistribution and target site accumulation of nanomedicines in mice using hybrid 3D CT-FMT imaging.

"Given the fact that optical imaging is ever more extensively used to monitor the *in vivo* behavior of nanomedicines, our findings therefore advance the field by enabling the visualization and quantification of probe accumulation of a number of deeper-seated healthy tissues," says Lammers.

Fluorescence molecular tomography is a technique that enables quantitative imaging of the fluorescent probe, or marker concentration, to the picomolar level with no tissue depth limitation. Lasers are used to excite near-infrared probes for which planar detectors such as CCD cameras record excitation and emission images of the diffuse light propagation. Computer software then uses advanced algorithms to volumetrically reconstruct the accumulation and concentration of the optical imaging agents.

So far, though, the major shortcoming of FMT has been its inability to accurately assign the reconstructed probe accumulation signal to a given organ of interest.

The team's hybrid imaging protocol, a combination of high-resolution microcomputed tomography and FMT, enables more meaningful and more quantitative *in vivo* analyses of the biodistribution of nanomedicines.

As Lammers points out, this initial proof-of-principle study solves one of the most important limitations that researchers working on the optical imaging of nanomedicine biodistribution have been confronted with, *i.e.* the inability to non-invasively and accurately assess probe accumulation in non-superficial organs.

Hybrid 3D CT-FMT imaging can be used at preclinical levels for a large number of *in vivo* applications, ranging from the molecular diagnosis of receptor expression or enzyme activity to the longitudinal visualization and quantification of drug delivery and drug release.

The researchers caution that there is still quite some room for improvement and optimization of this technique.

"Particularly," says Lammers, "the volumetric fluorescence reconstruction can be improved by including tissue-specific optical properties such as absorption and scattering in the reconstruction method. Furthermore, we are working on methods to perform automated organ segmentation to reduce the required effort per scan and achieve higher user independence."

Featured scientist: Dr Twan Lammers
Organization: Department of Experimental Molecular Imaging, Helm-holtz Institute for Biomedical Engineering, RWTH-Aachen University (Germany) (http://exmi.rwth-aachen.de/)
Relevant publication: S. Kunjachan, F. Gremse, B. Theek, P. Koczera, R. Pola and M. Pechar, *et al.*, Noninvasive Optical Imaging of Nanomedicine Biodistribution, *ACS Nano*, 2013, 7(1), 252, DOI: 10.1021/nn303955n

5.2.15 Measuring Femtoscale Displacement for Photoacoustic Spectroscopy

The development of nanoscale devices and applications requires ultra-sensitive sensing systems that not only can offer atomic resolution imaging but also sub-nanometer scale displacement detection, zeptogram level mass sensing, or single bio-molecular sensing.

Optical scanning methods—the optical transduction of a micro- or nano-cantilever's motion—are widely used to achieve high deflection sensitivity down into the sub-nanometer range.

Researchers have developed a novel sensor that addresses some of the shortcomings that have plagued existing optical scanning systems, namely size, complexity, and cost.

"Our sensing technology is completely electrical and capable of sensing very small displacement as low as in the femtometer range," says Abdul Talukdar, former PhD student from the University of South Carolina-Columbia (currently at Intel Corporation). "This sensitivity has enabled the detection of nanogram level explosives (RDX) and other analytes."

Talukdar and principal investigator Prof. Goutam Koley and the research team, which included scientists from the universities of South Carolina, Alberta, Calgary, and Clemson University, developed on III–V nitride based microcantilever embedded with a highly sensitive three-terminal device or heterojunction field effect transistor (HFET) (Figure 5.5).

In collaboration with Canada, the research team has demonstrated for the first time, real-time sensing of nanogram level explosives that is done photo-acoustically and with a completely electrical readout.

The researchers found that, utilizing the piezoelectric properties of III–V nitride materials, the piezoelectric polarization-induced changes in two-dimensional electron gas can be utilized to transduce displacement with very high sensitivity.

"Thus it is possible to detect femtoscale displacement with a microcantilever operating below 100 kHz without moving towards a nanocantilever operating at MHz, which is very difficult to fabricate, requires complex measurement systems, and often is very costly," says Talukdar.

He adds that, while current research trends in cantilever-based sensing appears to be moving toward nanoscale devices, this involves complex fabrication schemes that are sometimes difficult to replicate and their comparable performance can be difficult to confirm.

200 μm

Figure 5.5   SEM image of AlGaN/GaN HFET embedded GaN microcantilever array
(14 microcantilevers) for the integration of different sensing attributes.
(Image: Abdul Talukdar.)

By utilizing inherent piezoelectric material properties and a heterojunction transistor embedded at the base of a microcantilever—what the team calls a "piezotransistive microcantilevers"—the device offers very high sensitivity with a very simple measurement technique.

The demonstrated sensitivity of this device is the best among state-of-the-art electrical technologies, notes Talukdar.

The results show electrical sensing performance comparable to optical means, and thus advancing the possibility of replacing optical readouts with electrical readouts from relevant equipment to offer cheaper, lighter, smaller, and better devices.

Importantly, the new device can readily be implemented in any scanning probe based microscopy—such as atomic force microscopes—eliminating the need for optical readout; it can be used in array-based sensing in real time; as a strain, force, and displacement sensor.

The team points out that, as the materials (AlGaN/GaN) are optically transparent, they can easily be integrated with wearable technologies, for instance for health monitoring.

Aluminium nitride and gallium nitride, two materials with a wide bandgap and strong piezoelectric properties, are also often used in devices that need to operate in harsh environments. Therefore, these new sensors lend themselves to space and aircraft applications, and any other mission in remote areas under extreme pressure or temperature.

The next stage of this work is to integrate the complete sensing system into a single portable package. Another goal is to demonstrate array-based sensing and imaging to detect multiple analytes (*e.g.* different explosives) in real time.

Talukdar cautions that there are two key challenges for this novel sensor technology:

Firstly, AlGaN/GaN is less defective when grown on SiC or sapphire. But for cantilever-based MEMS/NEMS, a silicon substrate is desirable. However AlGaN/GaN on Si is more defective. Thus the challenge is defect-less material growth on silicon, a task which already has progressed significantly.

Another major challenge for this technology, when used in photoacoustic spectroscopy, is the generation of acoustic waves from an analyte, which eventually oscillate the microcantilever and the HFET detects the deflections.

"A laser system/light source is needed to excite the analytes with sufficient power, however these laser systems are bulky, costly, and power hungry," Talukdar explains. "Thus, even with replacing the optical readout with the piezotransistor, which significantly reduces the size, cost, and the power consumption, it still is the excitation laser source that hinders further miniaturization and simplification. Only with a better excitation source can the whole system be packaged into a hand-held portable spectroscopy system."

Featured scientist: Dr Abdul Talukdar (https://scholar.google.de/citations?user=UC2ahVcAAAAJ&hl=en); Prof. Goutam Koley's Nanoscale & Sensors Lab (https://www.ces.clemson.edu/nesl/)
Organization: Intel Corporation; Clemson University, Anderson, SC (USA)
Relevant publication: A. Talukdar, M. Faheem Khan, D. Lee, S. Kim, T. Thundat and G. Koley, Piezotransistive transduction of femtoscale displacement for photoacoustic spectroscopy, *Nat. Commun.*, 2015, **6**, 7885, DOI: 10.1038/ncomms8885

5.2.16 Reduced Graphene Oxide Platform Shows Extreme Sensitivity to Circulating Tumor Cells

The complexity of the microenvironment of a biological cell is influenced by many factors, including surface topography and chemistry; matrix stiffness; mechanical stress; molecular liquid composition and other physiochemical parameters. However, most artificial bio-interfaces are developed based on just a single chemical or physical factor to direct cell behaviors. The functions performed by these artificial bio-interfaces are far simpler than those performed in the natural cell microenvironment.

In an effort to more closely mimic a cell's natural environment, a research team in China has fabricated an antibody-modified reduced graphene oxide (rGO) platform and used it to significantly improve the efficiency for capturing circulating tumor cells.

This is an important issue in cancer medicine because metastasis—the process of cancer spreading from its original site to distant tissues—is caused by marauding tumor cells that break off from the primary tumor site and ride in the bloodstream to set up colonies in other parts of the body. These breakaway cancer cells in the peripheral blood are known as circulating tumor cells (CTCs). Detecting and analyzing these cells can provide critical information for managing the spread of cancer and monitoring the effectiveness of therapies.

The scientists note their platform's extremely sensitive CTC static capture performance without the need for complex microfluidic operations.

"Our antibody-modified rGO films possess several synergistic topographic interactions provided by their low stiffness which, compared to other rigid artificial substrates, more closely mimics the cellular microenvironment of living organisms, thus facilitating cell–environment interactions," says Dr Yingying Li, a researcher in Prof. Shutao Wang's research group at the Technical Institute of Physics and Chemistry, Chinese Academy of Sciences, in Beijing.

"For example," she explains, "the rough texture of rGO might enhance topographic interactions between CTCs and biointerfaces; the low stiffness nature of rGO might facilitate cell–matrix interactions, thereby contributing to highly efficient CTC capture. Moreover, the negative charge and superhydrophilicity of antibody-modified rGO renders the surface inert to nonspecific cell adhesion, which is helpful for decreasing the white blood cell background—an issue that remained a problem in previous studies."

To demonstrate the effectiveness of their rGO material, the scientists fabricated a test kit to perform sensitive CTC specific recognition *via* a simple static method, without the need for complex microfluidic devices or operations. They showed that the antibody-modified rGO films can efficiently capture CTCs from fresh whole blood (which contains less than 10 CTCs per mL).

The test kit, suitable for clinical use, requires a blood sample from the patient that can be as low as 1 mL; which is much smaller than in conventional testing methods.

"While we addressed four microenvironmental factors that have an impact on cells' properties, the real nature of cell interactions is much more complex with almost endless physical and chemical properties influencing cells," Li points out. "While we are confident that more and more complex and multi-matching platforms will be designed and fabricated, getting close to mimicking natural cell environments is a huge challenge."

Featured scientists: Prof. Shutao Wang's research group (http://159.226.64.162/web/30490/home)
Organization: Technical Institute of Physics and Chemistry, Chinese Academy of Sciences, Beijing (PR China)
Relevant publication: Y. Li, Q. Lu, H. Liu, J. Wang, P. Zhang and H. Liang, *et al.*, Antibody-Modified Reduced Graphene Oxide Films with Extreme Sensitivity to Circulating Tumor Cells, *Adv. Mater.*, 2015, **27**, 6848, DOI: 10.1002/adma.201502615

5.3 Analyzing and Manipulating Single Cells Becomes Possible

Researchers are very interested in investigating and manipulating the biomechanical properties of the inner and outer structure of cells due to their relevance in many important topics in biology such as intracellular and intercellular dynamics; tissue and organ formation and their homeostasis; but also in medicine as the formation and development of diseases such as inflammatory disorders or tumors.

5.3.1 Untethered Active Microgripper for Single-Cell Analysis

There are a wide range of passive devices such as beads, wells and tubes that can be used to capture and confine single cells. Active cell grippers with moving parts rely on electrical modalities, which can be challenging to implement off-chip and in a highly parallel manner.

To deal with this issue, researchers have developed an untethered active microgripper that can be used to capture and contain single cells. These single-cell grippers, made from biocompatible and bioresorbable materials, derive their actuation energy from thin-film stresses. They can be arrayed for high throughput *in vitro* assays and imaging.

Since their operation does not need wires or batteries, they can be released for use as untethered tools.

"We employed varying sizes of these grippers to capture individual fibroblasts and red blood cells," says David H. Gracias, a professor at Johns Hopkins University. "These cells were alive and could be assayed or fixed for imaging. Because these devices are fabricated in 2D and subsequently folded into 3D, future studies could explore patterned topography such as spikes, holes, nanoscale roughness, and biochemical surface functionalization in specific designs onto one or more device walls."

The researchers note that their approach is inspired by previous studies on the stress-based roll-up and self-folding of thin films and the energy required to enable gripper motion is derived from the differential residual stress in nanoscale bilayers.

"Our approach utilizes photolithography, which is a high-throughput technique capable of fabricating 500 000 to 10 million single grippers on a 3-inch wafer or potentially over 100 million on a 12-inch wafer which is the size of wafers currently used in CMOS fabrication facilities," explains Kate Malachowski, a graduate student in Gracias' group and lead author of a paper on this work. "Additionally, the grippers can be actuated to close around single cells *en masse*, creating devices with patterns in all three dimensions."

The team developed several gripper variants with three or four arms, varying in size from 10 to 70 μm in length (tip-to-tip when open), which is an appropriate size range to grasp a variety of individual cells.

Since the grippers can be manufactured *en masse* and also operate autonomously, they can be utilized to capture and analyze single cells *in vivo* in a high throughput manner.

Additionally, since the process is compatible with conventional microfabrication, the surfaces of the devices can be patterned with electrical or optical biosensors to enable real-time single cell analysis after capture.

On the *in vivo* side, the scientists anticipate that single cell grippers could be utilized to capture and possibly analyze single cells within hard to reach places in the human body. These capabilities are enabled by the composition of materials that can be bioresorbable allowing for clearance from the body by dissolution, a small size that enables access to small conduits, and the absence of any wires or tethers allowing them to be deployed and operated in a highly parallel manner.

Gracias cautions that there still remain numerous challenges before these grippers could see real-world applications: "On the *in vitro* side, we are still challenged by the integration of biosensing elements on the surfaces of the devices; less than perfect yields; and cell viability especially for real-time monitoring over prolonged durations. For *in vivo* applications, there are a huge number of challenges including guidance; introduction into the body; retrieval; site specific or targeted actuation; and patient safety."

Previously,[11] Gracias' has shown that larger untethered grippers could be used to biopsy gastrointestinal organs, suggesting that the realization of this concept is within reach.

Featured scientists: Prof. David Gracias' research group (http://engineering. jhu.edu/chembe/faculty/david-gracias-2/)
Organization: Whiting School of Engineering, Johns Hopkins University, Baltimore, MD (USA)
Relevant publication: K. Malachowski, M. Jamal, Q. Jin, B. Polat, C. Morris and D. Gracias, Self-Folding Single Cell Grippers, *Nano Lett.*, 2014, **14**(7), 4164, DOI: 10.1021/nl500136a

5.3.2 New Technique Precisely Determines Nanoparticle Uptake into Individual Cells

In conventional chemotherapy, the patient is pumped full of cytotoxic drugs that go everywhere in the body with the hope that enough of the drug molecules reach the cancer cells and target their nuclear DNA to damage it or destroy the cell. By contrast, nanomedicine attempts to pinpoint individual tumor cells and deliver nanoparticles with or without therapeutic payloads right into the nucleus where they destroy the malignant cell by chemical or mechanical action.

While nanoparticles are emerging as drug carriers for targeted nanomedicines, preclinical assays to test nanoparticle efficacy are hampered by the lack of methods to quantitatively determine internalized particles. A novel method, developed by scientists at the California NanoSystems Institute (CNSI), is suited to pave the way for preclinical testing of nanoparticles to establish dose-efficacy relationships and to optimize biophysical and biochemical parameters in order to make better drug delivery vehicles.

The team, led by Claudia Gottstein, Adjunct Assistant Professor, MCD Biology, and Director, Biological Nanostructures Lab at CNSI, demonstrated that it is possible to determine the exact number of nanoparticles inside a cell through a combination of three methods and a mathematical model which they developed to link the data from these three methods.

This a generic technique for precise determination of nanoparticle uptake into cells, making possible a readout of the average number of nanoparticles internalized per cell.

"This is of utmost importance for a number of biomedical applications," Gottstein explains. "The differentiation between only adhered and internalized particles has been a big unsolved problem in the field. For example, many drugs work intracellularly, and mere binding without internalization would render those drugs ineffective. We present a generic solution to this problem."

Furthermore, as she points out, comparison of data from different labs has so far been difficult at best. "Our method allows for standardization and facile side-by-side analysis of internalization data from different materials and different labs."

Specifically, the new method can solve three problems:

- Distinction of internalized and adsorbed particles in a quantitative fashion;
- Correction of errors induced by fluorescent tagging;
- Comparison of data between different materials and different labs.

Gottstein and her team used a high-throughput technique—fluorescence activated cell sorting/scanning (FACS)—combined with confocal laser scanning microscopy (CLSM). Already, researchers have been combining FACS and CLSM to obtain quantitative data on particle association with cells, and qualitative data on internalization—but the data are presented as independent sets. By contrast, the CNSI team measures internalization quantitatively by CSLM and integrates this data with high-throughput FACS data for a precise measurement of nanoparticle uptake. In addition, they introduce a correction factor—which has an impact on confocal microscopy as well as on FACS measurements—that accounts for a potential change of fluorescence intensity upon internalization.

The researchers first derived and validated their method using polystyrene particles and then applied it to measure the impact of different surface coatings of vesosomes on their uptake by cells of the reticuloendothelial system.

Apart from basic research on materials' properties and their influence on particle behavior in biological contexts, potential applications of this work include preclinical testing of drugs (correlation of internalized particles with drug effect) and the optimization of drug delivery vehicles (impact of biochemical and biophysical parameters on targeting performance).

"We see this method as a starting point for further refinements, such as automatization and further delineation of particles in distinct subcellular compartments," says Gottstein. "This will provide a wealth of information."

She points out that some challenges need to be overcome in the areas of developing efficient fluorescent labeling techniques for a wide array of nanoparticles, investigating the impact of the label itself on biological and biophysical functions such as binding, internalization, and aggregation.

Featured scientist: Prof. Claudia Gottstein
Organization: California NanoSystems Institute, Santa Barbara, CA (USA)
(http://www.cnsi.ucsb.edu/)
Relevant publication: C. Gottstein, G. Wu, B. Wong and J. Zasadzinski, Precise Quantification of Nanoparticle Internalization, *ACS Nano*, 2013, 7(6), 4933, DOI: 10.1021/nn400243d

5.3.3   Optical Sensor Detects Single Cancer Cells

The nanotechnology-enabled detection of a change in individual cells, for instance cell surface charge, presents a new alternative and complementary method for disease detection and diagnosis. Since diseased cells, such as cancer cells, frequently carry information that distinguishes them from normal cells, accurate probing of these cells is critical for early detection of a disease.

Scientists have already reported[12] on sensitive flow sensing of a single cell based on graphene field-effect transistors. In this work, it was demonstrated that an array of graphene transistors integrated with microfluidics can be employed to detect the change to the cell surface charge of a cell infected with a virus.

"Relative to the temporal and spatial resolutions of optical sensing, the temporal and spatial resolutions of transistor-based graphene flow sensing of a single cell are greatly restricted," Zhi-Bo Liu, an associate professor at the School of Physics, Nankai University, points out.

He notes that the main challenge in the optical flow sensing of single cells is to improve the sensitivity to achieve a fast response and a non-contact approach, and to sense low concentrations of diseased cells in a small-volume sample size.

To overcome this challenge, Liu and his collaborators have designed a graphene-based optical refractive index sensor.

"We obtained our ultrasensitive graphene optical sensor by controlling the thickness of high-temperature reduced graphene oxide," explains Liu. "With this method, the limits of sensitivity and resolution for refractive index sensing are increased to 4.3×10^7 mV per RIU and 1.7×10^{-8}, respectively. The resolution is the highest value reported so far for refractive index sensors."

The researchers found that, in order to achieve the most sensitive sensing, ~8-nm-thick high-temperature reduced graphene oxide (h-rGO) is the optimal choice as a sensing layer compared with other types of graphene.

"The refractive indices of cancer cells are significantly larger than those of normal cells," explains Liu. "In our experiments, we showed that our sensor in a microfluidic channel detects lymphocytes and cancer cells extracted from blood. We mixed a low percentage (1%) of cancer cells with normal lymphocytes and even at this very low proportion detected the signal changes in voltage with our sensor."

He adds that this graphene-based optical single-cell sensor can provide comprehensive information regarding each cell signal, which is extremely important for research on the cell submicroscopic structure.

There are still a few challenges to overcome before this type of sensor could be used in a clinical environment. Currently, this flow sensing method can only study blood cells that are suspended. Furthermore, the type of cell that can be analyzed is still very limited.

"In the future, we hope to improve the measurement mechanisms and methods to achieve the purpose of measuring various types of cells," says Liu. "Also, the sensor's ultrahigh resolution and sensitivity to refractive index measurements can be extended to other areas, such as drug discovery, environmental monitoring, and gas- and liquid-phase chemical sensing."

Featured scientist: Prof. Zhibo Liu (http://oldphysics.nankai.edu.cn/grzy/rainingstar/index.htm)
Organization: School of Physics, Nankai University (PR China)
Relevant publication: F. Xing, G. Meng, Q. Zhang, L. Pan, P. Wang and Z. Liu, *et al.*, Ultrasensitive Flow Sensing of a Single Cell Using Graphene-Based Optical Sensors, *Nano Lett.*, 2014, **14**(6), 3563, DOI: 10.1021/nl5012036

5.3.4 Catch and Release of Individual Cancer Cells

Fractals are structures built up from repeated sizings of a simple shape to make a complex one. A fractal is a geometric structure that can repeat itself towards infinity. Zooming in on a fragment of it, the original structure becomes visible again.

In biological systems, fractals can be found everywhere—bronchial trees, vasculature, and nerve cells. These amazing structures can provide a specific interfacial contact mode that is highly efficient for absorbing sunlight, transporting nutrition, exchanging oxygen and carbon dioxide, and signal transduction.

Due to their attributes, artificial fractal structures have attracted considerable scientific interest for their appealing applications in antennas, solar cells, and biosensors.

"Recently, fractal nanostructures have been discovered on the surface of cancer cells," explains Shutao Wang, a professor at the Institute of Chemistry,

Chinese Academy of Sciences. "They result in cancer cells having a higher fractal dimension than normal cells, which can be used to distinguish cancer cells from normal cells. This important finding inspired us to explore whether fractal nanostructures can be utilized to program interfacial materials for cancer cell recognition, which will be of great importance for clinical diagnosis."

Wang and his team have demonstrated the fabrication of programmable fractal gold nanostructured interfaces and their outstanding specific recognition of rare cancer cells from whole blood samples along with their effective release capability.

By engineering the fractal gold nanostructures to match the surface fractal features of cancer cells, the researchers achieved a high cell-capture efficiency of rare circulating cancer cells from whole blood samples without the assistance of complex fluidic manipulation.

The team generated three kinds of fractal gold nanostructures (FAuNSs) with controllable fractal dimensions using a one-step electrochemical deposition approach.

"The electrodepositing potential and supporting electrolyte are two major factors influencing the topography of the nanostructures," explains Wang. "In our experiment, relative smooth nanostructures were produced at low potentials, and nanostructures with more complex details were generated at higher potentials."

Exploring the potential of their FAuNSs, the scientists found that the fractal structured interfaces showed a much higher cell-capture efficiency compared to flat gold nanostructures (62% ± 13% for high FAuNS *versus* 3% ± 1%, for flat).

Wang notes that an average of 98% of the captured cells can be easily released with outstanding cell viability (*ca.* 95%) through an electrochemical cleavage of the Au–S bonds, which could provide undamaged cancer cells for further studies, such as phenotype assessments, genomic/mRNA analysis, and screening of anticancer drugs.

With increasing accuracy, "liquid biopsies"—where circulating tumor cells (CTCs) are isolated from blood samples—are becoming a viable complement or even alternative to invasive biopsies of metastatic tumors. CTCs are of great interest for evaluating cancer dissemination, predicting patient prognosis, and also for the evaluation of therapeutic treatments, representing a reliable potential alternative to invasive biopsies and subsequent proteomic and functional genetic analysis.

The challenge with liquid biopsies is to build a system that makes it possible to perform both the isolation and the molecular characterization of CTCs, preferably at a single-cell level.

"The biggest problem is that CTCs, in clinically relevant concentrations, are very rare: they are found in numbers in the order of 1 to 10 CTCs per mL of whole blood in patients with metastatic disease," explains Wang. "Existing technologies cannot reach the performance requirement of clinics, due to the low capture efficiency, low purity, and low specificity. However, advanced nanomaterials provide an opportunity to solve this problem and we hope that our three dimensional nanostructures can greatly improve the capture efficiency."

In particular, due to its high release rates, the FAuNSs assemblies provide an excellent platform for releasing captured CTCs undamaged for further studies, such as phenotype assessments, genomic/proteomic analysis, and screening of anticancer drugs.

Featured scientist: Prof. Shutao Wang' research group (http://wangshutao. iccas.ac.cn/)
Organization: Institute of Chemistry, Chinese Academy of Sciences, Beijing (PR China)
Relevant publication: P. Zhang, L. Chen, T. Xu, H. Liu, X. Liu and J. Meng, *et al.*, Programmable Fractal Nanostructured Interfaces for Specific Recognition and Electrochemical Release of Cancer Cells, *Adv. Mater.*, 2013, 25(26), 3566, DOI: 10.1002/adma.201300888

5.3.5 Sensing of Single Malaria-Infected Red Blood Cells

One of the graphene research areas that is seeing vast scientific interest is the biological interfacing of graphene, for instance for sensor applications. An example of these bio-applications is a graphene sensor that is integrated with microfluidics to sense malaria-infected red blood cells at the single-cell level.

"Graphene's attributes—good interface with living cells and biorecognition proteins; atomic thinness; optical transparency; current stability; and electrical sensitivity—enabled us to design and fabricate a selective flow-catch–release sensing platform, in which, for the first time, malaria-infected red blood cells can be optically differentiated without the need for fluorescence staining and simultaneously, electrically detected by graphene one cell at a time in a microfluidic channel," explains Professor Kian Ping Loh, a member of the Graphene Research Centre at National University of Singapore (NUS). The work was carried out by doctorate student Priscilla Ang, in collaboration with Professor Chwee Teck Lim's group at NUS' Division of Bioengineering.

Loh and his collaborators have demonstrated that this graphene sensor is able to generate dynamic disease diagnostic patterns in terms of conductance changes and characteristic dwell times (Figure 5.6).

The main motivation for designing such a device lies in the fact that a pathological cell–cell interaction is essentially mediated by a charge-based phenomenon. Any disease-induced structural changes will result in a noticeable change in the overall cell surface charge density, which can be used as an additional parameter for differentiation of disease state.

"Hitherto, most disease detection and diagnostic tools rely on microscopy and antibody staining, which require specialized training and may succumb

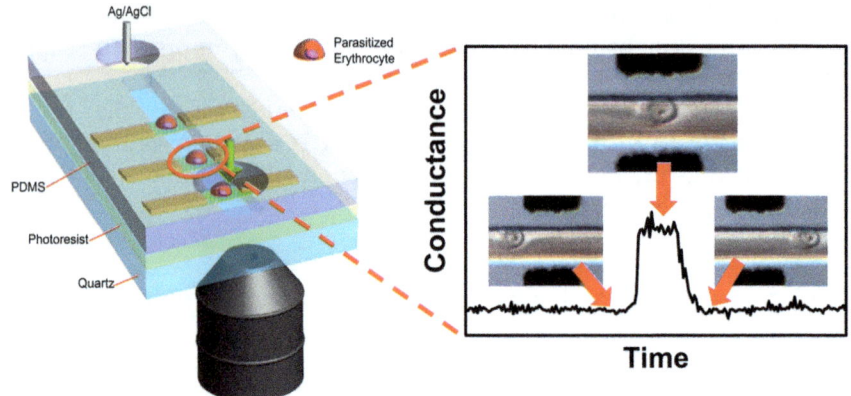

Figure 5.6   Schematic illustration of an array of graphene transistors in a microfluidic channel through which cells flow. Specific binding between ligands
on positively charged knobs of infected red blood cells and receptors
functionalized on graphene induces a distinct conductance change.
Conductance returns to baseline value when the infected cell exits the
graphene channel. (Image: Priscilla Ang, NUS.)

to misdiagnosis due to human errors," explains Loh. "While polymerase
chain reaction (PCR) remains the gold standard, it is unsuitable for routine
use due to its high operating cost. Besides these challenges, the heterogeneity of cellular behaviors in a large population of cells also makes such measurements insensitive to changes occurring in individual cells."

Therefore, as Loh points out, detecting a change in cell surface charge at
the single cell level presents a new alternative and complementary method
for disease detection and diagnosis. As the NUS team has successfully
demonstrated, this can be done by employing an array of graphene transistors in which any change to the cell surface charge translates to a distinct
change in the transconductance of the graphene. Moreover, easy integration
with microfluidics enables the manipulation, isolation and statistical analysis of individual diseased cells.

Malaria is among the most deadly infectious diseases on the planet. The
hallmarks of an infection by *P. falciparum*—one of the most virulent para-
site strains—include several irreversible structural modifications of the par-
asitized red blood cells. During an infection, these structural modifications
cause the cytoadherence of affected red blood cells to endothelial cells lining
the blood vessels and capillaries. CD36 receptor proteins on the endothelial
cells are involved in this specific interaction.

The researchers exploited this for their flow–catch–release scheme by
functionalizing graphene with endothelial CD36 receptors for the selective
capture of the malaria-infected cells when the diseased blood—consisting of
a mixture of healthy and infected red blood cells—flows through a microfluidic channel.

In addition to electrical sensing (a positively gated graphene will exhibit
an increase in its channel conductance upon the binding of the infected cell
because of electrostatic doping), the optical transparency of graphene allows

the simultaneous optical monitoring of binding events *via* differential interference contrast (DIC) microscopy.

"Analyzing the cell surface charge density will shed new insights onto cell–cell interactions governing pathophysiology in near physiological conditions," Loh outlines the possibilities of this sensing approach. "In addition, our device is robust and can be employed to study and detect any kind of diseased cell. The ability to do a statistical percentage count of the infected cells as the population of infected and healthy cells flow through the graphene transistors in a microfluidic channel exhibits great promise for clinical diagnostic applications."

There are many challenges ahead—for instance the economical production of high-quality graphene—before this device can be commercialized. However, Loh and his team can envision the incorporation of graphene transistors in a lab-on-a-chip device, which allows high-throughput flow sensing of infected cells with small volume consumption and automated electrical readout for disease detection and diagnosis.

Featured scientists: Prof. Kian Ping Loh's Carbon Convergence Technology Laboratory (http://staff.science.nus.edu.sg/~thecarbonlab/index.html); Prof. Chwee Teck Lim's research group (http://www.bioeng.nus.edu.sg/nanolab/nanolab.html)
Organizations: Department of Chemistry; Department of Biomedical Engineering, National University of Singapore
Relevant publication: P. Ang, A. Li, M. Jaiswal, Y. Wang, H. Hou and J. Thong, *et al.*, Flow Sensing of Single Cell by Graphene Transistor in a Microfluidic Channel, *Nano Lett.*, 2011, **11**(12), 5240, DOI: 10.1021/nl202579k

5.3.6 Novel Mechanobiological Tool for Probing the Inner Workings of a Cell

In order to study inner cell properties, researchers often use internalized particles as micro- or nanoprobes. However, almost all available techniques provide only two-dimensional manipulation and tracking of these particles.

"Besides the use of particle tracking methods, the decipherment of the complex dynamics at the single cell level also requires novel techniques that provide additional information about dynamical changes of the cell morphology," says Cornelia Denz, a professor at the Institute of Applied Physics, University of Münster. "Especially for the quantitative study of dynamic cellular processes, quantitative phase microscopy is more suitable than other qualitative imaging techniques."

Denz and her group, together with researchers from the university's Center for Biomedical Optics and Photonics, have demonstrated a biophotonic holographic workstation that combines the complementary features of holographic optical tweezers (HOT) and self-interference digital holographic microscopy (DHM), in order to investigate biomechanics properties at the single cell level.

"On the one hand, HOT enable the confinement of a microscopic object of nano- or micrometer size and move it precisely in three dimensions by means of tailored light fields," Álvaro Barroso Peña, a PhD student in Denz's group and first author of a paper on this work, explains. "On the other hand, DHM in off-axis configuration provides amplitude and quantitative phase information of the optical wave that propagates through a transparent sample. This allows simultaneous quantitative phase imaging of the cell morphology as well as 3D tracking of transparent objects without fluorescence labeling."

In their study, the researchers show that the combined implementation of HOT, DHM, and microspheres (with a diameter of 1 μm) as sensitive probes inside cells provides a novel mechanobiological tool for minimally-invasive quantitative 3D probing of the intracellular morphology without any fluorescence labels. The polystyrene particles, incorporated into cells by phagocytosis, can be moved at will with sub-micrometer accuracy within the cellular volume.

Barroso notes that the movement of the probe particles—which is restricted only by intracellular constraints—together with an accurate estimation of the exerted forces, provides key information about the inner cellular structure.

"In addition, our approach demonstrates to be minimally invasive as neither the trapping laser beam nor the imaging laser source induces serious damage to the investigated cells," he says.

In their proof-of-principle study, the scientists demonstrate the capability of their method by analyzing the inner cell structure changes during a very important biological process: cell swelling provoked by osmosis stimulation.

This paves the way to study the intracellular changes during other dynamic cellular processes.

Furthermore, this novel approach allows the determination of damage in living cells induced by the trapping laser. This can provide valuable information to be used in related studies where the effect of the trapping laser on the biological specimen is unknown.

Going forward, the team plans to extend their study to living organisms. In this respect, model organisms such as zebrafish (*Danio rerio*) are good candidates to explore *in vivo* the intracellular structure by sensing the movement of optically manipulated internalized particles.

Featured scientists: Prof. Cornelia Denz's research group (http://www.uni-muenster.de/Physik.AP/Denz/)
Organizations: Institut für Angewandte Physik; Centrum für Biomedizinische Optik und Photonik (http://campus.uni-muenster.de/index.php?id=529); Westfälische Wilhelms-Universität Münster (Germany)
Relevant publication: Á. Barroso, M. Woerdemann, A. Vollmer, G. von Bally, B. Kemper and C. Denz, Three-Dimensional Exploration and Mechano-Biophysical Analysis of the Inner Structure of Living Cells, *Small*, 2013, 9(6), 885, DOI: 10.1002/smll.201201851

5.3.7 Snail-Inspired Nanosensor Detects and Maps mRNA in Living Cells

Every cell is a busy messaging center, with thousands of genes churning out messenger RNA (mRNA) transcripts for translation into functional proteins. Cell fate, function, and phenotype are significantly dictated through the spatiotemporal control of mRNA expression. Therefore, determining the mRNA expression of an individual cell can reveal critical insights into that particular cell's health and physiological state.

"Presently, several techniques for detecting mRNAs are available, which include *in situ* hybridization and polymerase chain reaction (PCR)," says David T. Leong, an assistant professor in the Department of Chemical & Biomolecular Engineering at National University of Singapore (NUS). "However, these single-point and end-point techniques require the killing of the cells and are thus unable to capture the expression of mRNA in real time and locality with high precision."

Leong and his team describe a new way of preparing functional DNA nanostructures that can provide accurate quantification and visualization of mRNA transcripts in living cells.

"One key feature of our DNA-based mRNA sensor design is that it is highly resistant to non-specific enzymatic degradation, and thus was shown to significantly reduce the likelihood of generating false-positive signals," Leong explains. "This is made possible because of our snail-inspired design. Just like how the shell of the snail protects the snail from the harsh environment but does not completely restrict it from sensing its surroundings, we incorporated this 'protective-yet-accessible' design concept into our DNA nanodevice which we termed *nano-SNail-inspirEd Locator* (nano-SNEL)."

Nano-SNEL has two main components: a sensor/reporter to detect mRNA, and a DNA-pyramid shell. The latter not only serves as a carrier for the sensor to easily enter into a cell but also functions as a decoy or sacrificial shell to protect the sensor from enzymatic attack.

"We showed that our nano-SNEL could synergistically present itself as an excellent intracellular probe, without interfering with either specific targeting by or hybridization-induced fluorescence of the probes," says Leong. "When the cells were treated with nano-SNEL, we were able to attain robust *in situ* hybridized signals and the associated intracellular spatial distribution could be mapped since the cell itself is not destroyed."

A key feature of nano-SNEL is that it is very stable even under harsh physiological conditions. This opens up a myriad of possible applications where DNA-based nanomaterials can be applied directly.

Because the targets—DNA, mRNA, miRNA—are ubiquitous in all biological systems, nano-SNEL is highly versatile as a diagnostic tool. For example, defective expression of mRNA linked to certain heterogeneous diseases such as cancer might be detectable at the single cell resolution using nano-SNEL.

Outside the cell, nano-SNEL can also be used to detect viral RNA that may be found in blood serum such as the Ebola virus. This is not possible with mRNA sensors that are susceptible to non-specific enzymatic degradation.

"In our lab, we are particularly interested in the use of DNA-based nanomaterials for biomedical applications," notes Leong. "Compared to other types of classical organic/inorganic nanomaterials, DNA as a malleable putty for nanomedicine is still at its infancy and thus leaves some uncharted territory for exploration. We hope to share with the community a few interesting applications arising from the use of this exciting starting material."

"The use of DNA-based materials for nanomedicine applications has gained significant traction in recent years; we believe future studies will need to establish a better understanding of the critical relationship between DNA nanostructures and cell biology," concludes Leong. "With these critical insights, we hope to develop smarter and more sophisticated DNA nano-devices to meet some of the challenges of modern medicine. We cannot do all these on our own as the success of DNA nanomaterials as a functional material for biomedical applications definitely requires the cross pollination of ideas from colleagues of diverse backgrounds and expertise."

Featured scientists: Prof. David T. Leong's lab (http://blog.nus.edu.sg/leonglab/)
Organization: Department of Chemical & Biomolecular Engineering, National University of Singapore
Relevant publication: C. Tay, L. Yuan and D. Leong, Nature-Inspired DNA Nanosensor for Real-Time in Situ Detection of mRNA in Living Cells, *ACS Nano*, 2015, **9**(5), 5609, DOI: 10.1021/acsnano.5b01954

5.3.8 Silicon Chips Inserted into Living Cells Can Feel the Pressure

In 2010, a team of researchers in Spain produced silicon chips that can be internalized inside living cells to be used as intracellular sensors.[13] Following up on this work several years later, the same team, led by José Antonio Plaza, who heads the Micro and Nano Tool group at Instituto de Microelectrónica de Barcelona IMB-CNM (CSIC) and a group from the 3DLab (Development, Differentiation and Degeneration) at Centro de Investigaciones Biológicas in Madrid (CSIC) coordinated by Teresa Suárez, demonstrated a nanomechanical chip that can be internalized to detect intracellular pressure changes within living cells, enabling an interrogation method based on confocal laser scanning microscopy.

"Our goal was to fabricate a chip small enough to be inserted into a living cell and detect mechanical loads," says Rodrigo Gómez-Martínez, a member of the team. "As a result, we have been able to show that this nanostructured device can detect intracellular pressure changes. This is the first time that pressure can be detected by a sensor located inside an intact cell, preserving the integrity of the cell membrane."

Suárez points out that the presence of the nanochip does not seem to affect the cell's structure or viability—cells with the sensor chip inside were alive, healthy and able to divide (with the sensor remaining inside one of the cells).[14]

The challenge for the researchers was to produce a device with an integrated sensor system small enough to fit inside a cell.

The design that they used comprises a mechanical sensor defined by two membranes separated by a vacuum gap, and an optical reference area. The internalized sensors only represented 0.2% of the total volume of a typical HeLa cell.

"The membranes act as parallel reflecting mirrors, constituting a Fabry-Pérot resonator that is partially transparent for some wavelengths," explains Gómez-Martínez. "An external pressure deflects the membranes and changes the gap; which, in turn, modifies the intensity of the reflected light at the centre of the membranes."

He points out that the presence of a sensor inside a vacuole has several inherent advantages: "First, it can give information about how an external pressure is transmitted mechanically to organelles. Second, it prevents the eventual existence of mechanical cross-sensitivity on the devices because of other organelles or cytoskeletal filaments, which can induce small forces and displacements. Third, better-quality confocal laser scanning microscopy images are obtained when the sensors are immersed in a medium with a uniform refractive index (Figure 5.7)."

Extracellular pressure is a common load in many situations. For instance, human cells experience a pressure of 0.2 bar from throughout the body, which can increase during certain activities, and deep-sea animals can be exposed to 200 bar upon diving.

"Our experiments support the supposition that the cytoskeletons of human HeLa cells do not mechanically withstand extracellular pressures in the studied range and under our experimental cell culture conditions," says Suárez. "Thus, extracellular pressure is transmitted through the cytosol to the inner compartments. The implication is that intracellular transmission of fluid pressure follows Pascal's law."

Going forward, the team will try to improve their sensor by increasing its sensitivity and include thinner mechanical layers, autofocus and tilt-stage systems, and computer-assisted measurements.

Another objective is to design chips that can measure other intracellular parameters—biochemical, electrical, thermal *etc.*—in order to investigate more complex cellular processes.

"Intracellular mechanical sensors will provide information directly from inside the cellular environment about these cellular forces and will provide new opportunities," notes Plaza. "We believe that this is a first step towards a wide-ranging field of intracellular nanochips that will offer a different perspective on fundamental problems in cell biology."

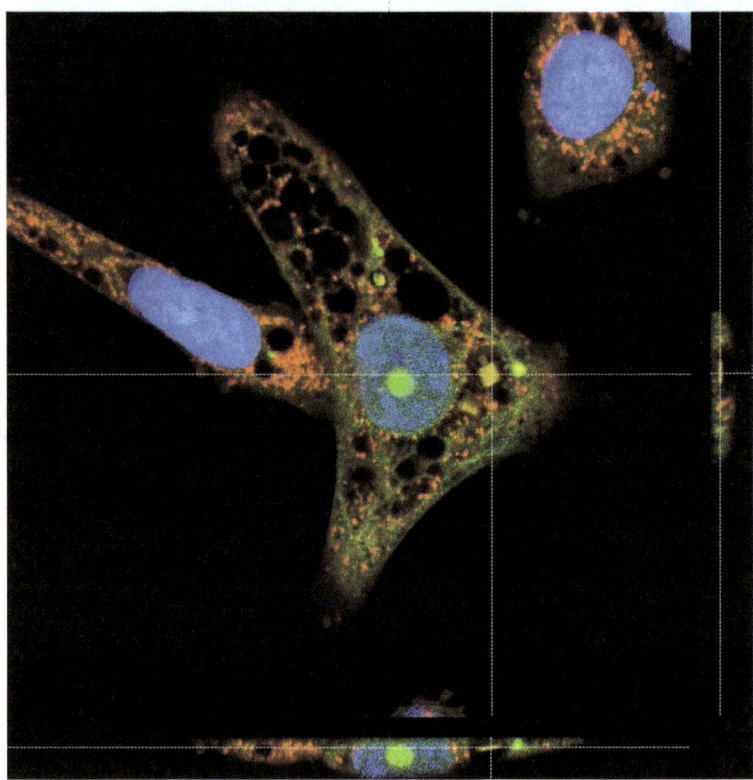

Figure 5.7   A HeLa cell displaying an internalized chip. Overlay of confocal images and an orthogonal projection of confocal images showing that the chip is inside the cell. The cells were loaded with vital dyes CellTracker Green and MitoTracker Red before fixation. (Image: Alberto Hernández-Pinto, Centro de Investigaciones Biológicas, CIB (CSIC).)

Featured scientists: Dr José A. Plaza (http://mnt.imb-cnm.csic.es/index. php); Teresa Suárez; Rodrigo Gómez-Martínez
Organization: Micro and Nano Tool Group, Instituto de Microelectrónica de Barcelona; 3DLab, Centro de Investigaciones Biológicas, Madrid (Spain) (http://www.cib.csic.es/en/grupo.php?idgrupo=20)
Relevant publication: R. Gómez-Martínez, A. Hernández-Pinto, M. Duch, P. Vázquez, K. Zinoviev and E. de la Rosa, *et al.*, Silicon chips detect intracellular pressure changes in living cells, *Nat. Nanotechnol.*, 2013, **8**(7), 517, DOI: 10.1038/nnano.2013.118

5.3.9 Direct Observation of How Nanoparticles Interact with the Nucleus of a Cancer Cell

Cancer researchers are looking to nanoparticles as a drug carrier capable of localizing and directly releasing drugs into the cell nucleus, thereby circumventing multi-drug-resistance and intracellular drug-resistance mechanisms to effectively deliver drugs to the vicinity of DNA, leading to a high therapeutic efficacy.

Nanoparticles have been explored for a wide range of applications, including drug delivery vehicles, imaging contrast probes, and therapeutic agents. However, although increased therapeutic efficacy has been realized—and evidence that nanoparticles can affect intracellular function has been seen *via* fluorescent labels—there have been no reports on visualizing at nanoscale dimensions how nanoparticles interact with specific organelles. How nanoparticles interact with the cell nucleus, for example, will have implications not only for the fundamentals of cancer biology but also for the design of translational therapeutic agents.

In a breakthrough for nanomedicine cancer research, scientists have demonstrated the direct visualization of interactions between drug-loaded nanoparticles and the nucleus of a cancer cell.

Teri W. Odom, Board of Lady Managers of the Columbian Exposition Professor of Chemistry and Professor of Materials Science and Engineering at Northwestern University, and her team, report two breakthrough results on the interaction of drug-loaded nano-constructs with the cancer cell nuclei:

(1) Nano-constructs trafficked to the nucleus *via* a shuttling protein produced invaginations in the nuclear envelope at the site of the drug-loaded nanoparticles; and

(2) Changes in nuclear phenotype—specifically increases in the number, depth, and shape of the nuclear envelope invaginations—were strongly correlated with changes in cell function.

"We have reported the first imaging results on how drug-loaded nanoparticles affect nuclear phenotype," says Odom. "The most surprising aspect of this work is that changes in nuclear phenotype were correlated with the biological response of the treated cancer cells."

The team's two-component nano-constructs consist of gold nanostars and the single-stranded DNA aptamer AS1411. AS1411 has been tested as a chemotherapeutic agent because of its ability to strongly bind to nucleolin, a protein that is over expressed in cancerous cells. By blocking several functions of nucleolin, AS1411 can ultimately result in tumor cell death.

Odom points out that, surprisingly, high-resolution TEM images showed major changes in nuclear phenotype near the site of the nano-construct; the nuclear envelope was extremely deformed in over 60% of the tumor (HeLa) cells with nano-constructs.

To test whether the concentrated release of aptamer from the constructs could increase nuclear envelope folding further, the researchers used ultra-fast (femtosecond), light pulses to detach AS1411 from the surface of the gold nanostars. They found that this increased the number of local deformations in the nucleus.

"That cancer cell function can be correlated with deformations in the nucleus suggests that major challenges in particle-based, nuclear-targeted therapy can be overcome," says Odom. "For example, complete internalization of nano-constructs inside the nucleus is not necessary if induced physical changes in nuclear phenotype can disrupt nuclear functions. Also, because morphological effects are induced by nanoparticles outside the nucleus, there may be no limitation in the size or shape of nanoparticles that can achieve a similar response. These factors should provide insight into the development of new strategies to design drug-loaded nanoparticles with increased therapeutic efficacy."

In subsequent research, the scientists are testing this nano-construct in a variety of carcinomas *in vitro*. Because the shuttling protein nucleolin is over-expressed on all proliferating cancer cells, they expect to observe high anti-cancer potency in a wide range of cancers.

Featured scientists: Prof. Teri W. Odom's research group (http://chemgroups.northwestern.edu/odom/index.htm)
Organization: Department of Chemistry, Northwestern University, Evanston, IL (USA)
Relevant publication: D. Dam, J. Lee, P. Sisco, D. Co, M. Zhang and M. Wasielewski, *et al.*, Direct Observation of Nanoparticle–Cancer Cell Nucleus Interactions, *ACS Nano*, 2012, **6**(4), 3318, DOI: 10.1021/nn300296p

5.3.10 A Precise Nanothermometer for Intracellular Temperature Mapping

Green fluorescent protein (GFP)—originally found in a jellyfish—has played a crucial role in life science research, providing insights into many fundamental questions that have paved the way to the biology and medicine of the future. Since the mid-1990s, when the protein was successfully cloned, GFP has been used in research laboratories worldwide as a visual marker of gene expression and protein localization, easily observed *via* light (optical) microscopy. GFP can be linked to other proteins and is primarily used to track dynamic changes in living cells. In 2008, biologists who discovered and developed the protein as a laboratory tool won a Nobel prize for their work.[15]

Researchers in Spain have demonstrated how GFP can also act as an efficient nano-thermometer inside cells.

There has been no reliable method to measure intracellular temperature. This novel approach is based on GFP, where the GFP is transfected by the cells. Using this approach, the cell mechanisms are not modified or altered during the measurement by the presence of GFP.

"Because temperature governs many vital cellular processes, its accurate and non-invasive monitoring should contribute, in parallel with other existing GFP-based techniques, to further unravel the complex machinery of cell mechanics where the temperature plays an important role," says Romain Quidant, ICREA Professor at The Institute of Photonic Sciences (IFCO).

Several microscopy techniques have been proposed to address the need for monitoring intracellular temperature in molecular biology. Most of them rely on introducing synthetic nano-objects into living cells, a fact that can cause stress in the cells and alter their behavior.

Quidant and his team wanted to find a method to measure temperature that does not harm the cell, but at the same time can give robust imaging opportunities—namely, fast and accurate measurements. Their approach relies on measuring fluorescence polarization anisotropy (FPA) of GFP and combines fast and accurate acquisition with confocal spatial resolution.

Jon Donner, a PhD student in Quidant's group, explains the—well-established—underlying physics which relates temperature and molecular FPA: "In general, a population of fluorophores illuminated by a linearly polarized light re-emits partially polarized fluorescence due to the random orientation of the molecular dipoles. When the temperature increases, the Brownian rotational motion of the fluorophores is accelerated. Hence, the molecules will rotate more during their fluorescence lifetime. The more the molecules rotate, the more the re-emitted photons will lose the memory of the incident light polarization. Consequently, a temperature increase leads to a decrease of the degree of polarization (anisotropy) of the fluorescence. Using a suitable calibration, FPA leads to an absolute temperature measurement. This is one of the main advantages of this technique."

In their temperature measurement experiments, the researchers used single living cells and dispersed gold nanoparticles in the extracellular medium. Then they locally heated the cells by focusing an infrared laser 50 μm aside from the studied cell (they chose not to shine on the cell directly to demonstrate that FPA variations measured in the cytosol are unambiguously due to temperature variations and not to a possible IR-assisted perturbation of the GFP fluorescence emission process). They were able to precisely measure the temperature inside the cell resulting from the heated nanoparticles outside it.

"We hope that our method will become a powerful tool to unravel intimate cellular processes or mechanics at the single cell level," says Sebastian Thompson, a postdoc in Quidant's group. "Benefiting from its full compatibility with widespread GFP cellular biology, it complements the existing toolbox for biologists and has the potential to extend our understanding of life science."

He notes that this approach could become particularly useful for temperature-based therapies such as photothermal cancer therapy, where heat is locally delivered to cancer tumor cells *via* photothermal conversion of nanoparticles. This is a field that is already quite advanced and where clinical trials are being conducted. To better calibrate the parameters of these processes, one needs to monitor the temperature of the cells by disturbing the cells as little as possible.

Going forward, the researchers will work on improving the sensitivity of their method.

"After that, we would love to see this research extended to *in vivo* applications," says Quidant.

Featured scientists: Prof. Romain Quidant's Plasmo Nano-Optics Group (https://www.icfo.eu/research/group_details.php?id=27)
Organization: The Institute of Photonic Sciences (ICFO), Barcelona (Spain)
Relevant publication: J. Donner, S. Thompson, M. Kreuzer, G. Baffou and R. Quidant, Mapping Intracellular Temperature Using Green Fluorescent Protein, *Nano Lett.*, 2012, **12**(4), 2107, DOI: 10.1021/nl300389y

5.3.11 Direct Observation of Drug Release from Carbon Nanotubes in Living Cells

Carbon nanotubes (CNTs) offer a number of advantages for delivering drugs to specific locations inside the body, which suggest that they may provide an improved result over nanoparticles. They have a larger inner volume, which allows more drug molecules to be encapsulated; this volume is more easily accessible because the end caps can be easily removed; and they have distinct inner and outer surfaces for functionalization. Already, research has shown the ability of CNTs to carry a variety of molecules such as drugs, DNA, proteins, peptides, targeting ligands *etc.* into cells—which makes them suitable candidates for targeted delivery applications.

Doxorubicin (DOX), a popular anticancer drug, is promising for use in controlled drug-release systems. Due to its aromatic structure, DOX can be attached to the sidewall of single-walled carbon nanotubes (SWCNTs) and then detached to yield free molecules in acidic conditions because of the pH-dependent stacking interaction between DOX and nanotubes. Researchers have already successfully applied a DOX-loaded carbon nanotube drug delivery system for both *in vitro*[16] and *in vivo*[17] cancer treatment.

"The precise process of drug release from its vehicle inside the living cell is still not clear," says Da Chen, a professor in the College of Material Science and Technology at Nanjing University of Aeronautics and Astronautics. "Understanding this process is quite important to better understand the subcellular mechanisms of drug delivery systems therapy and therefore to design new systems and controlled-release approaches. In the DOX-loaded carbon nanotube drug delivery system, DOX is supposed to release to intracellular lysosomes after being delivered into cells due to the acidic microenvironment in the lysosomes. However, thus far, the direct observation of this drug-release process inside the living cells has not been accomplished. Tracking this process in living cells requires the determination of the location and movement of the carbon nanotube carriers and the drug status—loading or release—simultaneously, which seems quite difficult to achieve."

In their work, Chen and his collaborators have developed a unique two-dye labeling method to directly track the release process of DOX from carbon nanotube carriers in living cells. The team demonstrated the direct observation of the time-dependent drug-release process using a laser scanning confocal microscope. Based on this observation, they propose a three-step process model for subcellular drug release from a carbon nanotube vehicle.

To perform their observations, the researchers covalently conjugated a dye that exhibits bright green fluorescence to chitosan-coated carbon nanotubes. This allowed them to track the nanotube location and movement inside cells. Then the drug DOX, which is also a kind of dye that exhibits bright red fluorescence, is attached by means of π-stacking (the attractive, noncovalent interactions between aromatic rings) onto the side-wall of the SWCNTs.

"This pH-sensitive interaction between DOX and SWCNTs should induce drug release in acidic conditions," says Chen. "Thus, from the performance of these two kinds of fluorescence—together or separated—we monitored the drug status on nanotube carriers, and also the pH change in specific subcellular organelles that triggered the release."

From their observations of the subcellular behavior of drug-loaded SWCNTs, the scientists concluded the mechanism of drug release inside the cell.

The team notes that these findings are partly consistent with the previously proposed mechanism for drug release. "Additionally, we noticed some exceptional phenomena in our experiments," says Chen. "For instance, after being internalized into the cells, SWCNTs are actively transported into a perinuclear area without obvious exocytosis, which is different from the traditional endocytosis pathway we observed previously. So the cellular metabolic pathway of SWCNTs still requires systematic study."

Nevertheless, the experimental establishment of drug-release mechanisms in living cells, as demonstrated in this work, may provide important insights to better understand the drug-release process and aid the future design of new drug-delivery systems.

Featured scientist: Prof. Da Chen
Organization: College of Materials Science and Technology, Nanjing University of Aeronautics and Astronautics (PR China)
Relevant publication: B. Kang, J. Li, S. Chang, M. Dai, C. Ren and Y. Dai, *et al.*, Subcellular Tracking of Drug Release from Carbon Nanotube Vehicles in Living Cells, *Small*, 2012, 8(5), 777, DOI: 10.1002/smll.201101714

5.3.12   Functionalizing Living Cells

Modifying living cells by coating them with a nanolayer of functional materials in order to provide them with new structural and functional features has developed into a popular research area for bionanotechnology researchers. In contrast to genetic manipulation techniques, this approach modifies the functionality of a cell simply by attaching polymers or nanoparticles to the cell's surface.

"In our work we have been concentrating on improving the ways of positioning of nanoparticles and nanotubes onto living cells," says Rawil F. Fakhrullin's, Senior Lecturer, Biomaterials and Nanomaterials Group, Department of Microbiology at Kazan Federal University. "Imagine a system where a large cell is coated with millions of nanoparticles which potentially can penetrate into the cytoplasm and eventually kill the cell. Early research reports were based on the direct, unmediated deposition of nanoparticles onto cells, and sometimes it was required to pre-treat cells in harsh conditions. Obviously, the cells did not survive, and could be used only as sacrificial templates. In contrast, our groups managed to utilize the popular layer-by-layer (LbL) technique in combination with cells and as a result, cells were still viable and functional after the treatment."

Fakhrullin has worked with Yuri M. Lvov, Professor and Tolbert Pipes Eminent Endowed Chair on Micro and Nanosystems at Louisiana Tech University, to describe the techniques for cell encapsulation with LbL self-assembly *via* sequential adsorption of oppositely charged components: polyelectrolytes, nanoparticles, and proteins (Figure 5.8).

Fakhrullin explains that this LbL method for encapsulation is based on the consecutive deposition of polycations and polyanions, bound together through electrostatic interactions and applied first for planar film production and later for encapsulation of colloids, including biological cells.

He notes that this simple encapsulation technique enables researchers to process many biological cells in parallel.

"The ability to design shells of any composition, containing nanosized layers of polymers, proteins, and nanoparticles in a predetermined order enables control of the capsules' properties, such as sensitivity to temperature and pH, permeability, and structural stability," says Fakhrullin. "These surface-functionalized cells thus have their intrinsic functions enhanced or altered."

Using this fabrication approach, scientists can easily assemble functionalized cells into larger 3D blocks—such as artificial multicellular structures or engineered tissues—or immobilize them onto patterned surfaces. Another direction might be the encapsulation of nutrients in the coatings. Yet another example is the protection of LbL-coated cells from the immune system for cell delivery and therapies.

Fakhrullin and Lvov outline the four main research directions in bioencapsulation:

- LbL functionalization of isolated microorganisms (*e.g.*, fungi, algae, and bacteria);
- functionalization of isolated mammalian cells, including human cells;

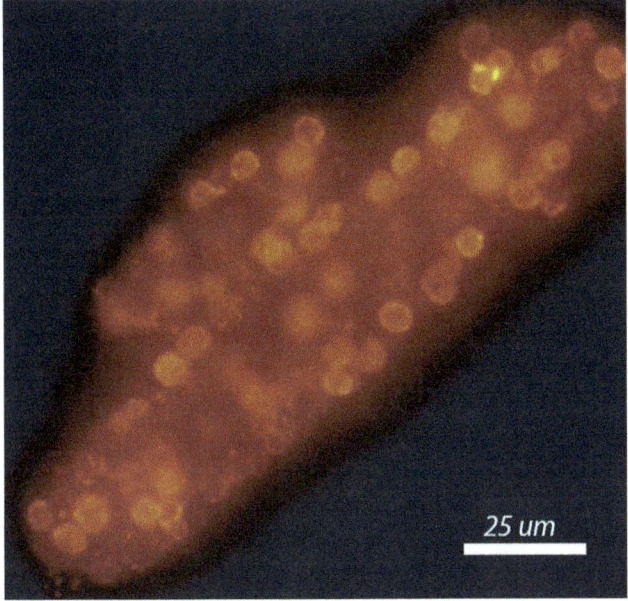

Figure 5.8 Artificial 3D multicellular clusters built from nanocoated yeast cells. (Image: Dr Fakhrullin, Kazan Federal University.)

- encapsulation of cell aggregates, tissue sections, and microorganisms (*e.g.*, microworms); and
- encapsulation of viruses.

Another important area concerns the LbL functionalization of living cells with nanoparticles.

"Polyelectrolytes facilitate adhesion of nanoparticles to biological cells, thus providing the stability of the sandwich-like polyelectrolyte/nanoparticle coating and suppressing nanoparticle internalization through the cell walls into the cytoplasm," explains Fakhrullin.

During the LbL process, a 20–60-nm-thick flexible, hydrogel-like coating is assembled onto a living cell, in effect producing an artificial structure that mimics a cell wall. Toxicological research has shown that the fragile cellular membrane of human cells (less so on more robust microbial cells) can be affected by the deposition of extracellular polymer multilayers.

Scientists hypothesize that the apparent toxicity of polycations can be explained by their electrostatic interaction with the cellular membrane, which causes pore formation and subsequent cell death. To resolve this issue, LbL assembly with natural biocompatible polyelectrolytes, polysaccharides, poly(amino acids), DNA, and polyphenols aims to improve encapsulated shell viability.

Functionalized cells may find a number of practical applications, including as biosorbents, biosensors, in spore formation, and in tissue engineering.

"The advantage of using polymer-modified cells is the almost unlimited quantity of different combinations (*i.e.* wall architectures) possible," says

Fakhrullin. "Functional nanoparticles and nanodevices can be embedded into wall multilayers, which, in turn, may further attenuate the functionality of the composite shell. The modified cells not only act as microtemplates but also exhibit intrinsic functionality, including proliferation."

"In my opinion," he continues, "the most important results can be obtained with magnetic cells. For instance, we have shown[18] how the direct magnetic functionalization of human cells can be applied to manipulate cells using a simple magnet, thus opening new avenues for cell therapy."

Fakhrullin thinks that we will see an exponential growth of research papers in the area of functionalized cells. He argues that there are many still unused nanocoatings and even more types of cells.

"We believe that the major challenge now is to control precisely the time when the coatings are still on the living cell, and also to prepare 'smarter' coatings, which would help control the transport to and from the cells, their communications with other cells, *etc.*"

Featured scientists: Dr Rawil F. Fakhrullin's Biomaterials and Nanomaterials Group (http://old.kpfu.ru/eng/departments/f1/k7/index.php?id=5&idm=3); Prof Yuri M. Lvov (http://www2.latech.edu/~ylvov/)
Organizations: Department of Microbiology, Kazan Federal University (Russian Federation); Institute for Micromanufacturing, Louisiana Tech University, Ruston, LA (USA)
Relevant publication: F. Rawil, Fakhrullin and M. Yuri, Lvov 'Face-Lifting' and 'Make-Up' for Microorganisms: Layer-by-Layer Polyelectrolyte Nano-coating, *ACS Nano*, 2012, **6**(6), 4557, DOI: 10.1021/nn301776y

5.4 A Glimpse at the Numerous Benefits that Nanomedicine Has in Store for Us

Many nanomedicine applications—especially early commercial products—are geared towards drug delivery. However, researchers are pursuing many other application areas that include nanostructure scaffolds for tissue replacement; nanostructures that allow transport across biological barriers; remote control of nanoprobes; integrated implantable nanoelectronic sensory systems; and multifunctional chemical structures for drug delivery and targeting of disease.

5.4.1 High-Tech Band-Aids

In order to fabricate stimuli-responsive materials, researchers have shown a lot of interest in asymmetric materials such as modulated gels, which consist of a controlled layer that is responsive to an environmental stimuli and a nonresponsive substrate layer. And while much effort has gone into creating free-standing films through layer-by-layer (LbL) assembly, relatively little attention has been paid to the asymmetric properties or functionalization of the two surfaces of such free-standing LbL films.

A research team, led by Catherine Picart, a professor at Grenoble INP in France, and Jian Ji, a professor in the Department of Polymer Science Engineering at Zhejiang University in China, has fabricated an asymmetric free-standing layer-by-layer film with asymmetric wettability—one surface is superhydrophobic and the other is hydrophilic. The superhydrophobic side is water-repellent while the hydrophilic side can absorb/desorb water easily.

The superhydrophobic surface was shown to be self-cleaning, anti-wetting and to limit bacterial adhesion, while the hydrophilic surface could extensively deliver bactericidal silver ions.

The team used an amplified LbL method to prepare a film of up to 20 micron thickness. When released from its substrate, this free-standing asymmetric membrane had one rough and superhydrophobic surface and the other surface was flat and hydrophilic.

In previous work, the team had already reported the stimuli responsiveness of their asymmetric, free-standing film.[19] In this follow-up work, they have added additional functionality to the superhydrophobic/hydrophilic asymmetric film.

They loaded the film with silver nanoparticles, which resulted in the hydrophilic surface becoming bactericidal by extensively releasing silver ions. Importantly, the film can be transferred to another interface, such as human skin.

"The superhydrophobic/hydrophilic asymmetric film may find application in wound healing," says Ji. "Imagine a medical dressing where the superhydrophobic surface on top protects the wound from dust and bacteria while the hydrophilic surface at the bottom delivers drugs that can accelerate wound healing or help avoid infections."

He adds that the excellent anti-adhesion ability of the top surface stems from a combination of micro–nano hierarchical structure and superhydrophobicity—which leads to anti bacteria adhesion. "Of note," he says, "the size of the microstructure on the film here was about 2–5 μm, close to the size of *E. coli*."

One issue the scientists mention is that this film is flexible in wet environments but brittle in a dry state. One solution would be to reinforce the film by introducing nanoparticles, such as nano-clay, into the multilayer film.

Overall, this concept of asymmetrically functionalizing multilayer free-standing films may open a new avenue for the LbL technique. Apart from wound dressings, asymmetric, multifunctional films may be of great potential for use in the areas of barrier films, separation, transportation or drug delivery.

Featured scientists: Prof. Catherine Picart (http://www.lmgp.grenoble-inp. fr/research/catherine-picart-261416.kjsp); Prof. Jian Ji (http://www.fe.zju. edu.cn/english/redir.php?catalog_id=6589&object_id=7151)
Organizations: Laboratoire des Matériaux et du Génie Physique, Grenoble IMP (France); Zhejiang University, Hangzhou (PR China)
Relevant publication: L. Shen, B. Wang, J. Wang, J. Fu, C. Picart and J. Ji, Asymmetric Free-Standing Film with Multifunctional Anti-Bacterial and Self-Cleaning Properties, *ACS Appl. Mater. Interfaces*, 2012, 4(9), 4476, DOI: 10.1021/am301118f

5.4.2   Surface-Modified Nanocellulose Hydrogels for Wound Dressing

Cellulose is a biopolymer consisting of long chains of glucose with unique structural properties, whose supply is practically inexhaustible. It is found in the cell walls of plants where it serves to provide a supporting framework, *i.e.* a sort of skeleton. Nanocellulose from wood—*i.e.* wood fibers broken down to the nanoscale—is a promising nanomaterial with potential applications as a substrate for printing electronics, filtration, or biomedicine.

Researchers have reported a method to control the surface chemistry of nanocellulose. Using a specific chemical pretreatment as an example (carboxymethylation and periodate oxidation), a team from the Paper and Fibre Research Institute (PFI) in Norway demonstrated that they could manufacture nanofibrils with a considerable amount of carboxyl groups and aldehyde groups, which could be applied to functionalizing the material.

Using their technique for the production of nanocellulose with tailor-made surface chemistry, the team fabricated hydrogels with pH-responsive characteristics. These nanocellulose gels have a significantly higher swelling degree in neutral and alkaline conditions, compared to an acid environment. This material could be of great interest for critical wound healing applications.

"The purpose of this study was to produce nanocellulose from wood with controlled surface chemistry and morphology, which could potentially be exploited to add functionality to the material," explains Gary Chinga-Carrasco, a senior research scientist at PFI (Figure 5.9).

Chinga-Carrasco is also project manager of the NanoHeal project[20] at PFI, a multi-disciplinary research program with the goal of developing novel material solutions for use in advanced wound healing based on nanofibrillated cellulose structures.

According to Chinga-Carrasco, there are two aspects to consider in this respect.

One is the production of nanocellulose. There are many efforts in Norway and abroad focusing on production of nanocellulose with low-energy consumption.

"In our case," says Chinga-Carrasco, "with a modification of the pre-treatment we can produce a tailor-made nanocellulose with controlled surface chemistry and a specific morphology."

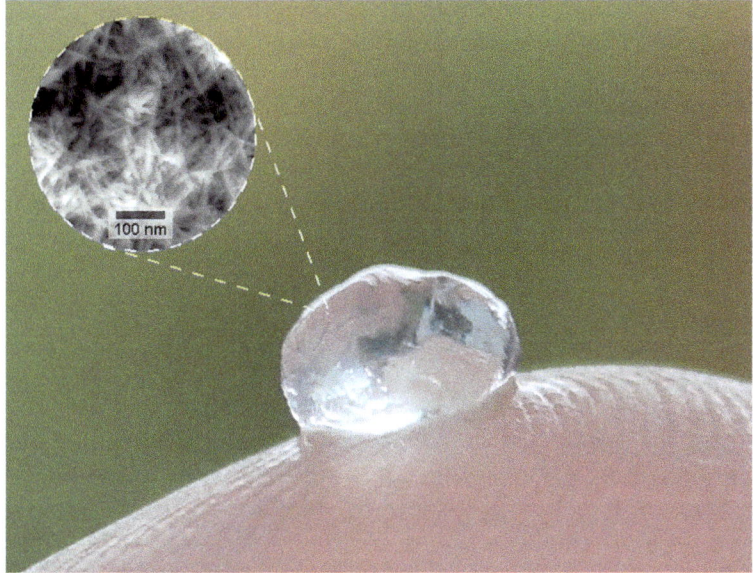

Figure 5.9   Cross-linked nanocellulose hydrogel. The hydrogel is composed of cellulose nanofibrils with widths less than 20 nm (inset). (Image: Gary Chinga-Carrasco, PFI Biocomposites.)

The other aspect is the application that Chinga-Carrasco and his team are researching extensively in the NanoHeal project: the production of nanocellulose qualities that are suitable for biomedical applications, in this specific case for wound healing.

"We have various groups working on assessing the suitability of nanocellulose as a barrier against wound bacteria and also with the assessment of the cytotoxicity and biocompatibility," he says. "However, as a first step we have intensified our work on the production of nanocellulose that we expect will be adequate for wound dressings."

A specific activity that the PFI researchers and collaborators are working on in the NanoHeal project is the production of an ultrapure nanocellulose, which is important for biomedical applications. Considering that the nanocellulose hydrogel material can be cross-linked and have a reactive surface chemistry, there are various potential applications.

A specific application is as a dressing for wound healing, another is scaffolds.

"Production of an ultrapure nanocellulose quality is an activity that we are intensifying together with our research partners at the Institute of Cancer Research and Molecular Medicine in Trondheim," notes Chinga-Carrasco. "The results look good and we expect to have a concrete protocol for production of ultrapure nanocellulose soon, for an adequate assessment of its biocompatibility."

The researchers caution that some points should be taken into consideration when considering nanocellulose materials for biomedical applications.

One is the thorough assessment of biocompatibility with the human body. Another aspect is the interaction between nanocellulose and wound bacteria, which the PFI team is researching with the Department of Tissue Engineering and Restorative Dentistry at Cardiff University.

"These are aspects not only important from a wound healing application point of view, but that also will help in the design of a biodegradable biocomposite material with antimicrobial properties, which can have a range of additional potential applications in the future," Chinga-Carrasco concludes.

Featured scientist: Dr Gary Chinga Carrasco (http://www.pfi.no/About-PFI/ PFI-Contacts/Gary-Chinga-Carrasco/)
Organization: Paper and Fibre Research Institute, Trondheim (Norway)
Relevant publication: G. Chinga-Carrasco and K. Syverud, Pretreatment-dependent surface chemistry of wood nanocellulose for pH-sensitive hydrogels, *J. Biomater. Appl.*, 2014, **29**, 3 423, DOI: 10.1177/0885328214531511

5.4.3 Curcumin Nanoparticles as Innovative Antimicrobial and Wound Healing Agents

Despite significant progress in the medical treatments of severe burn wounds, infection and subsequent sepsis persist as frequent causes of morbidity and mortality for burn victims. This is due not only to the extensive compromise of the protective barrier against microbial invasion, but also as a result of growing pathogen resistance to therapeutic options.

Innovative therapies are urgently needed that overcome mechanisms of pathogen resistance—not only for thermal injuries but in general—and are easily administered without concerning systemic side effects.

"Antimicrobial resistance continues to be a growing crisis, highlighted by the FDA's Generating Antibiotic Incentives Now (GAIN) program, through which three new antibiotics with the indication for acute bacterial skin and skin structure infections were rapidly approved in unprecedented succession. All three however are systemically administered, and we have yet to see new topical antimicrobials emerge," says Adam Friedman, Assistant Professor of Dermatology and Director of Dermatologic research at the Montefiore-Albert Einstein College of Medicine. "For me, this gap fuels innovation, serving as the inspiration for my research with broad-spectrum, multi-mechanistic antimicrobial nanomaterials."

Friedman and a team of researchers from Albert Einstein College of Medicine and Oregon State University have explored the use of curcumin nanoparticles for the treatment of infected burn wounds, an application that resulted in reduced bacterial load and enhanced wound healing.

Turmeric (*Curcuma longa* L.) is the shining star among the cornucopia of traditional medicinal plants. It has a long history of usage in traditional medicine in India and China. Ancient Indians have known the medicinal properties of turmeric—*i.e.* curcumin—for several millennia.

In the scientific literature, there is a large body of evidence showing that curcuminoids exhibit a broad spectrum of biological and pharmacological activities including anti-oxidant, anti-inflammatory, anti-bacterial, anti-fungal, anti-parasitic, anti-mutagen, anti-cancer and detox properties. Curcumin's unique ability to work through so many different pathways with its extraordinary antioxidant and anti-inflammatory attributes can have a positive influence in combating almost every known disease.

"There has been tremendous excitement regarding curcumin in multiple fields of medicine, most prominently in oncology," Friedman points out. "Here, for the first time, we demonstrated that curcumin nanoparticles were more effective at both accelerating thermal burn wound closure and clearing infection with methicillin-resistant *S. aureus* (MRSA) as compared to curcumin in its bulk size."

Friedman and his team utilized an innovative sol–gel-based polymerization technique to create silane composite nanoparticles that incorporate curcumin within a highly structured porous lattice. The versatility of the resulting nanoformulation allows for loading of different active ingredients, with therapeutic efficacy when applied topically, intradermally, and intravenously.

"While so much is known about curcumin's therapeutic potential, there have been numerous limitations with respect to clinical translation resulting from its poor solubility, instability at physiology pH and unsightly yellow-orange color," says Friedman. "Nanotechnology can and has overcome many of these impediments. At the nanoscale, the likelihood of curcumin interfacing with its intended target is much greater."

To sum it up, this work nicely demonstrates that curcumin nanoparticle technology circumvents the difficulties inherent in curcumin administration, enabling delivery of this therapeutic substance. Unlike currently used treatments, curcumin nanoparticles are less likely to select resistant bacterial strains or delay wound healing.

"We believe our technology has the potential to serve as a novel topical agent for burn wound infection and possibly other cutaneous injuries," Friedman concludes.

Featured scientist: Prof. Adam Friedman (http://www.einstein.yu.edu/departments/medicine/divisions/dermatology/faculty/profile.asp?id=12408)

Organization: Montefiore-Albert Einstein College of Medicine, New York, NY (USA)

Relevant publication: A. E. Krausz, *et al.*, Curcumin-encapsulated nanoparticles as innovative antimicrobial and wound healing agent, *Nanomedicine*, 2015, **11**(1), 195, DOI: 10.1016/j.nano.2014.09.004

5.4.4 Multifunctional RNA Nanoparticles to Combat Cancer and Viral Infections

Today's nanotechnology research often focuses on the search for new materials and techniques, and safety and biomolecular aspects are often postponed for future studies. To avoid erroneous conclusions stemming from endotoxin contamination and sterility issues from commercial starting materials or residual manufacturing components, nanomaterial fabrication for biomedical applications usually requires purification steps.

Another problem with some of these compounds is their bio incompatibility and accumulation in the human body, which may cause health complications later on.

To overcome these fabrication issues, the use of naturally occurring biological materials—such as RNA or DNA—for drug formulation may become the next big step in nanoparticulate therapy development.

"There is a significant need for new therapeutic approaches to combat diseases such as cancer and viral infections," says Bruce A. Shapiro, PhD, a senior investigator at the Center for Cancer Research, National Cancer Institute (NCI). "Using RNA as a therapeutic modality brings to bear an entirely new approach, which not only allows for the construction of uniform scaffolds for attachment of functional entities, but also permits the use of all the different types of functionalities that are inherent in natural RNAs."

Research, led by Shapiro, demonstrates that multifunctional RNA nanoparticles with a nanoring design allow the use of different types of functionalities inherent in natural RNAs. The combinatorial nature of these nanospheres promotes higher detection sensitivity of diseased cells and significant silencing efficiencies of targeted genes in cells and *in vivo*.

"Computationally designed RNA- and RNA–DNA-based nanorings presented in our work have multiple advantages in diagnostics and delivery of functional moieties to diseased cells," Shapiro points out. "These advantages include—but are not limited to—(1) tight control of structural homogeneity, processivity and targeting; (2) programmability; (3) precise control over folding and self-assembly; (4) simple conjugation with different natural nucleic-based functionalities—exemplified by different RNA aptamers, RNA interference triggers (siRNAs), and proteins, to name just a few; (5) thermal and chemical stability; (6) biocompatibility and biodegradability; (7) relatively low immune response; (8) relatively low cost of production; and (9) improved stability of complexes with cationic carriers."

He adds that nanoparticles of this type have the ability to incorporate a tremendous amount of versatility, while still maintaining control over stoichiometry, which is extremely important for any potential therapeutic agent.

Kirill Afonin, at the time of this work a research fellow at the NCI (and now an assistant professor of chemistry at UNC Charlotte) and first author of a paper on this work, recounts that the concept of nanorings was first introduced[21] through the computational design by Yaroslava Yingling and Bruce Shapiro and then was further experimentally tested in Luc Jaeger's laboratory at UCSB.

"Our current work shows the versatility of nanoscaffold applications that allows their controlled functionalization with different aptamers, fluorescent dyes, and proteins for biosensing purposes as well as simultaneous delivery of various siRNAs and RNA–DNA hybrids to diseased cells," says Afonin. "We carried out extensive *in vitro* and *in vivo* characterizations of the resulting functional RNA nanoparticles and performed detailed characterization and particularly challenging visualization of functional nanorings."

To demonstrate the combinatorial nature of the scaffolds, the scientists functionalized their nanorings with up to six RNA aptamers selected to bind malachite green dye and significantly increase its emission, which is otherwise undetectable in aqueous solutions.

The team visualized the RNA-containing nanorings using single-particle cryo-electron microscopy (cryo-EM). Dr Alexey Koyfman, a research scientist in the Department of Pharmacology and Toxicology at University of Texas Medical Branch and co-author of the paper, notes that this characterization revealed that siRNA arms do not point straight out as expected from the model. After reconstruction, it became apparent that siRNA arms in the nanoring point upward, creating a crown shape. Also, looking from the top, the siRNA arms are positioned in a pinwheel fashion around the ring.

To show the feasibility of the nanorings, the team developed a set of two nanorings constructs and tested it with HIV-1 infected cells.

"The approach that uses multiples targets within one nanoparticle is very interesting from the HIV-1 point of view," explains Angelica Martins, a researcher working in the HIV Drug Resistance Program at NCI. "Since HIV-1 has a high mutational rate, the virus evolves to escape and quickly becomes resistant to the drugs. Therefore, simultaneous targeting of multiple parts of the HIV-1 genome can increase the genetic barrier and decrease the chance of the viral escape."

She points out that using the nanorings, one would guarantee that all siRNAs will enter the same cell and silence several targets. HIV-1 was a good system to test the nanorings and the multiple targets approach.

"However," says Martins, "siRNAs have been shown to be not the best approach as a therapeutic measure for HIV-1. Nevertheless, some research groups demonstrated that the use of siRNA treatment of a virus like filovirus (*e.g.* Marburg, Ebola) with more acute infection works well as the RNAi therapy. I believe that the multiple target approach should be further tested in these viruses and the results of this combinatorial use of siRNAs should be even better as a therapy."

An important next step for the researchers is to expand the library of RNA-based nanoscaffolds and to determine potential advantages—or disadvantages—of their application to therapeutic nanotechnology including delivery against specific disease related targets (cancers, viruses, *etc.*).

As Shapiro concludes, "combining the different design strategies and utilizing the various computational approaches, for example to construct a user-friendly database of RNA nanoscaffolds, would greatly simplify the design approach and make the structures readily available for other scientists to use in their research. This ultimately can lead to a combined computational and experimental pipeline to develop such particles."

Featured scientist: Dr Bruce A. Shapiro (https://ccr.cancer.gov/Gene-Regulation-and-Chromosome-Biology-Laboratory/bruce-a-shapiro); Prof. Kirill Afonin (http://chemistry.uncc.edu/people/faculty/kirill-afonin-phd) *Organization*: Gene Regulation and Chromosome Biology Laboratory, Center for Cancer Research, National Cancer Institute, Frederick, MD (USA); Department of Chemistry, University of North Carolina at Charlotte, NC (USA)
Relevant publication: K. Afonin, M. Viard, A. Koyfman, A. Martins, W. Kasprzak and M. Panigaj, *et al.*, Multifunctional RNA Nanoparticles, *Nano Lett.*, 2014, **14**(10), 5662, DOI: 10.1021/nl502385k

5.4.5 Replacing Antibiotics with Graphene-Based Photothermal Agents

At a meeting of infectious disease experts in Copenhagen in March 2012, the Director General of the World Health Organization (WHO), Dr Margaret Chan, warned vividly that the growing threat of antibiotic-resistant bacterial strains may pose grave risks for society: "A post-antibiotic era means, in effect, an end to modern medicine as we know it. Things as common as strep throat or a child's scratched knee could once again kill."[22] Chan pointed out that there is a global crisis in antibiotics caused by rapidly evolving resistance among microbes responsible for common infections that threaten to turn them into untreatable diseases. Every antibiotic ever developed is at risk of becoming useless.

Tackling this topic, the WHO published a book, *The evolving threat of antimicrobial resistance—Options for action.*[23] It describes examples of policy activities that have addressed antimicrobial resistance in different parts of the world, with the aim of raising awareness and in particular to stimulate further coordinated efforts.

Chapter 1 of this book, "The evolving threat of antimicrobial resistance—Introduction", lays out facts and details that form the basis for Chan's warning: a bacterium that causes disease reacts to the antibiotics used as treatment by becoming resistant to them, sooner or later. This natural process of adaptation—antimicrobial resistance—means that the effective lifespan of antibiotics is limited. Unnecessary use and inappropriate use of antibiotics favors the emergence and spread of resistant bacteria. A crisis has been building up over decades, so that today many common and life-threatening infections are becoming difficult or even impossible to treat, sometimes turning a common infection into a life-threatening one. It is time to take much stronger action worldwide to avert a situation that entails an ever increasing health and economic burden.

"Our group took this issue seriously years ago when death rates associated with multiple organ failure caused by bacterial infection appeared to increase in the intensive care units of local hospitals," says Young-Chien Ling, a professor in the Department of Chemistry at National Tsing Hua University if Taiwan. "We started to explore treatment options using nanomaterials in the hope of finding antimicrobial mechanisms different to conventional antibiotics."

Ling's group has been working on antibacterial nanomaterials since the mid-2000s and first reported antibacterial paper[24] where they developed a simple, green, and cost-effective way of coating ZnO nanoparticles on a paper surface without the use of binders.

Following the same logic of capturing and killing bacteria, the researchers have replaced the ZnO nanoparticles with graphene.

This allowed the bacteria—both gram-positive *Staphylococcus aureus* and gram-negative *Escherichia coli*—to be efficiently captured by glutaraldehyde and concentrated (or immobilized) by the magnetic properties of magnetic reduced graphene oxide functionalized with glutaraldehyde (MRGOGA). The bacteria are rapidly killed by multiple means, including conventional oxidative stress as well as physical piercing and photothermal heating of graphene by near-infrared (NIR) laser irradiation (Figure 5.10).

"In this context, our proposed antimicrobial treatment system is a binary treatment system that allows the targeted bacteria being separated and concentrated from the healthy biofluids/cells/tissues/organs first and subjected to focused killing thereafter, minimizing adverse side effects," explains Ling.

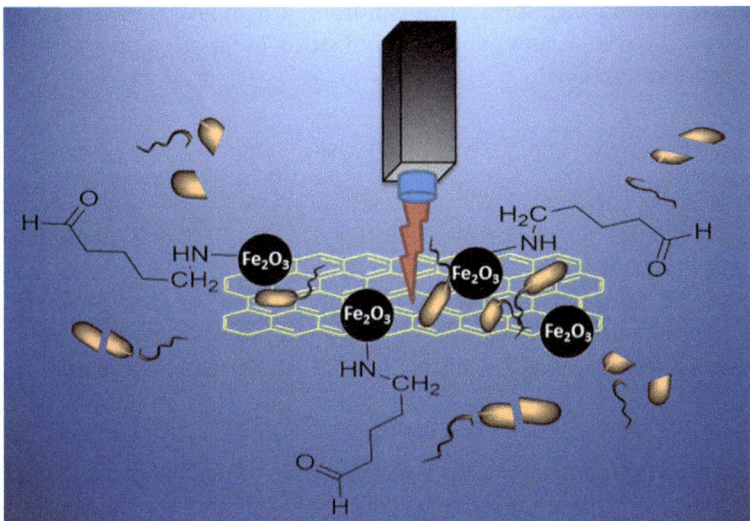

Figure 5.10 Schematic of graphene-based photothermal antibacterial therapy by the proposed binary treatment system of capturing and killing bacteria offering a promising solution against the global antibiotics crisis. (Image: Dr Ling, National Tsing Hua University.)

"The controllable and multiple killing mechanisms offer additional advantages of enhanced efficacy and reduced time."

Photothermal therapy has been widely developed by combining pulsed laser and strong light-absorbing materials such as gold nanoparticles, carbon nanotubes, and graphene. Their unique optical properties enable these nanomaterials to absorb light irradiation and release it as heat.

Graphene and graphene derivatives have been explored widely in biomedicine due to their exceptional chemical and physical properties. In previous work,[25] researchers have made the surprising finding that graphene-based nanomaterials—graphene oxide, graphene oxide and reduced graphene oxide—possess excellent antibacterial properties. They discovered that the toxicity mechanism of graphene-based nanomaterials includes oxidative stress, physical piercing through the sharp edges of graphene nanowalls, as well as photothermal heating.

In their subsequent research, the team combined the advantages of the excellent photothermal properties of reduced graphene oxide upon NIR laser irradiation and glutaraldehyde as efficient capturing agent towards both *S. aureus* and *E. coli*. The magnetic characteristic of this novel photothermal agent results in the captured bacteria being readily trapped into small volumes by an external magnetic force.

The researchers demonstrated that MRGOGA efficiently captured and effectively killed bacteria within 10 minutes upon NIR irradiation.

"Furthermore," says Ling, "we employed a microfluidic chip system to demonstrate the re-usability of MRGOGA, offering a biocompatible platform for photothermal sterilization. The magnetic and low cytotoxicity properties of MRGOGA also make it an ideal candidate for *in vivo* biomedical applications by taking advantages of easy mobilization at targeted position and minimum cellular damage."

Featured scientist: Prof. Young-Chien Ling (http://chem-en.web.nthu.edu.tw/files/13-1117-33470.php)
Organization: Department of Chemistry, National Tsing Hua, Hsinchu University (Taiwan)
Relevant publication: M. Wu, A. Deokar, J. Liao, P. Shih and Y. Ling, Graphene-Based Photothermal Agent for Rapid and Effective Killing of Bacteria, *ACS Nano*, 2013, 7(2), 1281, DOI: 10.1021/nn304782d

5.4.6   Nanotechnology Against Acne

Nanotechnology is finding applications in the treatment of infectious and inflammatory diseases, particularly skin disease. These applications work in one of two ways: utilizing nanomaterials that have inherent antimicrobial properties; or incorporating known therapeutics into nanoscale vehicles to enhance delivery and improve efficacy.

"An ideal approach to develop a topical therapy for microbial infection in skin would combine both," says Adam Friedman, Assistant Professor of Dermatology and Director of Dermatologic research at the Montefiore-Albert Einstein College of Medicine.

Researchers at the Albert Einstein College of Medicine and University of California Los Angeles School of Medicine have developed a nanoparticle platform using chitosan—a polysaccharide from the shells of crustaceans such as crabs and shrimp—for the treatment of inflammatory skin diseases such as acne vulgaris.

The team demonstrated that not only were these nanoparticles effective at killing *Propionibacterium acnes*, the gram positive bacteria associated with acne, they inhibited the damaging inflammation that results in the large, painful lesions associated with inflammatory acne.

Acne is one of the most common dermatologic diseases affecting between 40–50 million people each year. While best known as a bothersome part of puberty, affecting approximately 75% of teenagers, acne can persist or even first start during adulthood, causing emotional and physical distress as well as permanent disfigurement.

Topical therapies for acne, including benzoyl peroxide, salicylic acid, topical antibiotics such as clindamycin, and retinoids, all suffer from various related side effects including irritation, redness, peeling and scaling, bacterial resistance, and resulting discoloration from the associated irritation in patients with darker skin types. These adverse events often serve as major limiting factors influencing patient compliance and ultimately impacting efficacy.

"Nanotechnology has tremendous potential in the field of dermatology for both cosmetic and medical applications," says Friedman. "In fact, there are a number of over-the-counter products utilizing nano-vehicles to more effectively deliver active agents for skin rejuvenation and sun protection. Along these lines, we encapsulated benzoyl peroxide in our chitosan–alginate nanoparticles for the purpose of improving both its activity and importantly, tolerability, as an anti-acne agent."

He explains that, in order to synthesize nontoxic, biodegradable, and biocompatible nanoparticles (<50 nm in size) that can be used for the treatment of cutaneous infections, the team chose a derivative of the biostructural crustacean shell polymer chitin—chitosan—along with the well-known thermally stable gelling agent, alginate.

The antimicrobial activity of chitosan is well-established and it has been used as a preservative for food packaging to prevent spoilage. It has also been evaluated as a component for antimicrobial textiles used in clothing for health care workers.

"We have found that the nanoparticles themselves have antimicrobial activity not only against *P. acnes*, but also against methicillin-resistant *Staphylococcus aureus* and *Escherichia coli*, again highlighting the broad value of this platform," says Friedman.

"Our results demonstrate that chitosan–alginate nanoparticles can not only be utilized as a delivery tool but have inherent antimicrobial and anti-inflammatory properties which could be useful for other therapeutic applications," says Dr Jenny Kim, Associate Professor of Dermatology, Chief of Dermatology at the Veterans Affairs Greater Los Angeles Healthcare System. "This is an exciting finding with clinical relevance as development of novel therapeutics using chitosan–alginate nanoparticles could potentially be useful for the treatment of acne and other infectious and inflammatory skin diseases."

Featured scientists: Prof. Adam Friedman (http://www.einstein. yu.edu/departments/medicine/divisions/dermatology/faculty/profile. asp?id=12408); Prof. Jenny Kim (https://www.uclahealth.org/provider/ jenny-kim-md-phd)
Organizations: Montefiore-Albert Einstein College of Medicine, New York, NY (USA); UCLA Health, Los Angeles, CA (USA)
Relevant publication: A. Friedman, J. Phan, D. Schairer, J. Champer, M. Qin and A. Pirouz, *et al.*, Antimicrobial and Anti-Inflammatory Activity of Chitosan–Alginate Nanoparticles: A Targeted Therapy for Cutaneous Pathogens, *J. Invest. Dermatol.*, 2013, **133**(5), 1231, DOI: 10.1038/jid.2012.399

5.4.7 Biofunctionalized Silk Nanofibers Repair the Optic Nerve

A main difference between the central and peripheral nervous system is the lack of regeneration after a neurotrauma, leading to severe and irreversible handicaps. While biomaterials have been developed to aid the regeneration of peripheral nerves, the repair of central nerves, such as the optic nerve or nerve cells in the spinal cord, remains a major challenge for scientists.

The ability to regenerate central nerve cells in the body could reduce the effects of trauma and disease in a dramatic way and nanotechnologies offer promising routes for repair techniques.

"Since the first regenerative trial, achieved by J. F. Tello a hundred years ago,[26] our knowledge in the discrepancies between the two systems have dramatically increased, without, however, leading to realistic therapeutic strategies," says Thomas Claudepierre, at the time of this work a professor in the Faculty of Medicine at the University of Leipzig (he is now with the faculty of the University of Lorraine, Nancy, in France). "This is mainly due to the multifactorial cascade of events during the neurodegenerative process following a trauma in the central nervous system."

As Claudepierre explains, in the case of damage to the optic nerve, axons of neurons (namely the retinal ganglion cells or RGC) that form the optic nerve first degenerate following a nerve crush or cut. Surrounding glia cells will lose their orientation and form a glia scar at the site of the lesion, modifying profoundly the matrix deposition and the global structure orientation. Glia cells undergo gliosis, a proliferative state where they secrete

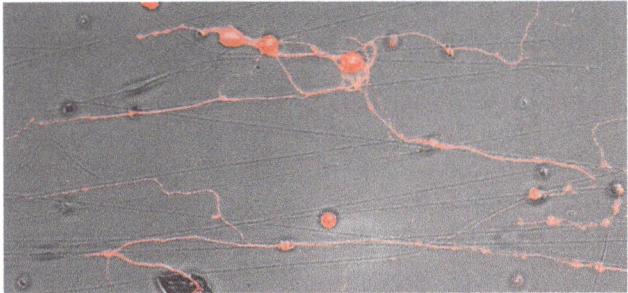

Figure 5.11 Retinal ganglion cells (labeled in red with beta III tubulin antibody) growing on electrospun silk fibers; the neurites are extending mainly in close contact with silk, which is seen in contrast phase. (Image: Dr Thomas Claudepierre, University of Leipzig.)

survival but also noxious factors that will affect the neuroretina. The RGC fail to re-grow their axons due to lack of orientation clues for the growth cone and massively die.

In their study, Claudepierre and a team from Tufts University and University of Leipzig attempt to rescue RGC death and enhance their regeneration using an electrospun material made of biofunctional nanofibers (Figure 5.11).

"The goal of our study was to test new biomaterials that could provide a permissive growth support that is orientated and can be functionalized with survival factors for RGC (growth factors, matrix molecules *etc.*), and can ultimately develop as a 3D nerve guide that can be implanted at a site of a nerve lesion to regenerate the traumatized axons," says Claudepierre.

He points out that, while progress has been made in generating artificial nerve-regrowth systems using different approaches, no nerve guide has yet emerged that can: (1) physically align the growth of regenerating nerves; (2) switch neurons to a regenerative state and promote nerve growth and axonal adhesion; (3) eliminate scar-tissue formation; (4) avoid rejection by the body; (5) not swell to impinge on the nerve; (6) degrade to form by-products that are nontoxic or non-inflammatory; and (7) be easily handled by surgeons.

The researchers selected silk, as it is a highly biocompatible material used for years in surgery. It can be prepared in purified orientated fibers using electrospinning methods. Last but not least, it can be biofunctionalized during the spinning process.

"We selected growth factors known to act on RGC survival and neurite elongation," says Claudepierre. "We used purified primary RGC isolated using the immunopanning method from newborn rodents. Our *in vitro* model focuses therefore on the first step of the trauma: can an axotomized RGC axon regenerate along a silk guide containing survival molecules?"

In their work, the research team demonstrated that electrospun silk fiber is fully biocompatible and allows RGC survival at their contact; that RGC growth cones preferentially follow the silk fibers and that RGCs extend robust neurites expressing the axonal marker GAP43.

"Moreover," as Claudepierre notes, "we found the same rate of survival, regeneration and axon differentiation when growth factors were embedded in the silk fibers and not added to the medium. The embedded process also conferred stability to growth factors that are otherwise easily degraded in the medium."

"In our study, we provide evidence that silk is a highly biocompatible material in regenerative neuroscience, it guides the growth cone and can topically provide surviving molecules that are protected from degradation and can act therefore over a long period of time, compatible with the duration of a regenerative process in the central nervous system," he sums up the results of the team's work.

The researchers' next goal is to demonstrate that a silk guide can also provide orientation clues for glia cells and can therefore reduce the amount of glia scarring and keep gliosis under control.

"We are actually testing the behavior of primary glia cells at silk contact and perform co-culture of glia and RGC on this biomaterial," Claudepierre describes the thrust of their follow-up work. "We will test the effect of various matrix molecules on the guidance of the growth cone and orientation of glia cells along the silk guide. Once we obtain the best combination of surviving factors and guidance signals, we will develop a 3D guide that we will implant in rodents at the site of an optic nerve lesion to promote a massive axon grow that will be able to cross the lesion site. The final challenge will be to reconnect the axons to their physiological target in the brain, restoring the normal cartography of retinal projections."

Such strategies, if successful, could lead to medical applications in the treatment of traumatic optic neuropathies and would also be useful to handle other central nervous system lesions.

Featured scientist: Thomas Claudepierre
Organization: Universitätsklinikum Leipzig (Germany)
Relevant publication: C. R. Wittmer, T. Claudepierre, M. Reber, P. Wiedemann, J. A. Garlick, D. Kaplan and C. Egles, Multifunctionalized Electrospun Silk Fibers Promote Axon Regeneration in the Central Nervous System, *Adv. Funct. Mater.*, 2011, **21**, 4232, DOI: 10.1002/adfm.201100755

5.4.8 Move Over Chips—Here Come Multifunctional Labs on a Single Fiber

Optical fibers have revolutionized telecommunications by providing higher performance, and more reliable telecommunication links with an ever decreasing bandwidth cost. Optoelectronic and fiber optic components are now manufactured on a large scale and the world is interconnected by an information superhighway built of glass.

In parallel with these developments, fiber-optic sensor technology has been a major user of technologies associated with the optoelectronic and fiber optic communications industry. Today, with the rapid advance of communications and especially sensing applications, there is an ever increasing need for advanced performance and additional functionalities. This, however, is difficult to achieve without addressing fundamental fabrication issues related to the integration onto optical fibers of advanced functional materials at the micro- and nanoscale.

The key aspect of suitable fabrication methodologies for lab-on-fiber devices is to adapt all the standard fabrication processes and tools in terms of material deposition (spin coating, dip coating, sputtering, evaporation, *etc.*) and sub-wavelength patterning and post-processing (FIB, EBL, RIE, *etc.*) to operate on the tip of an optical fiber. However, this approach is not straightforward since spin coating and etching procedures are very challenging to handle when operating on such a substrate.

Solving these technical problems, however, will open up the possibility of developing multifunctional labs integrated onto a single optical fiber, exchanging information and combining sensorial data.

"The realization of highly integrated optical fiber devices requires that several micro-and nanostructures be fabricated, embedded, and connected in order to achieve the necessary light–matter interaction and physical connection," explains Andrea Cusano, a researcher in the Optoelectronic Division of the Department of Engineering at the University of Sannio in Italy. "As a consequence, a critical issue that needs to be addressed consists of the definition of a reliable fabrication procedure able to integrate and process—at micro- and nanoscales—several materials with the desired physical, mechanical, magnetic, chemical, and biological properties onto unconventional substrates such as the optical fiber tip."

Cusano, together with Emanuela Esposito from the Istituto di Cibernetica "E. Caianiello", leads a team that is working on the development of "lab-on-fiber" technology, *i.e.* multifunctional optical nanoprobes within a single optical fiber.

Lab-on-fiber technology will allow the implementation of sophisticated, autonomous multifunction sensing and actuating systems—all integrated in individual optical fibers. Such a technology would ensure unique advantages in terms of miniaturization, light weight, cost-effectiveness, robustness and power consumption.

"Multifunctional labs integrated into a single optical fiber, exchanging information and combining sensorial data, could provide effective auto-diagnostic features as well as new photonic and electro-optic functionalities useful in many strategic sectors such as optical processing, environment, life science, safety and security," says Esposito. "Lab-on-fiber technology would also provide the basis for the exploitation of many novel and intriguing phenomena that are at the forefront of scientific optical research, involving light manipulation phenomena and excitations of guided resonances in photonic crystals and quasi-crystals, as well as combined plasmonic and photonic effects in hybrid metallo-dielectric structures."

The Italian research team proposes a reliable fabrication process that enables the integration of dielectric and metallic nanostructures on the tip of optical fibers, thus representing a further milestone in the lab-on-fiber technology roadmap.

"In our research, we propose and validate a novel fabrication process based on a 'direct writing' approach taking advantage of conventional deposition and nanopatterning techniques, typically used for planar substrates and suitably adapted to directly operate on an optical fiber tip," says Cusano.

In essence, the team's fabrication process consists of three main technological steps: (1) dielectric overlay deposition, with a flat surface and controlled thickness over the fiber core region, by means of a properly customized spin coating process; (2) nanoscale patterning of the deposited overlay by an electron-beam lithography (EBL) tool; and (3) superstrate overlay deposition, where different functional materials—metallic or non-metallic—can be deposited by various techniques (*e.g.* sputtering, thermal evaporation), properly customized to operate with optical fibers.

As the scientists point out, a distinguished feature of this process is the use of a customized spin-coater chuck, which allows a flat and reproducible resist layer deposited onto optical fiber tips, onto which nanostructured arrays are directly written using an EBL system. Importantly, this fabrication process follows almost ordinary lithographic techniques—here adapted to operate on fiber facets—allowing rapid prototyping with a 90% yield and the ability to produce robust and reusable devices.

To demonstrate the effectiveness of their proposed methodology, the team fabricated a miniaturized fiber tip device based on a 2D hybrid metallo-dielectric nanostructure supporting localized surface plasmon resonances (LSPR).

"We carried out both experimental and full-wave numerical analyses to characterize the resonant phenomena," says Cusano. "The measured Q-factors were higher than those observed in typical plasmonic crystal configurations. We have also shown that the LSPR can be easily tuned by adjusting the physical and geometrical parameters of the crystal nanostructure and can be designed to be very sensitive to modifications of the surrounding medium."

With a view towards possible applications, the researchers have also reported some preliminary results on the capability of their platform to be used for label-free chemical and biological sensing.

"Moreover, we have also demonstrated the surprising capability of our device to detect acoustic waves, taking advantage of the low elastic modulus of the patterned polymer," notes Cusano.

The research team, driven by the enormous potential of lab-on-fiber technology, is devoting their efforts to optimizing the fabrication process in terms of reliability and throughput as well as enlarging the set of functional materials to be integrated.

"Our dream is to get this technology to a point where it can compete with the already better established lab-on-chip technology—especially taking advantage of the higher versatility in integrating complex lighting systems as well as sophisticated optoelectronics components, devices and systems," says Cusano.

Featured scientist: Andrea Cusano
Organization: Optoelectronic Division, Department of Engineering, University of Sannio (Italy) (http://www.ing.unisannio.it/elettronica/indice.htm)
Relevant publication: M. Consales, A. Ricciardi, A. Crescitelli, E. Esposito, A. Cutolo and A. Cusano, Lab-on-Fiber Technology: Toward Multifunctional Optical Nanoprobes, *ACS Nano*, 2012, **6**(4), 3163, DOI: 10.1021/nn204953e

5.4.9 Nanoparticles Accelerate and Improve Healing of Burn Wounds

Skin thermal burns are a complex and major source of morbidity, mortality and healthcare expenditure in the United States, with 486 000 patients treated for burn injuries each year.[27]

Given the range of causes, from fire-associated injury to water scalding, patients commonly present with multiple and complex burns—wounds that often worsen and expand over the first few days due to the associated underlying inflammation and injury.

"Though the impact of these injuries is large, our current treatment armament falls short, with gold standard treatments lacking evidence to support their use, or even worse, delaying wound healing as was recently elucidated[28] with silver sulfadiazine," says Adam Friedman, Associate Professor of Dermatology and Director of Translational Research at the George Washington School of Medicine and Health Sciences.

To address this gap, Friedman and colleagues at the Albert Einstein College of Medicine utilized a unique nanotechnology that can both release the potent biomolecule nitric oxide (NO) over time, as well as facilitate nitrosation, the addition of an NO group to a biological molecule, which is central to many of NO's activities.

"The role of nitric oxide in wound healing is well established through all three phases," says Friedman. "Both NO itself and the act of nitrosation are exceedingly important in the transition from the inflammatory phase to the proliferative phase of wound healing. Therefore, we created a system that can do both."

The scientists evaluated a previously described *S*-nitroso-*N*-acetyl cysteine containing hydrogel-based nanoparticle platform (NAC-SNO-np) on *in vivo* burn wound closure, expansion and inflammation in a multi-burn model.

Burn wound expansion in a group treated with NAC-SNO-np in coconut oil was significantly attenuated compared to an untreated control, coconut oil, and control-np groups.

The clinically observed acceleration of wound healing was matched with the histologic evaluation of the burn wounds, with the NAC-SNO-np treated groups showing less persistent inflammation, more maturation, and healthier new tissue as compared to the other groups over time.

While still in its infancy, Friedman believes this technology will be brought from bench to bedside as it has been licensed by the company Nano Biomed Inc.

"The reality is, many impressive technologies fail to launch because the business and financial element aren't nurtured," says Friedman. "Nanotechnology in general has extraordinary potential to change the face of medicine, but it is unfortunately not enough to just have a good scientific basis or data."

"Along this theme, it is in fact only through the unique structure of the nanomaterial herein investigated that allows for the generation of nitric oxide and confers the nitrosation capacity offered by the NAC-SNO-np," he concludes.

Featured scientist: Prof. Adam Friedman (http://www.einstein.yu.edu/departments/medicine/divisions/dermatology/faculty/profile.asp?id=12408)
Organization: Montefiore-Albert Einstein College of Medicine, New York, NY (USA)
Relevant publication: A. Landriscina, T. Musaev, J. Rosen, A. Ray, P. Nacharaju, J. D. Nosanchuk and A. J. Friedman, N-acetylcysteine S-nitrosothiol nanoparticles prevent wound expansion and accelerate wound closure in a murine burn model, *J. Drugs Dermatol.*, 2015, **14**(7), 726.

5.4.10 A Nanoparticle-Based Alternative to Viagra

Here is another example of how nitric oxide nanoparticles could find applications in nanomedicine.

The majority of men who undergo radical prostatectomy for the treatment of prostate cancer will suffer from erectile dysfunction due to disruption of the cavernous nerve. This nerve has been identified as responsible for penile erection. The oral erectogenic PDE5 (phosphodiesterase-5) inhibitors such as Viagra™ rely on the functioning of this nerve to provide the initial burst of nitric oxide (NO) necessary to initiate an erection. Several other patient groups, such as diabetics, where endothelial nitric oxide production is impaired, are also refractory to orally administered PDE5 inhibitors.

The precise cause of this neuropraxia—the temporary loss of motor and sensory function due to blockage of nerve conduction—is unclear but it has been hypothesized to include direct trauma during surgery, damage from tissue electrocautery, disruption of the neural vasculature, and generalized local inflammation associated with the procedure.

In this condition, a nanoparticle delivery system may come to the rescue by targeting useful therapeutics for penile rehabilitation following radical prostatectomy.

Research by an interdisciplinary team of researchers at Albert Einstein College of Medicine and University of San Diego has demonstrated that nitric oxide nanoparticles (NO-NPs) could make them potentially useful agents in penile rehabilitation.

"Recent research has shown that NO-NPs may have several characteristics desirable for improvement of erectile function following radical prostatectomy," says Dr Adam Friedman. "We show in real time that NO-NP application in two different vehicles (lipophilic and hydrophilic) increase cutaneous blood flow for 90 minutes as compared to internalized controls."

Previously, the same group of researchers published a proof-of-principle article[29] demonstrating that topical application of NO-releasing nanoparticles can elicit an erectile response in an aging rat model of erectile dysfunction without stimulation of the cavernous nerve.

The hypothesis of this earlier investigation was that the NO-NPs could topically deliver sufficient nitric oxide to relax corporal smooth muscle tissue and elicit an erection.

"In the present study, we further tested this hypothesis to determine if topically applied NO-NPs would elicit an erectile response in an animal model of erectile dysfunction resulting from cavernous nerve transection," Friedman elaborates. "We hypothesized that the prolonged release of nitric oxide from the NO-NPs would lead to an overall increase in intracorporal blood pressure because of increased blood flow into the penis."

In a rodent model of radical prostatectomy, the researchers demonstrated the increase of microvascular blood flow by topical application of NO-NPs in combination with coconut oil or hyaluronic acid. This hydrogel was able to significantly increase blood flow through blood vessels 200 μm away from the site of NO-NP application to intact skin. The release of nitric oxide from the nanoparticles occurred over a period of seven hours, suggesting that their physiologic effects could last for at least this duration.

"Several technologies have been proposed for cutaneous delivery of nitric oxide, including but not limited to compressed gas cylinders that deliver it directly to wounds, acidified nitrite cream that releases nitric oxide, and diazeniumdiolates that give off nitric oxide spontaneously from a donor compound," says Friedman. "Unlike these and other nitric oxide delivery platforms, our NO-NPs do not rely on enzymatic conversion or donor compounds for nitric oxide delivery."

In the team's nanoparticle delivery platform, nitric oxide is spontaneously generated through the reduction of nitrite to nitric oxide, facilitated by the rich hydrogen bonding network provided by this unique nanoparticle technology.

Friedman points out that, as a result, neither depletion of host thiols nor toxicity as a result of residual parent compound, both of which have been previously reported, are of concern here.

Besides showing the *in vivo* mechanism of their nanoparticle-based system, the team's results also suggest efficacy of this delivery method for a broad range of diseases resulting from vascular dysfunction (*e.g.* Raynaud's, distal ulceration in scleroderma).

Featured scientist: Prof. Adam Friedman (http://www.einstein.yu.edu/departments/medicine/divisions/dermatology/faculty/profile.asp?id=12408)
Organization: Montefiore-Albert Einstein College of Medicine, New York, NY (USA)
Relevant publication: M. Tar, P. Cabrales, M. Navati, B. Adler, P. Nacharaju and A. Friedman, *et al.*, Topically Applied NO-Releasing Nanoparticles Can Increase Intracorporal Pressure and Elicit Spontaneous Erections in a Rat Model of Radical Prostatectomy, *J. Sex. Med.*, 2014, **11**(12), 2903, DOI: 10.1111/jsm.12705

5.4.11   Light-Triggered Local Anesthesia

Systemic pain medicine—which acts on the whole nervous system, rather than a specific area—can have side effects, such as nausea, feeling drowsy, or having trouble concentrating.

Besides pain relief, numerous serious medical conditions require medications that cannot be taken orally, but must be dosed intermittently, on an as-needed basis, and over a long period of time. Controlled and long-term drug release has been recognized as one of the most promising biomedical technologies for certain types of chronic diseases.

In order to avoid the inconvenience of repeated injections, remotely triggerable drug delivery systems have seen a lot of research interest. In these applications, a depot of drug is administered once, then repeatedly actuated *via* a safe external trigger such as an electrical impulse, a magnetic field, or near-infrared (NIR) light.

Researchers have also demonstrated a system that provides photo-triggered release of local anesthetics in a manner that could be adjusted by varying the irradiance and the duration of irradiation.

"We showed that gold nanoparticle-modified liposomes containing the local anesthetic Tetrodotoxin and the drug Dexmedetomidine could be used to provide adjustable, on-demand infiltration anesthesia," says Changyou Zhan, a researcher in the laboratory of Daniel Kohane, a professor of Anaesthesia at Harvard Medical School and director of the Laboratory for Biomaterials and Drug Delivery at Boston Children's Hospital. "Following the initial numbness after the injection wore off, irradiation of the injection site with near-infrared light led to the return of local anesthesia, once daily over five days."

"From a clinical point of view, this is important in that it demonstrates a method by which patients would be able to take control of relatively local pain, being able to deliver local analgesia on demand, for the duration and with the intensity desired," Kohane elaborates. "In the post-operative setting, technologies like these could make pain management easier for patients, and minimize the extent to which opioids and other systemic pain killers would have to be taken—and all their side effects."

In order to achieve repeated on-demand local anesthesia, the researchers chemically tethered gold nanorods—that are able to convert NIR light into heat—to liposomes containing Tetrodotoxin and Dexmedetomidine.

The gold nanorods would raise the temperature of the adjacent liposomal lipid bilayer above its transition temperature so that it changes from an ordered gel phase to a disordered liquid crystalline phase, and releases analgesic compounds.

The team tested the ability to provide a repeated sensory blockade triggered by remote NIR irradiation *in vivo* in rats.

"In theory, this sort of technology could be applicable to a range of excitable tissues—nerve, muscle, brain, heart, spinal cord," Kohane points out. "From the point of view of nanotechnology, it is an unusually clear demonstration of the ability of nanoscientifically-based triggering methods to induce local effects."

The challenge for the team is to make the formulations last longer and be triggerable at lower energies. The latter is important for increasing the depth in the body at which the particles can be triggered, and reducing the probability of tissue injury (burns), which could be caused by high irradiances and/ or prolonged irradiation times.

The scientists note that such devices could also be adapted to use in other excitable tissues, *e.g.* in the brain to prevent or treat seizures.

Featured scientist: Prof. Daniel Kohane's Laboratory for Biomaterials and Drug Delivery (http://kohane.tch.harvard.edu/our-team/prof-daniel-s-kohane/)
Organization: Harvard Medical School, Cambridge, MA (USA)
Relevant publication: C. Zhan, W. Wang, J. McAlvin, S. Guo, B. Timko and C. Santamaria, *et al.*, Phototriggered Local Anesthesia, *Nano Lett.*, 2016, **16**(1), 177, DOI: 10.1021/acs.nanolett.5b03440

5.4.12 Toward Next-Generation Nanomedicines for Cancer Therapy

The ultimate goal of drug delivery, especially with regard to cancer therapy, is to ferry most of the administered drug to the target, while eliminating or minimizing the accumulation of the drug at any non-target tissues. Nanomedicine applications with targeted nanoparticles are expected to revolutionize cancer therapy.

Already, more than 250 nanomedicine products are approved or in clinical study and the majority of current commercial applications of nanotechnology to medicine is geared towards drug delivery to enable new modes of action, as well as better targeting and bioavailability of existing medicinal substances.[30]

"However, the currently developed nanotechnology-based drug delivery systems, either 'passively targeted' or 'actively targeted', do not really improve the delivery of drugs to target tumors," says Hong Tan, a professor at the College of Polymer Science and Engineering at Sichuan University. "Therefore, we propose a next-generation nanocarrier that integrates various desired functions into a single nanosystem, which can harmonize with a complex physiological environment and display different properties sequentially, thus resulting in an excellent targeting effect and satisfactory bio-distribution of drugs."

Tan and his team, together with collaborators from the university's Laboratory Animal Center and the Ningbo Institute of Material Technology and Engineering, have designed and prepared an intelligent nanoscale delivery system based on multifunctional multiblock polyurethanes for targeted intracellular delivery of the anticancer drug Paclitaxel (PTX) into tumors.

Tan explains that these nanocarriers demonstrate a variety of attractive properties in a smart fashion, such as stealth character and long circulation, active targeting, pH-dependent shell detachment on arriving at the tumor site, improved cellular internalization, and triggered intracellular drug release in response to acidity within tumor cells.

According to the team, recent progress in nanoparticle engineering has improved drug targeting to some extent, but their use has not always translated into improved clinical outcomes. This is largely due to the fact that current target/multifunctional nanoparticles, regardless of how advanced they are, reach the target as a result of blood circulation, just like conventional drug delivery systems do.

"The traditional 'active targeting' approaches, using antibodies or other active ligands on the particle surface, do not really improve the accumulation of drugs into the target, because a large portion of the nanoparticles are taken up by the reticuloendothelial system (RES)," explains Tan. "Moreover, even if nanoparticles with prolonged circulation time can reach the tumor site by the enhanced permeation and retention (EPR) effect, the dense extracellular matrix and elevated interstitial pressure of tumors make drug penetration more difficult than in normal tissues, further leading to the clearance of nanoparticles from tumor sites."

Motivated by this rationale, Tan and his collaborators incorporated the targeting antibody molecules and tumor penetrating quaternary ammonium (GQA) cationic groups into the inner shell of multilayered polyurethane nanomicelles.

These active ligands can be screened and protected by a PEG outer corona during their circulation in blood in order to prevent RES uptake. When exposed to the acidic tumor environment, they detach to maximize the targeting effect and intracellular delivery of antitumor drugs.

"Such an intelligent nanovehicle can significantly improve the bio-distribution of drugs through which the drug concentration in tumor tissue can be greatly increased while those in other tissues are significantly decreased," notes Tan.

The team's work changes the traditional understanding of targeted drug delivery and provides a novel concept that nanoscale drug delivery systems need to harmonize with and respond to the highly complex biological environment inside the body and display specific required properties when necessary in order to achieve improved tumor accumulation and bio-distribution of therapeutics.

In addition, this work by the Chinese team has also broadened the area of polyurethane applications and opens a new chapter in the development of biodegradable polyurethanes for next-generation nanodelivery systems.

Going forward, the researchers plan to investigate the detailed mechanisms of cellular targeting and uptake of nanocarriers to understand whether each of the multiple functions will also work as well as expected *in vivo*. A next step then would be to introduce such multifunctional nanomedicines into the clinical domain. However, the additional complexity and cost, as well as regulatory hurdles, may make multifunctional nanomedicines difficult to be approved and commercially manufactured.

Featured scientist: Prof. Hong Tan
Organization: College of Polymer Science and Engineering, Sichuan University, Chengdu (PR China) (http://www.scu.edu.cn/cpse/index.htm)
Relevant publication: M. Ding, N. Song, X. He, J. Li, L. Zhou and H. Tan, *et al.*, Toward the Next-Generation Nanomedicines: Design of Multifunctional Multiblock Polyurethanes for Effective Cancer Treatment, *ACS Nano*, 2013, **7**(3), 1918, DOI: 10.1021/nn4002769

References

1. R. Gómez-Martínez, P. Vázquez, M. Duch, A. Muriano, D. Pinacho and N. Sanvicens, *et al.*, Intracellular Silicon Chips in Living Cells, *Small*, 2010, **6**(4), 499, DOI: 10.1002/smll.200901041.
2. R. Gómez-Martínez, A. Hernández-Pinto, M. Duch, P. Vázquez, K. Zinoviev and E. de la Rosa, *et al.*, Silicon chips detect intracellular pressure changes in living cells, *Nat. Nanotechnol.*, 2013, **8**(7), 517, DOI: 10.1038/nnano.2013.118.
3. B. Allen, P. Kichambare and A. Star, Synthesis, Characterization, and Manipulation of Nitrogen-Doped Carbon Nanotube Cups, *ACS Nano*, 2008, **2**(9), 1914, DOI: 10.1021/nn800355v.
4. M. B. Kadiiska, P. M. Hanna, L. Hernandez and R. P. Mason, In vivo evidence of hydroxyl radical formation after acute copper and ascorbic acid intake: electron spin resonance spin-trapping investigation, *Mol. Pharmacol.*, 1992, **42**(4), 723.
5. J. Shin, S. Choi, D. Youn and I. Kim, Exhaled VOCs sensing properties of WO_3 nanofibers functionalized by Pt and IrO_2 nanoparticles for diagnosis of diabetes and halitosis, *J. Electroceram.*, 2012, **29**(2), 106, DOI: 10.1007/s10832-012-9755-y.

6. F. Shen, M. Tan, Z. Wang, M. Yao, Z. Xu and Y. Wu, *et al.*, Integrating Silicon Nanowire Field Effect Transistor, Microfluidics and Air Sampling Techniques For Real-Time Monitoring Biological Aerosols, *Environ. Sci. Technol.*, 2011, **45**(17), 7473, DOI: 10.1021/es1043547.

7. E. Wallace, D. Stoddart, A. Heron, E. Mikhailova, G. Maglia and T. Donohoe, *et al.*, Identification of epigenetic DNA modifications with a protein nanopore, *Chem. Commun.*, 2010, **46**(43), 8195, DOI: 10.1039/C0CC02864A.

8. S. Kim, A. Gottfried, R. Lin, T. Dertinger, A. Kim and S. Chung, *et al.*, Enzymatically Incorporated Genomic Tags for Optical Mapping of DNA-Binding Proteins, *Angew. Chem., Int. Ed.*, 2012, **51**(15), 3578, DOI: 10.1002/anie.201107714.

9. Y. Ebenstein, N. Gassman, S. Kim, J. Antelman, Y. Kim and S. Ho, *et al.*, Lighting Up Individual DNA Binding Proteins with Quantum Dots, *Nano Lett.*, 2009, **9**(4), 1598, DOI: 10.1021/nl803820b.

10. H. Tao, J. Amsden, A. Strikwerda, K. Fan, D. Kaplan and X. Zhang, *et al.*, Metamaterial Silk Composites at Terahertz Frequencies, *Adv. Mater.*, 2010, **22**(32), 3527, DOI: 10.1002/adma.201000412.

11. E. Gultepe, J. Randhawa, S. Kadam, S. Yamanaka, F. Selaru and E. Shin, *et al.*, Biopsy with Thermally-Responsive Untethered Microtools, *Adv. Mater.*, 2013, **25**(4), 514, DOI: 10.1002/adma.201203348.

12. P. Ang, A. Li, M. Jaiswal, Y. Wang, H. Hou and J. Thong, *et al.*, Flow Sensing of Single Cell by Graphene Transistor in a Microfluidic Channel, *Nano Lett.*, 2011, **11**(12), 5240, DOI: 10.1021/nl202579k.

13. R. Gómez-Martínez, P. Vázquez, M. Duch, A. Muriano, D. Pinacho and N. Sanvicens, *et al.*, Intracellular Silicon Chips in Living Cells, *Small*, 2010, **6**(4), 499, DOI: 10.1002/smll.200901041.

14. This video shows a cell division of a HeLa cell enclosing a chip device: https://youtu.be/pOo26pc9vqw.

15. http://www.nobelprize.org/nobel_prizes/chemistry/laureates/2008/.

16. Z. Liu, X. Sun, N. Nakayama-Ratchford and H. Dai, Supramolecular Chemistry on Water-Soluble Carbon Nanotubes for Drug Loading and Delivery, *ACS Nano*, 2007, **1**(1), 50, DOI: 10.1021/nn700040t.

17. Z. Liu, A. Fan, K. Rakhra, S. Sherlock, A. Goodwin and X. Chen, *et al.*, Supramolecular Stacking of Doxorubicin on Carbon Nanotubes for In Vivo Cancer Therapy, *Angew. Chem., Int. Ed.*, 2009, **48**(41), 7668, DOI: 10.1002/anie.200902612.

18. M. Dzamukova, A. Zamaleeva, D. Ishmuchametova, Y. Osin, A. Kiyasov and D. Nurgaliev, *et al.*, A Direct Technique for Magnetic Functionalization of Living Human Cells, *Langmuir*, 2011, **27**(23), 14386, DOI: 10.1021/la203839v.

19. L. Shen, J. Fu, K. Fu, C. Picart and J. Ji, Humidity Responsive Asymmetric Free-Standing Multilayered Film, *Langmuir*, 2010, **26**(22), 16634, DOI: 10.1021/la102928g.

20. http://www.pfi.no/New-Biomaterials/Projects/NanoHeal/.

21. Y. Yingling and B. Shapiro, Computational Design of an RNA Hexagonal Nanoring and an RNA Nanotube, *Nano Lett.*, 2007, 7(8), 2328, DOI: 10.1021/nl070984r.

22. "Antimicrobial resistance in the European Union and the world (http://www.who.int/dg/speeches/2012/amr_20120314/en/).

23. Free download: http://apps.who.int/iris/bitstream/10665/44812/1/97892 41503181_eng.pdf.

24. K. Ghule, A. Ghule, B. Chen and Y. Ling, Preparation and characterization of ZnO nanoparticles coated paper and its antibacterial activity study, *Green Chem.*, 2006, 8(12), 1034, DOI: 10.1039/B605623G.

25. see for instance: W. Hu, C. Peng, W. Luo, M. Lv, X. Li and D. Li, *et al.*, Graphene-Based Antibacterial Paper, *ACS Nano*, 2010, 4(7), 4317, DOI: 10.1021/nn101097v.

26. F. Tello, La influencia del neurotropismo en la regeneracion de los centros nerviosos, *Trab. Lab. Invest. Biol.*, 1911, 9, 123.

27. Burn Incidence and Treatment in the United States: 2015 (http://www.ameriburn.org/resources_factsheet.php).

28. J. Rosen, A. Landriscina, A. Kutner, B. Adler, A. Krausz and J. Nosanchuk, *et al.*, Silver Sulfadiazine Retards Wound Healing in Mice via Alterations in Cytokine Expression, *J. Invest. Dermatol.*, 2015, 135(5), 1459, DOI: 10.1038/jid.2015.21.

29. G. Han, M. Tar, D. S. R. Kuppam, A. Friedman, A. Melman, J. Friedman and K. P. Davies, Nanoparticles as a Novel Delivery Vehicle for Therapeutics Targeting Erectile Dysfunction, *J. Sex. Med.*, 2010, 7, 224, DOI: 10.1111/j.1743-6109.2009.01507.x.

30. M. Etheridge, S. Campbell, A. Erdman, C. Haynes, S. Wolf and J. McCullough, The big picture on nanomedicine: the state of investigational and approved nanomedicine products, *Nanomedicine*, 2013, 9(1), 1, DOI: 10.1016/j.nano.2012.05.013.

CHAPTER 6

A Foray into the Multifaceted World of Nanotechnologies

Nanotechnology has many different facets. With the earlier examples, we have barely scratched the surface. There are many more areas where nanotechnology materials and applications will change the way products are manufactured or allow all kinds of multifunctional features to be combined in a product.

Metamaterials are an excellent example. They are precisely engineered composite materials that gain their properties from their structure rather than directly from their composition. Metamaterials' properties are not found in naturally occurring materials and could be used to make light-bending invisibility cloaks, flat lenses and other otherwise impossible devices.

With our technical capabilities today, the most advanced bottom-up nanotechnologies are a combination of chemical synthesis and self-assembly. But they already allow us to perform atomically precise manufacturing on a modest scale and this will lead to vastly improved materials, much more efficient manufacturing processes, and entirely new medical procedures.

And we haven't yet seen the end of where nanotechnology "doing it Nature's way" will lead. Recent developments in DNA-based nanotechnology have shown the suitability of this novel assembly method for constructing useful nanostructures. DNA molecules can serve as precisely controllable and programmable scaffolds for organizing functional nanomaterials in the design, fabrication, and characterization of nanometer scale electronic devices and sensors.

Another area that will be impacted in a huge way by nanoscale devices and applications is sensing, the buzzword here being *smart dust*. Smart dust refers to miniscule, even nanoscale, autonomous sensors, analytical systems

Nanotechnology: The Future is Tiny
By Michael Berger
© Michael Berger 2016
Published by the Royal Society of Chemistry, www.rsc.org

(lab-on-chip, lab-on-fiber), or even robots, that could be linked into entire sensor networks. Smart dust detects data about light, temperature, chemicals, pressure, vibrations or any environmental parameters you could think of, and is capable of transmitting this data remotely.

6.1 Nanorobotics—Motors and Machines at the Nanoscale

Science fiction style robots like Star Wars' R2D2 or the NS-5 model in *I, Robot* firmly belong in the realm of Hollywood—and so do "nanobots" à la Michael Crichton's *Prey*. Staying with both feet firmly on scientific ground, robotics can be defined as the theory and application of robots, completely self-contained electronic, electric, or mechanical devices, to activities such as manufacturing. Scale a robot down to a few billionth of a meter and you are talking nanotechnology robotics; nanorobotics in short.

The field of nanorobotics brings together several disciplines, including nanofabrication processes used for producing nanoscale devices, nanoactuators, nanosensors, and physical modeling at nanoscales. Robotic manipulation technologies, including the assembly of nanometer-sized parts, the manipulation of biological cells or molecules, and the types of robots used to perform these tasks also form a component of nanorobotics.

6.1.1 A Nanorobotics Platform for Nanomanufacturing

The rapid miniaturization of devices and machines has fueled the evolution of advanced fabrication techniques. However, the complexity and high cost of state-of-the-art high-resolution lithographic systems are prompting unconventional routes for nanoscale manufacturing.

Inspired by natural nanomachines, synthetic nanorobots have demonstrated remarkable performance and functionality. Nanoengineers at the University of California San Diego, have invented a new nano-patterning approach, named *Nanomotor Lithography*, which translates the autonomous movement trajectories of nanomotors into controlled surface features.

The research team, led by Professor Joe Wang, Chair of Nanoengineering, introduced a simple and efficient nanomotor-based nano-patterning technique based on self-propelled nanomasks and nanolenses that bring a twist to conventional static optical fabrication systems.

"Our nanomotor-fabrication strategy combines controlled movement of nanorobots with unique light focusing or blocking abilities for direct surface writing and provides researchers with considerable freedom for creating diverse features with different shapes and sizes," says Wang. "It allows us to generate spatially defined surface patterns, corresponding to the predetermined path of the nanorobots. We expect that directional propulsion and variation of nanorobot design can be used to generate more elaborate functional features."

For example, the shape and number of the assembled patches of the nano-motors can be customized to further increase the pattern complexity. Higher order nanorobot organization and modular motor design should lead to further improvements in pattern intricacy.

Diverse pattern morphologies can thus be achieved by rationally designing each component in the nanorobot assembly (Figure 6.1).

In this technique, the two types of nanomotors—self-propelled platinum nanowires as nanomasks and Janus spheres as near-field nanolenses—"swim" in water over photoresist surfaces. The nanowires replace the tradi-tional photomask while the Janus spherical motors concentrate the process-ing light and harness near-field optical effects for direct writing.

"Once light reaches the moving nanomotors, they can effectively perform near-field lithography: the opaque metallic nanowire motors allow for the nanoscale blocking of light while the transparent Janus sphere motors allow for efficient near-field concentration of light," explains Wang. "Due to the small dimension of the nanowire motors and the near-field focusing effect of the Janus sphere motors, sub-wavelength resolution can be achieved."

The team points out that the self-propelled motion, along with magnet-ically guided control, makes the new nanomotor approach simple, cost-effective and obviates the requirements for elaborate control systems used in common surface patterning techniques.

As proof-of-concept, the researchers demonstrated that their nanorobots can effectively manipulate the processing light beams for optical-based nanopatterning.

"By combining advanced nanorobot designs with diverse functionalities, we can realize future research and development of the nanorobot pattern-ing method," says Wang. "For example, it could become possible to use ther-mal, mechanical, electrical, plasmonic, and chemical effects for this kind of dynamic nanorobot lithographic technique by incorporating specific physi-cal functionality or surface chemistry into the nanorobot."

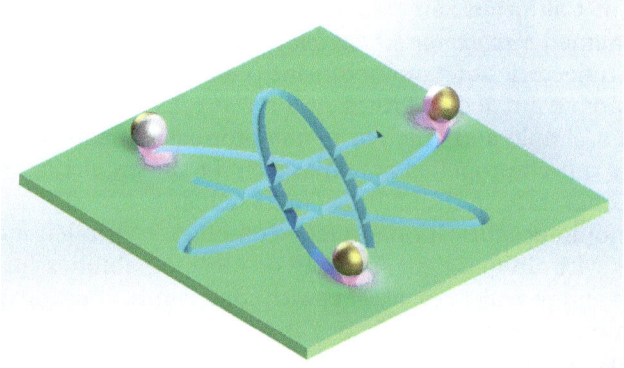

Figure 6.1   Schematic of nanomotor lithography by using a Janus sphere motor as a self-propelled nanolens. (Image: Jinxing Li, UC San Diego.)

He concludes that the current capabilities and future possibilities offered by motile optically active nanorobots will lead to a new "on-the-fly" nano-manufacturing platform and create new opportunities for surface science.

Featured scientists: Prof. Joseph Wang's laboratory for nanobioelectronics (http://joewang.ucsd.edu/)
Organization: University California San Diego, CA (USA)
Relevant publication: J. Li, W. Gao, R. Dong, A. Pei, S. Sattayasamitsathit and J. Wang, Nanomotor lithography, *Nat. Commun.*, 2014, **5**, 5026, DOI: 10.1038/ncomms6026

6.1.2 Graphene-Based Biomimetic Soft Robotics Platform

Soft robotics represents an exciting new paradigm in engineering. It challenges researchers to re-examine the materials and mechanisms that they use to make conventional hard robots so that they are more versatile, lifelike, and compatible for human interaction.

Among the various actuation mechanisms driven by different stimuli, light-driven systems have garnered more and more attention due to their advantages in wireless/remote control, localized rather than whole-field driven capabilities, and electrical/mechanical decoupling.

"While graphene shows decreasing absorption from visible to near-infra-red (nIR), there is brilliant photothermal conversion efficiency in the band of nIR, which has caused tremendous interest in biomedical applications, such as drug delivery and photothermal and photo-dynamic therapies," explains Dr Weitao Jiang, from the State Key Laboratory for Manufacturing Systems Engineering, Xi'an Jiaotong University in China. "Graphene can be incorporated into different polymer matrices to improve relevant properties, namely mechanical, thermal and electrical. Because of its photothermal effect and high thermal conductivity, graphene and its composites show promising photoresponsive properties."

Inspired by the photothermal effect of graphene in biomedical applications, Jiang's team proposed an easily fabricated and remote/wireless controlled light-driven approach to an actuation mechanism based on graphene nanocomposites.

"What has motivated us to develop an actuation mechanism based on graphene nanocomposites is its potential application in soft robotics applicable in clinical medicine, *i.e.*, implantable surgery robotics, or drug delivery devices," Jiang notes.

The researchers' soft and light-driven robotic bilayer platform is facile and scalable, and its fabrication only involves scraping coating and spin coating processing for the bottom and upper layer respectively.

As the key part in soft robotics platforms, stimuli-responsive materials have drawn enormous attention due to their brilliant intriguing shape or volume recovery properties under different external stimuli, which are helpful for creating mechanical motion rapidly and precisely.

In contrast to the actuation mechanism driven by electrical and electrochemical stimuli based on graphene nanocomposites, this photoresponsive soft platform can work both in air and water. The bilayer design combines soft matter PDMS and graphene. PDMS has already been widely used and studied in numerous fields and its good biocompatibility is a great advantage for applications *in vivo* or *in vitro*.

Meanwhile, graphene has been incorporated into different polymer matrices to improve mechanical, thermal and electrical properties. In previous work, polymer composites consisting of graphene nanoplatelets (GNP) and PDMS have been shown to exhibit a large light-induced reversible and elastic response.

In their work, Jiang and his team demonstrated an effective method for the fabrication of a polymeric bilayer biomimetic platform, which can be light-actuated both in air and water. The bilayer platform is composed of a pure PDMS layer and a PDMS/GNPs composited layer, in which each layer has a different coefficient of thermal expansion (CTE) and Young's modulus due to the existence of the GNPs.

The polymeric bilayer can be reversibly deflected at millimeter scale in response to nIR irradiation, which can be attributed to the photothermal effect of graphene. The deflection performances, *i.e.*, deflection magnitude and response time, are determined by the light intensity and GNP concentration.

To demonstrate the capabilities of their platform, the team designed biomimetic microfish, which can move forward, backward, and turn around in water under nIR irradiation, to mimic fish swimming in Nature. The moving directions and velocities can be remotely adjusted by light.

"Soft robotics is a fairly young sub-category of robotics, which combines classical principles of robot design with the study of fluids, gels, soft polymers, and other easily deformable matter," says Jiang. "It is learning from animals and plants in Nature that are composed primarily of soft, elastic structures which are capable of complex movement as well as adaptation to their environment. It represents an exciting new paradigm in engineering that challenges us to reexamine the materials and mechanisms that we use to make conventional hard robots so that they are more versatile, lifelike, and compatible with human interaction."

As Jiang points out, the results may not only be promising for developing light-driven drug-delivery platforms but also bio-robotic microgripper applications *in vivo* and *in vitro*.

"We believe that this soft robotic platform can further be explored for many other applications, such as biomimetic research, microcantilevers, micro/nanorobotics, drug delivery, minimally invasive medicine applications, implanting medical robots, *etc.*," he says. "Due to its excellent penetration ability in biological tissues, near-infrared light provides a promising approach to remotely actuate robotic devices within the body."

The researchers are already working on the next step, where they will integrate functional devices, such as a camera and stimuli tips, into the design and explore their applications in novel endoscopy and internal stimuli systems.

"We will also try to develop a new style of implantable flexible device which is either charged with near-infrared light or powered by the photothermal bilayer effect of soft robotics," Jiang concludes.

Featured scientist: Dr Weitao Jiang
Organization: State Key Laboratory for Manufacturing Systems Engineering, Xi'an Jiaotong University (PR China) (http://mail.sei.xjtu.edu.cn/tpl/sei/seienglish/index.html)
Relevant publication: W. Jiang, D. Niu, H. Liu, C. Wang, T. Zhao and L. Yin, *et al.*, Photoresponsive Soft-Robotic Platform: Biomimetic Fabrication and Remote Actuation, *Adv. Funct. Mater.*, 2014, **24**(48), 7598, DOI: 10.1002/adfm.201402070

6.1.3 How to Switch a Nanomachine On and Off

Over the past few years, researchers have demonstrated that microtubules driven by kinesin—a motor protein that converts chemical energy, derived from the hydrolysis of adenosine triphosphate (ATP), into mechanical work—make flexible, responsive and effective molecular shuttles for nanotransport applications.

In order to fully control these nanomotors, it has to be possible to switch them on, switch them off, and regulate the speed and direction of their movements—achievements that researchers haven't fully attained yet.

Several research labs have been working on building a fully controllable on/off switch mechanism of the motile properties of molecular machines such as kinesin-microtubules. Almost all of the methods show non-reversible and incomplete switching, namely, just one-time switching from a "fast to slow" or "slow to fast" state. Some research efforts accomplished reversible switching but in these cases the change in velocity was quite small. Most importantly, it has been impossible to stop the movement completely in such reversible switching methods.

In new work carried out in Professor Nobuyuki Tamaoki's laboratory at Hokkaido University, it has become possible, for the first time, to achieve complete control over on/off switching of the movement of a nanomachine.

In previous work,[1] Tamaoki's group already reported on their ongoing work on achieving reversible and repeated control over the motile properties of kinesin.

Following up on these results, the researchers investigated the photoresponsive inhibition properties of azobenzene-tethered peptides (azo-peptides) for the regulation of kinesin-driven microtubule motility.

"We synthesized various compounds and finally discovered a new peptide-azobenzene—containing a peptide and a terminal azobenzene unit—that has a reverse order of the amino acids of the peptide from the kinesin's tail," says Tamaoki. "This compound completely stops and starts the motility of kinesin-microtubules. It also has a reasonably good on/off switchability with UV or visible photoirradiation for one second each."

He notes that researchers have known that peptides whose amino acid sequences are taken from the tail part of kinesin show a moderate inhibition. "We thought about a covalent linkage between such a peptide and photoisomerizable azobenzene, which we have been studying for a long time, to switch various molecular functions."

The resulting photoresponsive azopeptide inhibitors can reversibly regulate microtubule motility over many cycles. If the concentration of the inhibitor is sufficiently high, the velocity of the microtubule can be stopped completely with photoirradiation.

"Our photoresponsive molecular motor system enables us to make an active spot allowing cargo-attached microtubules to move by irradiating with UV light selectively at any desired region," says Tamaoki. "Such an active spot could be moved freely just by moving the position of UV light in an inactivated background irradiated with visible light. In such a manner, with a focused UV light, we would select one specific microtubule attached with cargo and guide it to a desired point. As a consequence, all kinds of transportation of nano-objects for separation, mixing, concentration would be possible."

Research like this will not only lead to new applications of motor proteins in artificial nano-transportation systems but also contribute to a better understanding of the exact mechanism of natural molecular machines (today, even the mechanism of the conversion of the chemical energy to work is not fully understood yet).

"One of our goals is to change the energy source for the motility of bio-nanomachines from chemical energy to light energy in order to induce cycles of change in the molecular structure of kinesin," concludes Tamaoki. "With the demonstration of our novel photoswitching compound we may have reached an important stage in our thinking about the mechanism of the motility of nanobiomachines."

Featured scientist: Professor Nobuyuki Tamaoki's laboratory (http://tamaoki. es.hokudai.ac.jp/english/index.html)
Organization: Hokkaido University (Japan)
Relevant publication: W. Jiang, D. Niu, H. Liu, C. Wang, T. Zhao and L. Yin, *et al.*, Photoresponsive Soft-Robotic Platform: Biomimetic Fabrication and Remote Actuation, *Adv. Funct. Mater.*, 2014, **24**(48), 7598, DOI: 10.1002/adfm.201402070

6.1.4 Understanding Springs at the Nanoscale

Inspired by Nature's ingenious biological designs, researchers have persistently attempted to mimic these biofunctionalities to achieve technological breakthroughs. One of these morphologies—the unique shape of a helical coil—is not only interesting from a scientific standpoint but also pivotal, offering DNA its distinctive properties and propelling flagella in viscous fluids, to name a few.

Helically coiled springs are an integral element in many mechanical systems. Researchers have proposed nanoscale helically coiled morphology for enabling elastic memory devices, flexible electronics, impact protection, nanoinductors, and efficient electromagnetic shielding.

With the advent of personalized medicine on the horizon, researchers are now trying to use tiny springs made of carbon nanotubes, *i.e.* nanocoils, to propel nanorobots to perform microsurgeries.

A major challenge, however, is the lack of a detailed understanding of nanocoil mechanical properties. Indeed, the dynamic mechanical response of helically coiled structures is not fully understood due to the difficulties involved in exciting purely longitudinal/transverse resonances. This mathematical complexity has puzzled researchers for many years.

Professor Apparao M. Rao's group at Clemson Nanomaterials Center revisited this age-old problem and developed a three-pronged methodology, which entails experimental, analytical and computational techniques.

This study offers a better understanding of the shear and tensile contribution to the response when a helical coil is subjected to a transverse force.

The researchers, who had previously synthesized helically coiled carbon nanowires and nanotubes,[2] isolated and clamped a single helically coiled nanowire akin to a diving board or cantilever. This nanosized cantilever was then electrically resonated and detected *via* a method invented at Clemson called the "harmonic detection of resonance" (HDR) under a scanning electron microscope.

"Nanocoils are very special since they have multiple applications ranging from simple shock absorbers to diverse nanorobotic tools," says Rao. As his collaborators Herbert Behlow and Professor Malcolm Skove note: "Our protocol can also be used as a non-destructive probe for determining the material properties of not only nanocoils but any helically coiled material, in general."

The team successfully derived a much-needed closed form solution or a formula to predict transverse resonance frequencies of not just the first but also the second mode of any singly clamped helically coiled cantilever.

"The analytical solution was built upon the classical model for mechanical springs, and importantly, our protocol is applicable to any size of the coil," says Deepika Saini, the lead author of a paper on this work. "Hence, it will prove beneficial across many engineering and research fields allowing more accurate designs and early prediction of mechanical failures."

In addition, the team also observed fascinating mechanical resonance modes—non-planar as well as asymmetrical.

Professor Ramakrishna Podila, who initiated and participated actively in the project adds: "Through this project, the Clemson team has now shown that geometrically non-linear morphologies encompass rich physics which can be examined in greater detail, and this possibility opens doors to a better understanding of the mechanical properties of helically coiled cousins of DNA for designing futuristic building blocks for the materials world."

Featured scientists: Prof. Apparao M. Rao's nanomaterials research laboratory; Prof. Ramakrishna Podila
Organization: Clemson University SC (USA)
Relevant publication: D. Saini, *et al.*, Mechanical Resonances Of Helically Coiled Carbon Nanowires, *Sci. Rep.*, 2014, **4**, 5542, DOI: 10.1038/srep05542

6.1.5   Fast Molecular Cargo Transport by Diffusion

Engineered molecular motors are a prime example of nanotechnology's efforts to imitate biology in order to build artificial chemical systems with power and capabilities similar to living cells. For the more visionary goals of nanotechnology, functional and perhaps autonomous molecular motors will play an essential part, just like electric motors can be found in many appliances today. These nanomachines could perform functions similar to the biological molecular motors found in living cells, things like transporting and assembling molecules, or facilitating chemical reactions by pumping protons through membranes.

One group of researchers has explored a transport strategy that is different from these nanomotors: they utilized diffusion as an effective transport mechanism for DNA nanotechnology. These findings contribute a new aspect to be considered for the design of future DNA motors, molecular machines, and nanorobots, as they provide a simple way to transport molecules over distances of potentially several hundred nanometers; which is much faster than conventional DNA walkers or motors, which make many small and slow steps.

"In contrast to earlier work on DNA walkers, which sought to precisely control the transport path of a DNA walker on a DNA 'track', we give up this precise control and simply 'fly' from the start to the target position," says Friedrich C. Simmel, a professor for systems biophysics and bionanotechnology at the Technische Universität München. "As a benefit, the effective transport speed is much faster."

"Our work was motivated by the observation that small biological organisms such as bacteria do not have molecular motors—they are not needed as diffusion within a bacterial cell is very fast anyway," explains Simmel. "Molecular cargo is simply targeted by providing appropriate binding sites for the cargo. As our system is not compartmentalized—such as a bacterial cell—we tether the cargo to our transport platform, which effectively also creates some type of localization or compartmentalization."

He adds that another motivation is a general lesson learned from molecular machines and motors in biology: they utilize ("rectify") Brownian motion rather than work against it.

On the length scale of typical DNA nanostructures (10–100 nm), diffusion is an extremely fast process and thus can be used to transfer a molecule from one position on a nanostructure to another. In order to avoid escape of the cargo, it has to be tethered to the DNA structure with a "transporter arm".

If DNA molecules are transported (as in this case), they have to unbind from a starting position and later bind to a target position *via* DNA hybridization *i.e.*, the formation of DNA duplexes between sequence-complementary single strands of DNA.

However, hybridization is a relatively slow process. It turns out that on the scale of the DNA nanostructures used in this work, diffusion is much faster than these binding and unbinding reactions.

"As a consequence, the time required for transport from the starting to the end position is in fact independent from the distance," Simmel points out. "Thus the best—*i.e.* fastest—strategy to transport DNA molecules in this context is to simply make very large steps."

Using the scaffolded DNA origami technique,[3] the team created a rectangular platform with dimensions 90 nm × 60 nm, which was equipped with a 30-nm-long molecular tether attached to a flexible hinge in the center of the rectangle.

The tip of the arm can be fixed to an initial position or one of several target positions on the platform, and thus facilitate local transport of molecular cargoes attached to the arm. The role of the tether is to prevent diffusive loss of the cargo, whereas transport itself is driven by thermal motion.

In a typical experiment, the tether is released from its initial position by the addition of a trigger oligonucleotide. The tether then diffusively searches for an alternative binding position on the platform and, finally, attaches to the desired target site.

"Our kinetic studies indicate that this mechanism provides fast transport over distances of up to 52 nm with effective velocities of at least several nanometers per second and, thus, is much faster than transport with DNA walkers," notes Simmel.

Potential applications of this work are found in the field of DNA-templated synthesis: bringing compounds together to increase their local concentration and let them react or interact with each other. In combination with "mechanical" degrees of freedom, this results in systems that resemble "molecular assembly lines".

According to the researchers, another application can be found in DNA computing, where localization of DNA computational gates is supposed to speed up computation processes, reduce crosstalk between competing reactions and thus improve modularity of DNA circuits.

"One could also think of linking many systems such as ours together and let a molecular signal run through a network of DNA transporters," says Simmel. "Applications for these types of systems could be found in biosensing, where a biological input signal would be directly and autonomously processed by a biomolecular computation—rather than having to translate it into electrons or photons with a conventional sensor."

Scientists are beginning to learn how to utilize spatial organization and compartmentalization to control the speed and efficiency of (bio)chemical processes, and this work is just one example of that. "In the long run, we hope to create artificial, cell-like systems which have many of the capabilities of biological cells. We are currently exploring components and concepts for such systems," concludes Simmel.

He cautions, though, that one of the biggest challenges for the field is interfacing with the "outside world" and the combination with other chemistries: "While DNA is a great material to build programmable structures and develop artificial reaction networks, in applications one typically has to deal with other molecules. So the question is whether concepts developed for DNA-based model systems can be translated to other molecular systems, or how to utilize DNA components within hybrid systems."

Featured scientists: Prof. Friedrich C. Simmel's lab (http://www.e14.ph. tum.de/en/home/)
Organization: Technische Universität München (Germany)
Relevant publication: E. Kopperger, T. Pirzer and F. C. Simmel, Diffusive Transport Of Molecular Cargo Tethered To A DNA Origami Platform, *Nano Lett.*, 2015, **15**(4), 2693, DOI: 10.1021/acs.nanolett.5b00351

6.1.6   Micro- and Nanomotors Powered Solely by Water

Man-made micro- and nanoscale motors have received tremendous research interest owing to their great potential for diverse potential applications, ranging from targeted drug delivery, microchip diagnostics to environmental remediation. Particular attention has been given to self-propelled chemically-powered micro/nanoscale motors, such as catalytic nanowires, microtube engines or spherical Janus microparticles.

"Although significant progress over the past 10 years has greatly advanced the capabilities of these tiny man-made machines, catalytic motors have predominantly relied on an external hydrogen peroxide fuel that impedes many practical applications," notes Joseph Wang, Distinguished Professor, Department of Nanoengineering at the University California San Diego. "Extending the scope of chemically-powered nanomotors to diverse operations and new environments requires the identification of new *in situ* fuels in connection to new catalytic materials and reactions. Obviously, water is the ideal choice as fuel for the majority of practical nanomachine applications."

Wang and his team have demonstrated the first example of a water-driven bubble-propelled micromotor that eliminates the requirement for the common hydrogen peroxide fuel.

Although the water-splitting reaction has been used before to drive bipolar-electrochemical macroscale motors under an external electrical field,[4] it has not been used for the locomotion of self-propelled chemically powered micromotors.

By presenting the first example of a chemically powered micromotor propelled autonomously using water as the sole fuel source, the team greatly expands the scope of applications and environments of chemically powered nanomachines (in previous work, Wang and his team already demonstrated hydrogen-bubble-propelled zinc-based microrockets that rely on strong acids as fuels[5]).

In particular, as Wang explains, "these water-driven micromotors could have a profound impact on diverse biomedical or industrial applications of micromotors, ranging from targeted drug delivery to microchip diagnostics, where the use of the common peroxide fuel is not desirable or possible."

To accomplish the goal of using water for generating the bubble-thrust, essential for an efficient autonomous movement, the researchers have focused on the use of aluminium to reduce water to hydrogen. For this purpose, they identified the most appropriate aluminium-alloy composition. They designed a new bubble-propelled spherical Janus micromotor based on coating one side of an Al–Ga microparticle with a titanium layer and developed a new microcontact route for fabricating such water-propelled Al–Ga/Ti microparticles (Figure 6.2).

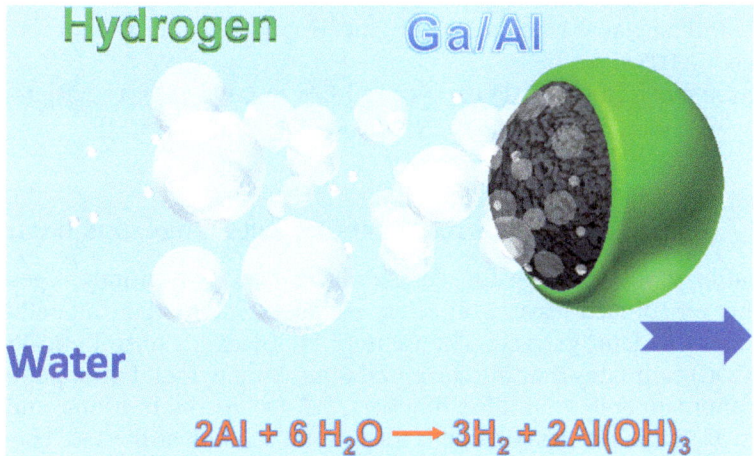

Figure 6.2   The water-driven Janus micromotor is composed of a partially coated Al–Ga binary alloy microsphere. The dark hemisphere (left) represents the Al–Ga alloy, while the green area (right) corresponds to the asymmetric Ti coating on one side of the sphere. The ejection of hydrogen bubbles from the exposed Al–Ga alloy hemisphere side, upon its contact with water, provides a powerful directional propulsion thrust. (Image: Wang Group, Department of Nanoengineering, University of California, San Diego.)

The fabrication process involves these major steps: aluminium particles (average size 20 μm) and liquid gallium are spread onto separate glass slides at a 1:1 mass ratio. The two slides are then pressed together until an Al–Ga alloy forms *via* microcontact mixing. During this period, gallium penetrates into the aluminium particles to form the outer alloy layer. Subsequently, one side of the Al–Ga microparticles is coated with a titanium layer *via* e-beam evaporation to form the asymmetric Janus microstructure. Brief sonication allows the separation of the micromotors from the glass slides.

Once the microparticles come in contact with water—or other fluids such as biological media—a rapid reaction between the aluminium alloy and water leads to a spontaneous generation of hydrogen bubbles. This ejection of hydrogen bubbles from the exposed Al–Ga alloy hemisphere side provides a powerful directional propulsion thrust.

The strong momentum of the bubble ejection propels the micromotor at a remarkable speed of 3 millimeters per second, which corresponds to 150 body lengths per second.[6]

Wang says that future efforts will aim at further improving the propulsion behavior of water-driven micromotors by enhancing the alloy reactivity *via* a judicious control of its microstructure and composition.

"We will also investigate new water-splitting catalytic reactions and materials that would offer extended motor lifetimes and/or higher efficiency, and will demonstrate practical applications of these water-powered micromotors."

Featured scientists: Prof. Joseph Wang's laboratory for nanobioelectronics (http://joewang.ucsd.edu/)
Organization: Jacobs School of Engineering, University California San Diego, CA (USA)
Relevant publication: W. Gao, A. Pei and J. Wang, Water-Driven Micromotors, *ACS Nano*, 2012, **6**(9), 8432, DOI: 10.1021/nn303309z

6.1.7 Self-Propelled Microrockets Detect Dangerous Bacteria

No matter if you are into big, fat hamburgers or eat entirely vegetarian, nibbling on spinach leaves and celery stalks—some food-borne pathogens will sooner or later get you. The Centers for Disease Control and Prevention (CDC) estimates that in the United States alone, food-borne pathogens cause approximately 76 million illnesses, 325 000 hospitalizations, and 5000 deaths. If that is not scary enough for you, take a look at the U.S. Food and Drug Administration's *Bad Bug Book*[7] that lists food-borne pathogenic microorganisms and natural toxins.

Early detection of food-borne pathogenic bacteria is critical to prevent disease outbreaks and preserve public health. This has led to urgent demands to develop highly efficient strategies for isolating and detecting this microorganism in connection to food safety, medical diagnostics, water quality, and counter-terrorism.

E. coli and other pathogenic bacteria are commonly detected using traditional culture techniques: microscopy, luminescence, enzyme-linked immunosorbent assay (ELISA), biochemical tests and/or the polymerase chain reaction (PCR). These techniques, however, are time-consuming, labor-intensive, and inadequate as they lack the ability to detect bacteria in real time. Thus, there is an urgent need for alternative platforms for the rapid, sensitive, reliable and simple isolation and detection of *E. coli* and other pathogens.

Taking a novel approach to isolating pathogenic bacteria from complex clinical, environmental and food samples, Wang's group has also developed a nanomotor strategy that involves the movement of lectin-functionalized microengines. Receptor-functionalized nanoswimmers offer direct and rapid target isolation from raw biological samples without preparatory and washing steps.

"Previously, we have demonstrated the use of new synthetic template-prepared microrockets,[8] functionalized with lectin receptors, for the efficient isolation of target bacteria from diverse real samples," explains Wang. "These modified self-propelled microengines offer very attractive capabilities for autonomous loading, directional transport and bacterial unloading—'catch, transport and release'—towards subsequent re-use, along with efficient and simultaneous transport of drug nanocarriers. The new smaller microengines we subsequently developed allow convenient label-free real-time visualization of the binding event and differentiation against non-target cells without the need for additional tagging."

In separate work,[9] the motion and power of self-propelled synthetic and natural nano/microscale motors have been exploited by the UCSD team as an attractive route for transporting target biomaterials, such as cancer cells or nucleic acids, but not for the capture and transport of pathogenic bacteria.

Wang's team also demonstrated, for the first time, the ability to capture and transport simultaneously the target bacteria along with drug-carrier polymeric spheres (towards a theranostics operation), as well as a chemically-triggered unloading (release) operation.

The efficient bacterial isolation platform that the team developed relies on the attractive behavior of its microrockets along with its functionalization with lectin receptors.

"Lectins are readily available sugar-binding proteins that offer an attractive route for recognizing carbohydrate constituents of a bacterial surface, *via* selective binding to cell-wall mono- and oligosaccharide components," explains Wang. "For example, ConA, the lectin extracted from *Canavalia ensiformis* that we used in this work, is a mannose- and glucose-binding protein that is capable of recognizing specific terminal carbohydrates of Gram-negative bacteria such as the *E. coli* surface polysaccharides."

Although lectins have been used as biosensor recognition elements for bacterial detection, their use in connection to nanomachines or nanoscale motion-based isolation is a novel approach.

As illustrated in Figure 6.3, the team's nanoscale bacteria isolation strategy utilizes the movement of ConA-functionalized microengines to scour, interact with and isolate pathogenic bacteria from distinct complex samples.

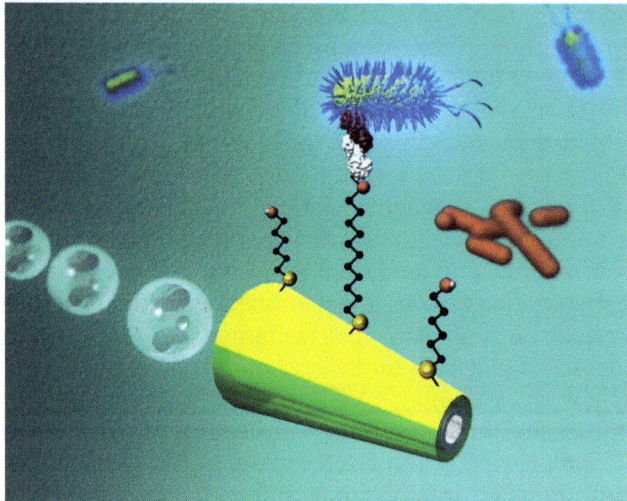

Figure 6.3   Micromachine-based isolation of bacterial targets from complex samples involving capture and release abilities of lectin-modified microengines. (Image: Wang Group, Department of Nanoengineering, University of California, San Diego.)

After bacteria have been captured, they then can be released in a controlled fashion by moving the microrocket through a low-pH glycine solution (which dissociates the lectin–bacteria complex).

Finally, but equally importantly, the microrockets are dual action, *i.e.* besides capturing and transporting target bacteria, they can simultaneously transport polymeric drug-carrier spheres to provide "on-the-spot" therapeutic action.

"The diverse capabilities of our lectin-modified microrockets make them extremely attractive for a wide range of fields, including food and water safety, infectious disease diagnostics, biodefense, and clinical therapy treatments," says Wang. "The incorporation of such a microengine-based bacterial isolation protocol into microchannel networks could lead to microchip operations involving real-time isolation of specific bacteria, its lysis and unequivocally identification (by 16S rRNA gene analysis)."

Overall, the new microengine "catch–transport–release" platform presents a unique approach for meeting the need for rapid, direct and real-time isolation of biological agents.

Featured scientists: Prof. Joseph Wang's laboratory for nanobioelectronics (http://joewang.ucsd.edu/)
Organization: Jacobs School of Engineering, University California San Diego, CA (USA)
Relevant publication: S. Campuzano, J. Orozco, D. Kagan, M. Guix, W. Gao and S. Sattayasamitsathit, *et al.*, Bacterial Isolation by Lectin-Modified Microengines, *Nano Lett.*, 2012, **12**(1), 396, DOI: 10.1021/nl203717q

6.1.8 Repair Nanobots on Damage Patrol

You cut yourself in the finger—and a few days later your skin has completely healed all by itself. Biological organisms have an amazing ability to automatically initiate self-healing and self-repair when they sustain damage. Materials engineers are dreaming about making materials that could do the same thing.

Inspired by the intrinsic self-repairing ability of biological systems, researchers have developed a class of artificial "smart" materials—called "self-healing materials"—which can repair internal or external damage;[10] more recently, they even developed an approach to self-healing electronic devices.[11]

Many of these approaches depend on the healing agents being incorporated into the material or on external stimuli to initiate the repair process. Complicating the matter are damages that originate at the micro- or even nanoscale, such as cracks and tears in thin-films, membranes and electronic circuits.

An ideal material and device self-healing system would be autonomous: damages would be detected, localized and repair initiated independent from outside interference and without the need for external control.

"For a system to perform autonomous repair, it must be capable of delivering the healing agents precisely to specific defect sites," says Professor Joseph Wang. "Such healing agents must carry out the following three key functions: convert environmental energy into mechanical work for directed motion; autonomously sense and detect the damage; and possess engineered repairing capabilities."

Wang's team, in cooperation with Professor Anna Balazs' group at the University of Pittsburgh, has engineered synthetic nanomotors that self-propel and autonomously detect surface cracks in electronic devices and rapidly restore the conductive pathway.

The team's nanomotors were inspired by the chemotaxis of neutrophils toward inflammation sites and the aggregation of platelets at the collagen fibers of a wound to stop bleeding.

The catalytic nanomotors are composed of conductive gold/platinum spherical Janus particles that self-propel efficiently in the presence of hydrogen peroxide fuel. They convert this fuel into directed motion to autonomously seek the surface cracks on the substrate.

"The presence of surface cracks introduces obstructions and gaps, which present both energetic barriers and potential wells to the random walk trajectories of the nanomotors," Wang explains the nanomotors' sensing mechanism. "The surface cracks act as potential wells, which confine and localize the nanomotors."

He notes that these nanomotors can also spontaneously self-assemble into clusters that can travel as groups toward the damage location.

By accumulating inside the cracks, the conductive nanomotors provide a "patch" that restores the electrical current in the broken circuit.[12]

The clustering of nanomotors is primarily due to the hydrophobic coating on their hemispheric gold surfaces (a self-assembled monolayer of octadecanethiol).

"Although the repair functionality involves the selective localization of conductive nanoparticles at cracked electrodes, the self-propelled nanomotor concept can be extended to repair the biological, mechanical, optical, or electronic properties of a wide range of damaged materials," says Wang.

"The use of self-propelled nanomotors to probe nano- and microscale environmental changes and to promote autonomous and precise localization at desired sites opens the door for artificial responsive nanosystems with advanced biomimetic functionalities for a wide variety of applications ranging from targeted drug delivery to self-healing nanodevices," he concludes.

Featured scientists: Prof. Joseph Wang's laboratory for nanobioelectronics (http://joewang.ucsd.edu/); Prof. Anna Christina Balazs (http://www.engineeringx.pitt.edu/AnnaBalazs/)
Organization: Jacobs School of Engineering, University California San Diego, CA (USA); Swanson School of Engineering, University of Pittsburgh, PA (USA)
Relevant publication: J. Li, O. Shklyaev, T. Li, W. Liu, H. Shum and I. Rozen, *et al.*, Self-Propelled Nanomotors Autonomously Seek and Repair Cracks, *Nano Lett.*, 2015, **15**(10), 7077, DOI: 10.1021/acs.nanolett.5b03140

6.2 Inspired by Nature, the Greatest Nanotechnologist of All

An increasing number of scientists apply methods and systems found in Nature to the study and design of engineering systems and modern technology—down to micro- and nanoscale systems. They believe that learning from natural designs is more likely to provide the cues for designing practical nanodevices than simply adopting a "trial and error" approach. The basic idea is that natural materials and systems can be adopted for human use beyond their original purpose in Nature.

6.2.1 Smart Materials Become "Alive" with Living Bacteria in Supramolecular Assemblies

Supramolecular chemistry deals with molecular building blocks that interact with each other in a dynamic manner, similar to what we see in Nature. Taking advantage of this, several "smart" materials have been developed for biomedical applications by carefully designing their building blocks. These materials have especially interesting properties such as self-healing and responsiveness to light and electricity.

"So far, smart materials have mostly been developed using artificially syn-thesized molecules and some simple proteins, which effectively limits the properties that can be endowed upon such materials," says Pascal Jonk-heijm, a professor at the MESA+ Institute for Nanotechnology and the Uni-versity of Twente. "In our work, we explored the possibility of developing a bacterial strain with the ability to interact dynamically with one such pop-ular supramolecular building-block—a pumpkin-shaped hollow molecule named cucurbit[8]uril (CB[8])."

By doing so, Jonkheijm, who heads the Bioinspired Molecular Engineering group, and his team, were able to show that living cells can also be used as components in supramolecular smart materials. Their novel strategy intro-duces specific, dynamic and reversible supramolecular functionality on the bacterial cell surface by adopting a bacterial display system that has been used before, exclusively to identify high affinity peptides for various proteins.

Although this study focused specifically on CB[8], the techniques demon-strated by the researchers can also be used to develop and study other strains for different host molecules.

"This could essentially allow us to give these materials exciting new properties like motility, growth and resistance against external agents," says Jonkheijm.

While groups made advances by introducing supramolecular elements on cell membranes by merging artificially made membranes, this is the first time the cells were made to produce and display supramolecular elements on their surfaces through genetic engineering.

"Since we have done this in bacterial cells, in principle, we can grow an infinite amount of cells with these supramolecular properties," Jonkheijm points out. "This laboratory bacterial strain was originally derived from the flora in human intestines, which means that our supramolecular strain could be incorporated in materials for biomedical applications."

Supramolecular materials have been designed with various different types of building blocks to have many exciting properties. Jonkheijm's group has also been actively developing dynamic and responsive materials to address biomolecules, cells and even tissues.

"The logical next step in the evolution of these materials was to incorpo-rate living cells as active components," he notes. "However, this has proven to be extremely challenging and has been approached by only a few research groups. We felt that the power of bacterial genetic engineering could help to address this challenge in a robust and sustainable manner and so we attempted this project."

This work addresses the issue of incorporating living cells in supramolec-ular materials only at a fundamental level. According to the researchers, one could imagine that these bacteria can carry within the bloodstream micro-scopic cargo, such as drugs in a microcapsule, and when triggered, release this cargo at a particular site, such as a cancerous or infected cell.

Living cells incorporated in supramolecular materials could possibly also be designed to secrete other components of the material, allowing them to heal the material or make it grow. Several such possibilities could be achieved to expand upon the impressive properties already available in existing supramolecular materials.

"The future of this research field is very bright and is limited only by our imagination and understanding," says Jonkheijm "These new discoveries and others that are being made every day will allow us to make smart materials with properties that mimic and surpass those found in Nature. Such materials can have an extremely wide range of applications from biosensors and implant coatings to even electronics."

He cautions, though, that researchers are facing several challenges, largely regarding the understanding of these systems. Living systems are extremely difficult to predict and a massive amount of trials and tedious analysis is required to understand the various processes involved.

"Large strides cannot be hastily made and an immense amount of knowledge needs to be gathered and connected from various field," Jonkheijm concludes. "However, in the past few decades, impressive advances in supramolecular materials have been made and with the same amount of effort, we will soon see several of these materials being incorporated in real world applications."

Featured scientist: Prof. Pascal Jonkheijm (https://www.utwente.nl/tnw/mnf/People/academic_staff/pascal_jonkheijm/)
Organization: MESA+ Institute for Nanotechnology, University of Twente (The Netherlands)
Relevant publication: S. Sankaran, M. Kiren and P. Jonkheijm, Incorporating Bacteria as a Living Component in Supramolecular Self-Assembled Monolayers through Dynamic Nanoscale Interactions, *ACS Nano*, 2015, 9(4), 3579, DOI: 10.1021/acsnano.5b00694

6.2.2 From Squid Protein to Bioelectronic Applications

Electrical devices, from light bulbs to smartphones, send information using electrons. All living things, on the other hand, send signals and perform work using ions or protons. Protons activate "on" and "off" switches and are key players in biological energy transfer. Proton-conducting materials have become important for a wide range of technologies, such as fuel cells, batteries, and biosensors. A great deal of research has been devoted to developing improved and application-specific proton conducting materials. A team of researchers that included Professor Pascal Jonkheijm even developed a proton-based transistor that could let machines communicate with living things.[13]

In separate work, scientists discovered and characterized novel electrical properties for the cephalopod structural protein reflectin.

"The figures of merit for reflectin's protonic conductivity compare favorably to those found for many artificial proton-conducting materials, such as ceramic oxides, solid acids, polymers and metal–organic frameworks," says Alon A. Gorodetsky, assistant professor at the Henry Samueli School of Engineering at the University of California, Irvine. "We believe that a better understanding of reflectin's proton conducting properties could inform the design and engineering of artificial proton-conducting materials."

However, reflectin's electrical properties had never been previously studied.

"We were inspired to pursue our studies in part by the Hanlon Group's work on electrical triggering of iridophores in squid skin," says Gorodetsky. "We postulated that reflectins might have unique electrical properties."

The team interrogated the protein by humidity-dependent direct current electrical measurements with both proton-blocking and proton-injecting contacts; alternating current electrical measurements in the presence of water and deuterium oxide; rationally guided mutagenesis experiments; and temperature-dependent electrochemical impedance spectroscopy. Together, these experiments indicate that reflectin functions as an efficient proton-conduction medium.

"Based on our measurements, we infer that reflectin exhibits the characteristics of a dilute acid," notes David Ordinario, a researcher in Gorodetsky's lab. "Bulk reflectin is quite unique in this regard; as far as we are aware, no other protein has been shown to mimic a dilute acidic solution so closely. Moreover, reflectin's maximum conductivity of 2.6×10^{-3} S cm^{-1} at 65 °C is among the largest values found for any naturally occurring protein. Within the context of other biological (and even artificial) proton-conducting materials, reflectin's figures of merit are impressive and may represent new benchmarks for proteins in the solid state."

Gorodetsky notes that these findings indicate that reflectins could represent a promising new class of modular proton-conducting materials for bioelectronics and other applications.

"We believe that reflectin-based transistors and devices might be especially useful for bioelectronic applications that require an inherently biocompatible conductive material," he says. "For example, we envision using reflectin-based devices to interface with neural cells and to read out ionic and protonic fluxes from these cells."

The team's future research directions will revolve around engineering and evolving reflectin for not only improved properties but also for specific applications outside of bioelectronics.

"One of the key challenges facing our work is that we do not know reflectin's precise structure, either in thin-films or in solution," Gorodetsky points out. "In our future work, by gaining insight into reflectin's structure, we hope to engineer the protein for optimum functionality in different types of electrical devices."

Featured scientists: Prof. Alon A. Gorodetsky's biomolecular electronics group (http://www.gorodetskygroup.org/)

Organization: Henry Samueli School of Engineering, University of California – Irvine, CA (USA)

Relevant publication: D. Ordinario, L. Phan, W. Walkup IV, J. Jocson, E. Karshalev and N. Hüsken, *et al.*, Bulk protonic conductivity in a cephalopod structural protein, *Nat. Chem.*, 2015, **6**(7), 596, DOI: 10.1021/acsnano.5b00694

6.2.3   An Octopus Might Point the Way to Stealth Coatings

As we have seen in the previous section, the protein reflectin has been explored for its optical properties. The same team of scientists, led by Gorodetsky, has explored how squid and octopuses might point the way to nanotechnology-based stealth coatings.

For a long time, scientists have been fascinated by the dramatic changes in color used by marine creatures, but they never quite understood the mechanism responsible for this. It was only recently that they found out that a neurotransmitter, acetylcholine, sets in motion a cascade of events that culminate in the addition of phosphate groups to a family of unique proteins called reflectins. This process allows the proteins to condense, driving the animal's color-changing process. These findings also revealed that there is a nanoscale mechanism behind cephalopods' ability to change color.[14]

Having begun to unravel the natural mechanisms behind these amazing abilities, researchers are trying to use this knowledge to make artificial camouflage coatings. Gorodetsky's lab addresses the challenge of making something appear and disappear when visualized with standard infrared detection equipment.

The team demonstrated graphene-templated, biomimetic camouflage coatings that possess several important advantages.

"We used reflectin, a protein that is important for cephalopod structural coloration, as a functional optical material," Gorodetsky explains. "We fabricated thin-films from this protein, whose reflectance—and coloration—could be dynamically tuned over a range of over 600 nm and even into the infrared. Our approach is environmentally friendly and compatible with a wide range of surfaces, potentially allowing many simple objects to acquire camouflage capabilities."

The novelty of these findings lies in the functionality of the team's thin-films within the infrared region of the electromagnetic spectrum, roughly 700 nm to 1200 nm, which matches the standard imaging range of infrared visualization equipment. This region is not commonly accessible to biologically derived materials. Gorodetsky notes that reflectin's tunable optical properties compare favorably to those of artificial polymeric materials.

"Given these advantages, our dynamically tunable, infrared-reflective films represent a crucial first step towards the development of reconfigurable and disposable biomimetic camouflage technologies for stealth applications," says Gorodetsky. "I could also imagine applications in energy efficient reflective coatings and biologically inspired optics."

The team began their studies by developing a protocol for the production of the histidine-tagged reflectin A1 (RfA1). Experimenting with a variety of substrates and surface treatments for the reliable formation of RfA1 thin-films, they achieved the best results by spincasting 5 to 10 nm films of graphene oxide on glass substrates. They then spread RfA1 onto the graphene oxide-coated substrates, yielding smooth films over centimeter areas.

These films showed a distinct coloration, depending on their thickness. For instance, a 125-nm-thick film was blue and a 207-nm-thick film was orange.

"Inspired by the dynamic optical properties of reflectin nanostructures, we sought to shift the reflectance of our RfA1 films into the infrared region of the electromagnetic spectrum," says Gorodetsky. "Given that some squid can dynamically modulate their skin reflectance across the entire visible spectrum and even out to near-infrared wavelengths of ~800 nm, we postulated that it should also be possible to tune the reflectance of our RfA1 thin films across a similar, or even larger, wavelength range. Thus, we sought conditions that would significantly increase the thickness of our RfA1 films and, consequently, shift their reflectance spectra toward the infrared."

To that end, the researchers explored the response of their RfA1 coatings to a variety of chemical stimuli. They discovered that exposing the films to vapor from a concentrated acetic acid solution induced a large, reversible shift in the reflectance spectra, caused by the acid-induced swelling of the closely packed RfA1 nanoparticles in the film.

"With the goal of fabricating dynamically tunable camouflage materials, which will self-reconfigure in response to an external signal, we are planning to develop alternative, milder strategies for triggering coloration changes in our material," Gorodetsky describes the team's future research plans.

Featured scientists: Prof. Alon A. Gorodetsky's biomolecular electronics group (http://www.gorodetskygroup.org/)
Organization: Henry Samueli School of Engineering, University of California – Irvine, CA (USA)
Relevant publication: L. Phan, W. Walkup, D. Ordinario, E. Karshalev, J. Jocson and A. Burke, *et al.*, Reconfigurable Infrared Camouflage Coatings from a Cephalopod Protein, *Adv. Mater.*, 2013, **25**(39), 5621, DOI: 10.1002/adma.201301472

6.2.4 Battery Parts Grown on a Rice Field

The list of nanotechnology research that gets inspired by Nature gets longer and longer—whether it is to mimic natural structures (*e.g.* peacock feathers, butterfly wings, gecko feet, lotus leaves) for nanoscale patterns and devices, or use naturally occurring nanoparticles instead of synthetic ones for the fabrication of high-tech materials and products.

Rice has also made it onto this list. Researchers in Korea have found that rice husks—the outer, protective covering of a rice kernel—can be a source of silicon that can be used for high-capacity lithium battery anodes.

Most of today's lithium-ion batteries rely on anodes made from graphite. There are several candidate electrode materials to replace graphite as the anode for lithium-ion batteries and silicon has been recognized as a favorable anode material because its capacity is 3–5 times larger than that of existing graphite anodes.

However, so far the cycle life performance of silicon anodes has been challenging due to large volume expansions. The problem was that silicon anodes broke down from repeated swelling and shrinking during charging/discharging cycles and quickly became useless.

This work demonstrates that rice husks, a major by-product from rice harvesting, can be used to produce silicon with an ideal porous nanostructure for use in high-capacity lithium-ion battery (LIB) anodes.

"The interconnected nanoporous structure of rice husks, developed by natural evolution for efficient cultivation of rice, can resolve important issues in silicon anode operation, enabling excellent cycling and power performance," says Jang Wook Choi, an associate professor at the Korea Advanced Institute of Science and Technology (KAIST), who led the work. "Given that annual global rice production reaches 700 hundred million tons, the promising battery data in our work shows how a part of the waste from rice husks—which reaches 20 wt% of the entire rice kernel—can be a massive resource to meet the ever-increasing demand for silicon in advanced batteries."

One of the research areas of Choi's energy nanomaterials group lies in bio-inspired materials for lithium-ion batteries. In previous work, for instance, the team developed a highly functional binder for lithium-ion batteries.[15]

Choi's team found a naturally existing nanoporous structure in rice husks with pores the size of tens of nanometers. Various purification treatments of ordinary rice husks, which preserved their unique 3D porous structure, resulted in over 99% pure silica. A further thermic reduction then transferred the 3D porous structure of the rice husk silica to silicon.

To evaluate the rice husk silicon as a LIB anode, the researchers prepared coin-type half-cells.

"For our tests, we coated the rice husk silicon with carbon layers (3–10 nm) to compensate for the intrinsically low electronic conductivity of silicon as shown in previous studies," explains Choi. "We found that this carbon-coated rice husk silicon anode material shows excellent discharge capacity retention during cycling using the stable interconnected porous structure. Its cycling performance is clearly better than that of bare silicon nanoparticles and carbonized polydopamine-coated silicon nanoparticles."

These results led him to conclude that the nanoporous structure found in rice husks is more suitable for solving the volume expansion problem in LIBs than any other artificially developed nanostructure. In particular, most of the artificially developed nanostructures have an issue of irregular particle sizes, but the structure from rice husks holds very uniform particles and pores.

As the silicon technology in the battery industry gets more mature, it could be possible that one of your next smartphones contains a part that grew on a rice field.

Featured scientists: Prof. Jang Wook Choi's energy nanomaterials group (http://nest.kaist.ac.kr/)
Organization: Korea Advanced Institute of Science and Technology (KAIST), Daejeon (Republic of Korea)
Relevant publication: D. Jung, M. Ryou, Y. Sung, S. Park and J. Choi, Recycling rice husks for high-capacity lithium battery anodes, *Proc. Natl. Acad. Sci. U. S. A.*, 2013, **110**(30), 12229, DOI: 10.1073/pnas.1305025110

6.2.5 Turning Trash into Treasure—Bioinspired Colorimetric Assays

Colorimetric sensing techniques require only the naked eye or ordinary visible color photography and are attractive because of their low cost, use of inexpensive equipment, and above all, their simple-to-understand results. Colorimetric sensors can be used for both qualitative analytic identification as well as quantitative analysis and find a wide range of applications in chemical and biomolecular analysis as well as integration into portable microfluidics lab-on-chip devices.

"Generally, a colorimetric probe consists of a recognition moiety and a signal moiety," explains Xiaodong Chen, an assistant professor in the School of Materials Science & Engineering at Nanyang Technological University. "The sensing strategy is that the recognition moieties exhibit selective response to targets depending on either coordination or chemical reactions between targets and ligands. The signal moiety—usually an organic chromophoric probe—then translates these detecting behaviors into color changes discernible by the eye. Both the components are indispensable and can affect colorimetric performance in terms of selectivity, sensitivity, response time, and so on."

Chen points out that for signal moieties, plasmonic nanoparticles, especially gold or silver ones, are preferable to organic chromophores due to their superior properties. Plasmonic nanoparticle-based colorimetric detection has drawn increasing attention due to the unique optical properties provided by surface plasmon resonance.

"For the recognition moiety, most probes are usually prepared by functionalizing chemically synthesized ligands on various chromophores or nanostructures for their special coordination chemistry with targets," says Chen. "For instance, thymine in DNA strands can be used for mercury ion (Hg^{2+}) detection through a T–Hg^{2+}–T complex, where Hg^{2+} selectively bridges two thymine molecules in a stable manner. However, in order to avoid this complicated synthesis and develop environmentally-friendly green methods, we think utilizing natural products is one of the most straightforward, simple, and preferred strategies, which is also consistent with the idea of green health and practical economy."

Chen and his team developed a plasmonic colorimetric assay to detect mercuric ions based on urine. Compared to other gold-nanoparticle-based colorimetric systems, it showed excellent selectivity to Hg^{2+}, especially without interference from Pb^{2+}, and good sensitivity for Hg^{2+} detection in industrial wastewater.

Specifically, the scientists carried out the strategy of loading "green" candidates on plasmonic nanoparticles for colorimetric assays, which is based on the synergistic effect of small molecules on the particle surface for target recognition. As a proof of concept, they realized their idea by fabricating a urine–nanoparticle system by just mixing urine and gold nanoparticles together directly with the proper ratio for colorimetric determination of Hg^{2+}.

The assay hinges on the stark red-to-blue color change of the gold nanoparticle solution upon contact with Hg^{2+} and its good sensitivity, owing to the large surface-to-volume ratio and much higher absorption extinction coefficient as against traditional organic chromophoric probes.

"The synergetic effect of uric acid and creatinine, existing in urine and decorated on gold nanoparticles, is the reason for selectively binding Hg^{2+}, leading to the aggregation of gold nanoparticles and thereby causing a visual color change," explains Chen.

In addition, he notes, the novelty of such an assay lies in its simplicity, inexpensiveness, and portability—without ligand synthesis, just using urine directly—which is very useful in remote and less industrialized areas. Finally, this assay development may serve as a good example of how to exploit natural products in exploring nanotechnology.

"In my view, bio-inspired applications are some of the most possible and practical directions for plasmonic nanoparticle-based assays and are a kind of green chemistry that is very considerate of health, environmental, and sustainability issues," says Chen. "Learning from Nature never ends and is so intriguing because it offers many possible solutions to our everyday problems, for instance: (1) bio-inspired mechanisms, such as hydrogenase for hydrogen; (2) bio-interface-inspired multifunctional structures designed and fabricated for different applications inspired by lotus leaves, butterfly wings, water strider legs and so on; (3) using the active constituent in natural products directly, such as functionalizing enzymes on nanomaterials for catalyzing/sensing as well as synthesizing gold nanoparticles depending on polyphenols inside tea leaves."

Featured scientists: Prof. Xiaodong Chen
Organization: School of Materials Science & Engineering, Nanyang Technological University (Singapore)
Relevant publication: J. Du, B. Zhu and X. Chen, Urine for Plasmonic Nanoparticle-Based Colorimetric Detection of Mercury Ion, *Small*, 2013, 9(24), 4104, DOI: 10.1002/smll.201300593

6.2.6 Flesh-Eating Fungus Produces Cancer-Fighting Nanoparticles

Naturally occurring nanomaterials can be found everywhere in Nature—in soil, ground and surface waters, volcanic ash, ocean spray, mineral composites, smoke (fullerenes and graphene have even been discovered in space[16])—and only with recent advances in instrumentation and metrology equipment are researchers beginning to locate, isolate, characterize and classify the vast range of their structural and chemical varieties.

Biogenic magnetite nanoparticles have been discovered in various organisms, ranging from bacteria to human brains, with various biological functions.

"Most studies on naturally occurring organic nanoparticles have focused on higher organisms," says Mingjun Zhang, a professor of biomedical engineering at the Ohio State University (at the time of this research he was an associate professor at the University of Tennessee, Knoxville). "Given the earth's rich biological diversity, it is reasonable to hypothesize that naturally occurring nanoparticles, of various forms and functions, may be produced by a wide range of organisms from microbes to metazoans."

In his research, Zhang looks at Nature for inspirations for solutions to challenges in engineering and medicine, especially on a small scale, such as bioinspired nanomaterials, bioinspired energy-efficient propulsive systems, and bioinspired nanobio systems for interfacing with cellular systems. For instance, he has explored ways to replace metal-based nanoparticles in sunscreen with natural nanoparticles[17] and explored the use of nanoparticles secreted from ivy for adhesive applications.[18]

In their work, Zhang's team also turned their focus on *Arthrobotrys oligospora*, a representative flesh eater with a predatory life stage in the fungal kingdom. They discovered that nanoparticles produced by *A. oligospora* hold promise for stimulating the immune system and killing tumors.

"It is really exciting to use a natural microbe system to produce nanoparticles for potential cancer therapy," says Zhang. "Originally, we were trying to understand how the fungus secretes an adhesive trap that can capture, penetrate, and digest free-living nematodes in diverse environments. By doing that we almost accidentally discovered the nanoparticles produced."

Zhang's team investigated the fungal nanoparticles' potential as a stimulant for the immune system, and found through an *in vitro* study that the nanoparticles activate secretion of an immune-system stimulant within a white blood cell line. They also investigated the nanoparticles' potential as an antitumor agent by testing *in vitro* the toxicity to cells using two tumor cell lines, and discovered that these nanoparticles do indeed kill cancer cells.

The researchers used a new culture method (fungal sitting drop culture method) that allowed them to observe fungal growth, as well as both secreted and surface-bound nanomaterials using light microscopy, scanning electron microscopy (SEM), and atomic force microscopy (AFM), without any disturbance from agar components present in solid media. Using this culture method, they were able to observe secreted nanoparticles, and characterize them.

Zhang notes that this discovery is the first step towards the development of natural nanoparticle-based therapeutics for cancer treatment and demonstrates the importance of looking to Nature for innovation in disease treatment.

Featured scientists: Prof. Mingjun Zhang's bio-inspired nanoparticles and bio-inspired robotics lab (http://mjzhanglab.org.ohio-state.edu/index.html)
Organization: Ohio State University, Columbus, OH (USA)
Relevant publication: Y. Wang, L. Sun, S. Yi, Y. Huang, S. Lenaghan and M. Zhang, Naturally Occurring Nanoparticles from *Arthrobotrys oligospora* as a Potential Immunostimulatory and Antitumor Agent, *Adv. Funct. Mater.*, 2013, **23**(17), 2175, DOI: 10.1002/adfm.201202619

6.2.7 Upconverting Synthetic Leaf Takes Its Cues from Nature

A large part of low-energy photons, such as in the deep-red and infrared, are lost during conventional photovoltaic or photochemical processes. However, about half of all the solar energy reaching the Earth's surface can be found in these wavelengths.

Harvesting this light more efficiently is possible thanks to a process called *photon energy upconversion* where two or more weak captured photons are converted into a single strong, energy-rich photon.

So far, triplet–triplet annihilation photon energy upconversion (TTA-UC) is the only upconversion method that has been experimentally demonstrated to operate with noncoherent low intensity illumination such as sunlight.[19]

However, the solid TTA-UC materials reported so far have focused mainly on elastomeric matrices with low barrier properties because the TTA-UC efficiency generally drops significantly in glassy and semi-crystalline matrices.

"To fully exploit the TTA-UC process in different applications, the importance of the development of solid-state-like TTA-UC materials for better device integration cannot be neglected since these materials are easily processable," says Katharina Landfester, a professor at the Max Planck Institute for Polymer Research (MPI-P) in Mainz, Germany.

Landfester and her group have successfully synthesized a bio-inspired upconverting solid-state-like film using nanocellulose.

As Landfester explains, the TTA-UC process uses a dense populated sensitizer triplet ensemble, created by optical excitation, as an energy reservoir for further energy transfer. In oxygen-rich environments, a major part of this stored energy is lost during the quenching process stimulated by the presence of molecular oxygen.

Therefore, in order to keep a high upconversion efficiency, the molecular system for this process has to be protected against quenching. For their design, the MPI-P researchers followed a bio-inspired strategy by using the same protection material that Nature uses in the structural buildup of leaves: cellulose.

By using pristine cellulose nanofibers, the team achieved a "green" functional nanomaterial with ultrahigh barrier properties. Since the cellulosic nanofibers are sustainable, biodegradable, and derived from an inexhaustible plant source, these results could pave the way for the creation of a fully recyclable and inexpensive TTA-UC solid-state-like film.

The semi-crystalline cellulose used in this work consists of amorphous cellulose—ensuring mechanical flexibility and processability of the mew material, using conventional printing techniques—and crystalline nanorods—ensuring high oxygen barrier properties.

The researchers fabricated their synthetic leaves by covering a layer of nanocellulose-based capsules with a diameter of 1.2 ± 0.3 μm and a wall thickness of about 30 nm, containing chromophores in the liquid core, with a high oxygen barrier protective layer of cellulose nanofibers. By altering the thickness of this protective layer, the rate of gas transport to the capsules can be regulated.

"Using the same material for the polymer film, delivering distinct mechanical properties, and for the shell of the upconversion nanocontainer, drastically reduces the free volume of the new materials, thus substantially reducing the oxygen permeability," notes Landfester.

She cautions that the analogy with the protection strategy used by Nature is incomplete: "Besides low oxygen permeability—the 'passive protection'—photosynthetic plants use a complicated mechanism for chemically deactivating singlet oxygen (created during the quenching process) and various radicals—*i.e.* 'active protection'. Our next step is to combine low permeability with active scavenging of the created molecular oxygen."

Featured scientists: Prof. Katharina Landfester's physical chemistry of polymers group (http://www.mpip-mainz.mpg.de/physical_chem_of_polymers)
Organization: Max Planck Institute for Polymer Research, Mainz (Germany)
Relevant publication: A. Svagan, D. Busko, Y. Avlasevich, G. Glasser, S. Baluschev and K. Landfester, Photon Energy Upconverting Nanopaper: A Bioinspired Oxygen Protection Strategy, *ACS Nano*, 2014, 8(8), 8198, DOI: 10.1021/nn502496a

6.2.8 Replicating Nacre Through Nanomimetics

Materials that have to perform under extreme conditions—jet engines, power turbines, catalytic heat exchangers, military armors, aircraft, spacecraft—require exceptional mechanical properties.

The remarkable properties of some natural materials have motivated researchers to synthesize biomimetic nanocomposites that attempt to reproduce Nature's achievements and to understand the toughening and deformation mechanisms of naturally occurring nanocomposite materials. One of the best examples is nacre, the pearly internal layer of many mollusc shells. It has evolved through millions of years to a level of optimization currently achieved in very few engineered composites.

Researchers have approached the preparation of artificial analogs of nacre by using various methods and the resulting materials have captured some of the characteristics of the natural composite—but so far have never fully replicated it.

Nacre has a layered structure composed of approximately 95% calcium carbonate ($CaCO_3$) nanoscale building blocks interfaced by about 5% organics. The typical characteristic of nacre is the "bricks and mortar" arrangement of these materials, which is thought to be crucial to the mechanical and other outstanding properties that nacre possesses.

Over the past decades, researchers have devoted significant efforts to investigate and mimic these brick and mortar structures, hoping that the understanding on formation of nacreous structure and biological mineralization would lead to new advances in materials technology and related applications.

Work performed at Imperial College London, led by professors Alexander Bismarck and Milo Shaffer, represents the first successful attempt to mimic the structure of nacre while maintaining the same characteristic geometry, aspect ratio and phase proportions.

"Our main insight was to adjust the absolute length scale of the constituent platelets to an intermediate range," says Shaffer. "Previously, large platelets—the same dimensions as the natural system—have been used, but are difficult to assemble correctly since the packing is poor; too little inorganic reinforcement is included. On the other hand, very thin nanosheets are often used, but they are too similar to the dimensions of the polymer molecules, so that too much polymer is necessarily included. Our intermediate size retains the same shape and aspect ratio as the natural platelets, but scaled to about 20 times smaller size."

He points out that the smaller size makes it possible to form a well-packed, aligned structure with a large inorganic fraction, using self-assembly techniques. In addition to replicating many of the excellent properties of nacre, the smaller size appears to increase plasticity.

"Most of the research in the field of artificial nacre has been based on the use of nanosheets reinforcement with graphene oxide, LDH and nanoclay, limiting the proportion of the inorganic phase to only 50–70 wt%—which are the wrong phase proportions," explains Francois De Luca, a PhD student in Shaffer's NanoHAC group. "Our work shows that good mimics of nacre can be achieved by maintaining the same geometry and proportions, even at a different scale."

The team used 10–20-nm-thick, layered double hydroxide (LDH) platelets with an aspect ratio similar to the aragonite platelets in nacre and "glued" them together with a simple organic "mortar" (PSS - poly(sodium 4-styrenesulfonate)). This soft polyelectrolyte is around ten times thinner than the natural biopolymer in order to retain the correct dimensional ratios and phase proportions as natural nacre.

They then used the layer-by-layer (LbL) assembly method to deposit well-controlled alternating layers, allowing the alignment of anisotropic nanoplatelets by simple sequential dipping.

The researchers caution that scaling up their process, or at least speeding it up, to manufacture a large quantity of materials in a short period of time, still needs to be investigated.

"Currently, the deposition of our nanostructure is quite time consuming in terms of thickness (15 hours for a micrometer), which we believe could easily be improved," says De Luca. "However, the great advantage of the layer-by-layer assembly is that very large surface areas can be coated—our simple dipping procedure can easily be scaled up."

Beyond the demonstration of achieving nanomaterials with the combination of high strength and stiffness with large plastic deformation, biomimetic nanocomposites like these could be used as protective and/or energy absorbing coatings, potentially also as a barrier coating against gas.

The next step for the team will be to design this Nature-inspired nanostructure over multiple levels of hierarchy—just like natural nacre.

"Indeed, the platelets contained in nacre are actually nanocomposites—nanograins of aragonite glued together in a soft biopolymer," De Luca points out. "Therefore, porous nano-reinforcements could be used and filled with a polymer afterwards to achieve another level of hierarchy below the platelet level."

However, the nanoscale of the platelet might already be limiting this research direction. Also, the use of higher aspect ratio platelets could be investigated in order to further improve the mechanical properties of the nanocomposites.

Featured scientists: Prof. Alexander Bismarck's polymer and composite engineering (PaCE) group (http://www3.imperial.ac.uk/polymersandcompositesengineering); Prof. Milo Shaffer's NanoHAC group (http://www3.imperial.ac.uk/nanostructuresandcomposites)
Organization: Imperial College London (UK)
Relevant publication: F. De Luca, R. Menzel, J. Blaker, J. Birkbeck, A. Bismarck and M. Shaffer, Nacre-nanomimetics: Strong, Stiff, and Plastic, *ACS Appl. Mater. Interfaces*, 2015, 7(48), 26783, DOI: 10.1021/acsami.5b08838

6.3 DNA Nanotechnology

DNA, the fundamental building block of our genetic makeup, has become an intense nanotechnology research field. DNA molecules can serve as precisely controllable and programmable scaffolds for organizing functional nanomaterials in the design, fabrication, and characterization of nanometer-scale electronic devices and sensors. The reason why DNA could be useful for the design of nanoelectric circuits is the fact that it actually is the best nanowire in existence—it self-assembles, it self-replicates and it can adopt various states and conformations.

6.3.1 DNA-Templated Nanoantenna Captures and Emits Light One Photon at a Time

The emission of light by a single molecule is a cornerstone of nano-optics that will enable applications in quantum information processing or single-molecule spectroscopy. However, a key challenge in nano-optics is to bring light to and collect light from nano-scale systems.

In conventional electronics, the interconnect between locally stored and radiated signals, for example radio broadcasts or mobile phone transmissions, is formed by antennas. For an antenna to work at the wavelength of light, it is necessary to downscale the structure by the same factor as the wavelength or the frequency of the wave, *i.e.* roughly by a factor of 10 million. So, at the nanoscale, a simple antenna like a dipolar TV antenna would be about 100 nm in size and made of two polarizable elements (the metallic rods of a TV antenna) and a feed element (an emitter–receiver).

Once the nanofabrication issues are sorted out, nano-optical antennas could become ubiquitous in all applications based on light–matter interactions such as sensing, light emission (*e.g.* LEDs) and detection, as well as light harvesting, *i.e.* for solar cell applications.

Back in 2005, in an article in *Science*, it already was proposed to make optical antennas to amplify the interaction between light and matter.[20] Although optical cavities could be used for this, they only work at a very specific wavelength and can only interact with emitter–receivers at very low temperatures (at room temperature, typical emitters such as molecules and semiconductors are broadband). On the other hand, antennas can function for a broad range of frequencies.

In the above-mentioned *Science* paper, it was proposed to build a basic optical antenna by putting a quantum emitter—the optical equivalent of the antenna feed element—between two gold or silver nanoparticles that replace the metallic rods of the radio wave antenna.

"But to do this," says Sébastien Bidault, a CNRS research scientist at the Institut Langevin's Optical Antennas and Sensing group, "there is a very complex technical issue: the position of the feed element and the particles must be controlled at the 1 nm scale and a lot of theoretical papers have since discussed this. This task is far beyond what conventional top-down lithography techniques, used in the fabrication of transistors, can do."

To solve this problem, Bidault and his team developed new bottom-up techniques to make dimers of gold nanoparticles linked by a single DNA strand where the distance can be tuned at the nanometer scale. They already demonstrated this in a paper[21] in 2011 where they describe purified suspensions of symmetric or asymmetric dimers linked by one DNA strand.

Following up on this earlier work, Bidault's team together with CNRS scientists from the Fresnel Institute used a DNA template to introduce a single dye molecule in gold particle dimers that act as antennas for light.

Through self-assembly, they grafted the gold nanoparticles and a fluorescent organic dye onto short synthetic DNA strands, which are only 10–15 nm long. These nanostructures are the equivalent of a dipolar TV antenna (rabbit ear antenna) downscaled by a factor of 10 million. The fluorescent molecule acts as a quantum source, supplying the antenna with photons, while the gold nanoparticles amplify the interaction between the emitter and far-field light (Figure 6.4).

The researchers produced in parallel several billion copies of these pairs of particles (in a purified water suspension) by controlling the position of the fluorescent molecule with nanometer precision, thanks to the DNA backbone.

"The need to have only one linking DNA molecule is to make sure we will only add one fluorescent molecule to drive the antenna," Bidault explains. "We measured several hundred molecule-driven antennas—using single molecule fluorescence measurements—and looked at their fluorescence lifetime. Comparing these measurements to electromagnetic theory, we obtained a quantitative agreement and were able to show that the position of the molecule is indeed controlled at the 1 nm scale. Also, since the molecule is a quantum emitter and the dimer is a passive element, the overall antenna is also a quantum emitter—it only emits one photon at a time. Both results have never been obtained experimentally before."

"We chose to study the simplest antenna geometry (dipolar antenna) to fully characterize its optical response and compare it to classical electromagnetic theory," says Bidault. "Since our systems are quantum emitters, they could be useful for ultrafast quantum communication, thanks to the short fluorescence lifetimes. An even more interesting aspect of our antennas is that they are made with DNA and are obtained purified as several billion copies

Figure 6.4 Schematic representation of a nanoantenna formed of two gold nanoparticles linked by a DNA double strand and supplied by a single quantum emitter. (Image: Mickaël P. Busson, Brice Rolly, Brian Stout, Nicolas Bonod, Sébastien Bidault.)

in water. Since the antenna response strongly depends on the DNA linker, we can develop new types of biochemical sensors with these nanostructures."

He points out that there are still numerous issues with these systems: "In particular, our antennas have a lot of ohmic (heat) losses, so the antenna efficiencies require optimization, for instance by using silver instead of gold or rods instead of spheres for the particles. More importantly, the antennas we made are the simplest (dipolar) geometry."

Other groups have already described more complex antennas in which the position of the feed element was less controlled but that had other desirable properties: in particular directivity.[22] So, Bidault and his collaborators have their work cut out for them to make bright directional antennas for light.

Featured scientist: Dr Sébastien Bidault's optical antennas and sensing group (https://www.institut-langevin.espci.fr/optical_antennas_and_sensing) *Organization*: Institut-Langevin, Paris (France) *Relevant publication*: M. Busson, B. Rolly, B. Stout, N. Bonod and S. Bidault, Accelerated single photon emission from dye molecule-driven nanoantennas assembled on DNA, *Nat. Commun.*, 2012, 3, 962, DOI: 10.1038/ncomms1964

6.3.2 DNA Nanopyramids Detect and Combat Bacterial Infections

Today's nanomedicine applications are dominated by platforms of synthetically manufactured materials—mostly nanoparticles of various elements and compositions—which require complex chemical manipulation. In addition, there is the lingering matter of potential toxicity to healthy cells during therapeutic application, which could occur due to the synthetic nature of these nanomaterials.

One way to eliminate the toxicity issue is by using truly biocompatible natural carriers for sensing and drug delivery applications. The emerging field of DNA nanotechnology may provide a solution. In addition to acting as a scaffold to deliver therapeutic molecules, DNA nanostructures allow precise modification of the structure and presentation of these components.

"Not only is DNA biocompatible, it also offers simplicity in its manipulation, high controllability in the end product size and shape, and modification versatility," explains David T. Leong, an assistant professor at the National University of Singapore (NUS)'s Department of Chemical and Biomolecular Engineering. "However, to date, there are very few studies that use DNA nanostructures as drug carriers and most of these studies are in anti-cancer applications."

Motivated by the lack of an efficient theranostic platform for bacterial infection, the groups of David Leong and Jianping Xie developed a novel theranostic platform which is made by utilizing a self-assembled DNA nanopyramid (DP) as a scaffold for incorporation of both detection and therapeutic moieties to combat bacterial infection.

The researchers utilized DNA as their starting block to develop nanoparticles with theranostic capability, in which the nanoparticle possesses the capability to detect the presence of a bacterial infection and eradicate infectious bacteria, all in one package.

"Commonly known as a genetic information repository, DNA is inherently a highly regulated polymer which could be tuned to have a precise size and shape by changing the DNA sequence," says Leong. "It is literally shape/size changing genetic engineering."

In their study, the NUS team exploited this malleability of DNA to form self-assembled pyramidal DNA nanoparticles.

They then loaded an active antimicrobial agent, actinomycin D (AMD), by making use of the drug's innate properties to intercalate between DNA bases, as agents packed into the struts of the DNA pyramids. Based on the number of bases on each strut, the nanopyramids measure around 10 nm.

The scientists also coupled the DNA vertices with fluorescent gold nanoclusters, which allowed them to detect dead bacteria due to AMD that they delivered with this construct (DPAu/AMD).

"Our DNA pyramid serves to be a simple, controllable, and biocompatible theranostic platform which can be used to combat bacterial infections," Leong concludes.

He notes that this work offers new possibilities in the field of nanomedicine. As DNA is naturally available in every living creature, nanoparticles formed with DNA as the base material are truly biocompatible, which could eliminate any possible toxicology that plagues several other nanomedicine platforms based on inorganic materials and especially those containing heavy metals such as cadmium.

Moreover, by exploiting Watson–Crick base pairing, theoretically, nanoparticles of any shape and size could be realized. Leong points out that this level of control is unique amongst polymeric material.

"This control also allows us to design nanoparticles that are small enough (6–20 nm) to be taken up by bacterial cells, which are considerably smaller (approx. 1 μm) than mammalian cells, ensuring efficient drug delivery to the infectious bacteria in order to eradicate them."

"Lastly," says Leong, "DNA bases offer many chemistry connections, allowing drug, protein, ligand, antibody, short oligonucleotide sequences or other small functional compounds to be specifically added onto the DNA nanoparticle; our creativity is the limit here."

To assess the drug delivery ability of their actinomycin D loaded DNA nanopyramids, they chose two bacterial strains, *E. coli* and *S. aureus*, which represent Gram-negative and Gram-positive groups, respectively.

"Our results show that both the nanopyramids alone as well as the nanopyramids coupled with the gold clusters did not exert any killing effect on the tested bacterial strains," says Leong. "Instead using the DPAu/AMD structure brought about a dose dependent killing effect on both *E. coli* and *S. aureus* cells. In addition, our result indicates that AMD packaged in DPAu show a significant killing effect of the bacteria when compared to the free AMD treatment."

This method could be applied to deliver drugs to combat various infections and illnesses, as the DNA pyramid is versatile enough to carry various drugs, for example through intercalation (as shown in the case of actinomycin D here) and sequence-specific-bound drugs. The geometric shapes also allow embedding of diagnostic materials that could be used to detect dead cells or bacteria, or to sense biomarkers specific to certain diseases.

Leong and his group see DNA as an important base material for the next generation of nanomedicine systems with a high level of achievable control of the major parameters that dictate nanoparticulate performance: "DNA-based nanomedicine platforms will be efficient in terms of drug loading and delivery as well as with lesser side effects. In addition, the ability to modify more modalities on a nanomedicine platform allows higher targeting specificities and detection."

Featured scientists: Prof. David T. Leong (http://blog.nus.edu.sg/leonglab/); Prof Jianping Xie (http://cheed.nus.edu.sg/stf/chexiej/)
Organization: National University of Singapore
Relevant publication: M. Setyawati, R. Kutty, C. Tay, X. Yuan, J. Xie and D. Leong, Novel Theranostic DNA Nanoscaffolds for the Simultaneous Detection and Killing of *Escherichia coli* and *Staphylococcus aureus*, *ACS Appl. Mater. Interfaces*, 2014, **6**(24), 21822, DOI: 10.1021/am502591c

6.3.3   3D-Printed "Smart Glue" Leverages DNA Assembly at the Macroscale

Designing systems that build themselves is one of the great dreams of nanotechnology researchers and they are taking great strides towards developing such "bottom-up" nanotechnology fabrication techniques. All biological objects are created from the bottom up—an approach in which the order is imposed from within the object being made, so that it "grows" according to some built-in design—but so far, this approach hasn't played as significant a role yet in technology.

Molecular assembly is quite prominently featured in popular science fiction; remember the matter compiler in Neal Stephenson's *The Diamond Age* or the cornucopia machine in Charles Stross' *Singularity Sky*? By contrast, today's examples of bottom-up technologies are specific chemical processes that work well for a particular task, but don't easily generalize for constructing more complex structures.

Fabrication processes based on DNA might change this: DNA origami—tiny shapes and patterns self-assembled from DNA—have been heralded as a potential breakthrough for the creation of nanoscale devices.

DNA origami is a method for folding long strands of DNA into whatever shape or pattern is desired. Using a computer-aided design program, a scientist can plot the desired nanoscale shape and the computer designs a set of short DNA strands. These get mixed with long DNA strands, heated up to nearly boiling, and cooled to room temperature over the course of a couple of hours. In a single drop of water, one then has 100 billion copies of the desired shape. The first DNA origami made were shapes such as triangles and smiley faces, and patterns such as maps of the western hemisphere, snowflakes, *etc.*

The interesting thing about DNA origami is that it allows researchers to make such nanoscale shapes and patterns purely by self-assembly. Because random arrangements wouldn't be very useful, these patterns then need to be arranged into meaningful structures, *e.g.* to wire up electronic circuits or to fabricate synthetic membrane channels.

All these groundbreaking accomplishments have been strictly at the nanoscale, though, and scientists are starting to leverage the power of DNA self-assembly at a much larger scale.

Work performed at the Ellington lab at the University of Texas at Austin has demonstrated that researchers indeed can build a macroscopic object held together solely by DNA interactions.

"We have developed methods to assemble DNA-functionalized microparticles into a colloidal gel, and to extrude this gel with a 3D printer at centimeter size scales," says Peter Allen, formerly a researcher in the Ellington lab and now an assistant professor at the University of Idaho.

He explains that the new process produces materials with several unique properties:

"Firstly, unlike conventional 3D printed objects, the extruded semi-solids are assembled solely by DNA–DNA interactions that are strong enough to support the object at the macroscale;

secondly, these objects have internal, microscale properties that are programmed by the nanoscale DNA interactions; by controlling the assembly of materials from the molecular to the macroscale, one of the challenges for self-assembling materials has been realized;

thirdly, the size of these objects can be large—up to centimeters—and the cost for this material is reasonable as the bulk of its volume is an inexpensive polymer rather than expensive DNA;

and finally, this bulk material is assembled under conditions in which cells can survive and grow. This material can be 'seeded' with cells during extrusion and these cells will proliferate within the colloidal gel matrix."

He points out that the ability to apply molecular intelligence to a substrate that can make a macroscopic object is a step toward a truly rationally programmed material.

The team's colloidal gels are programmable at three distinct scales: (1) at the nanometer scale, specific DNA interactions mediate individual microparticle-to-microparticle interactions; (2) at the micrometer scale, microparticle clusters form DNA dependent substructures. This microscale topology can be further controlled by using different sizes and stoichiometries of oligonucleotide-derivatized microparticles; (3) finally, the shape of the object can be patterned at the centimeter scale by 3D printing, and its material properties, such as porosity, can be altered through control over the composition of the colloidal mixtures.

Allen notes that these 3D printed, programmed, self-assembled materials present many opportunities for synthetic biology: "We assembled a colloidal gel from hydrogel particles. That was useful for microscopy as the hydrogel particles are very transparent. Hydrogel particles can be made with additional useful properties. I've started my new lab at the University of Idaho with this purpose: generate new types of functional particles that can be integrated into the self-assembled material."

For example, hydrogel particles can be created that release biological growth factors and that sense biological signals. The DNA-based programming of these materials can be expanded into both their assembly and their interactions with growing, developing tissue.

The key challenge is to retain and expand the "programmability" of these DNA-bearing microparticles while simultaneously increasing the physical strength of the assembled material.

Allen cites another example: "Cells in bio-printed tissue need to have a very specific microenvironment. Generating this optimal microenvironment in a manner that scales to a usable volume is a very exciting challenge. A technology such as DNA adhesives might eventually allow this microenvironment to be self-assembled – rather than serially sculpted into some other material. The control of DNA will allow for specific structures and the cheap hydrogel will keep the overall cost low."

"As we get better at programming DNA, our 'circuits' may someday mimic true biological development," he concludes.

Featured scientists: Prof. Peter B. Allen (http://www.uidaho.edu/sci/chem/faculty/peter-b-allen); the Ellington lab (http://ellingtonlab.org/)
Organization: The University of Texas at Austin; University of Idaho, Moscow, ID (USA)
Relevant publication: P. Allen, Z. Khaing, C. Schmidt and A. Ellington, 3D Printing with Nucleic Acid Adhesives, *ACS Biomater. Sci. Eng.*, 2015, **1**(1), 19, DOI: 10.1021/ab500026f

6.3.4   DNA Origami Nanorobot with a Switchable Flap

One of the most important developments in structural DNA nanotechnology—where DNA molecules are assembled into organized and programmable structures—has been the use of a scaffold strand and hundreds of short staple strands for the assembly of three-dimensional DNA origami objects. These objects are similar in size to virus capsids and have gained a lot of interest as nanocontainers for drug delivery. Since they can be programed to carry out specific tasks (such as transporting molecules and releasing them at a target site), these assemblies act like a robotic system—a nanorobot.

Not to be confused with the nanorobots of science fiction, for medical nanotechnology researchers a nanorobot, or nanobot, is a popular term for molecules with a unique property that enables them to be programmed to carry out a specific task.

In what is the smallest 3D DNA origami box so far, researchers in Italy have now fabricated a nanorobot with a switchable flap that, when instructed with a freely defined molecular message, can perform a specifically programmed duty.

Slightly larger nanocontainers with a controllable lid have already been demonstrated by others to be suitable for the delivery of drugs or molecular signals, but this new cylindrical nanobot has an innovative opening mechanism.

"While in previously developed objects the lids had a lock that can be removed, our container is equipped with a flap moved by an actuator that physically pulls it in the open position," notes Giuseppe Firrao, a professor in the Department of Agricultural and Environmental Sciences at the University of Udine. "It is like opening a window on a windy day: you can just unlock it and allow the wind to do the job or you can use your arm to pull the window, keep it firmly open, and eventually close it (Figure 6.5)."

The reversibility of this actuation mechanism has been demonstrated for another nano object in previous research.[23] Here, the team designed, constructed, and imaged a 100-nm-sized round DNA origami actuator capable of reversible motion in response to an external chemical stimulus.

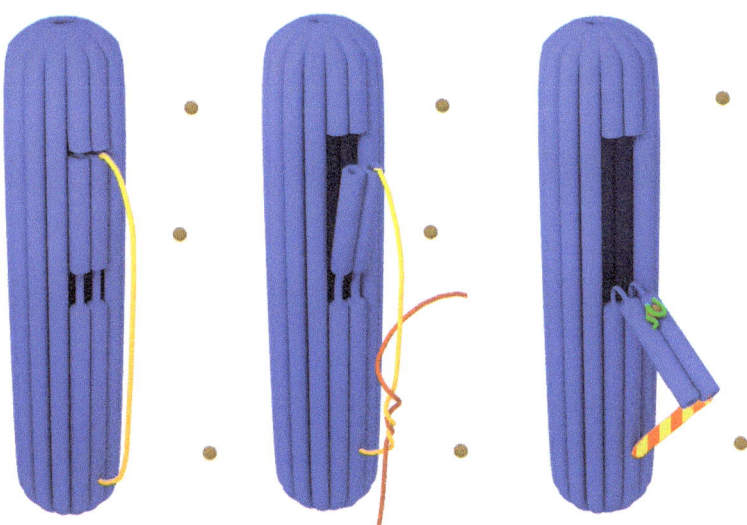

Figure 6.5   Schematic model of the 3D DNA origami nanorobot with a switchable flap. Closed 3D DNA origami (blue) with the unreacted probe (yellow) (left). The 3D DNA origami during the hybridization probe (yellow)/target (red) (middle) and an actuated 3D DNA origami armed with hemin/G-quadruplex DNAzyme complex (right); the probe (yellow) hybridized to target (red), the flap opened (blue) and a DNAzyme (green) pulled out of the cylinder and exposed to hemin (brown). (Image: Tomaso Firrao.)

For their new work, the team used the flap mechanism developed in the previous work and reduced it in size to be included in a 3D origami of cylindrical shape. The resulting nanorobot had estimated dimensions of 14 nm × 14 nm × 48 nm. The volume of the inside cavity (8 nm × 8 nm × 44 nm) was estimated to be 3 zeptoliters (10^{-21} liters, or 1 sextillionth of a liter). The flap dimensions were 9 nm × 5 nm.

"As far as our nanorobot is concerned, the lumen is large enough to accommodate a single-stranded nucleic acid, while the nano-object is small enough to be entirely contained within the capsid of viruses," says Firrao. "Our work therefore opens perspectives for the construction of nanorobots that can be hosted within cellular delivery vectors and can host, concurrently, selectively accessible molecular payloads in their internal cavity."

The next steps in the team's research will be to monitor the flap opening with plasmonics and to operate the nanorobot *in vivo*.

"By 'operate' we mean to control with external (*e.g.* light) or internal events the reversible movement of the flap, in order to hide or display a DNAzyme, an enzyme, or a functional nucleic acid," explains Firrao. "This will be done in the perspective of manipulating cell responses to environmental stimuli, including disease."

To gain precise control of opening and closing, nanocontainers will be essential in the development of robust systems for the delivery of molecular signals into living organisms. This work takes another step in this direction.

Featured scientist: Prof. Giuseppe Firrao (http://people.uniud.it/page/giuseppe.firrao)
Organization: Department of Agricultural and Environmental Sciences, University of Udine (Italy)
Relevant publication: E. Torelli, M. Marini, S. Palmano, L. Piantanida, C. Polano and A. Scarpellini, *et al.*, A DNA Origami Nanorobot Controlled by Nucleic Acid Hybridization, *Small*, 2014, **10**(14), 2918, DOI: 10.1002/smll.201400245

6.3.5 Fuzzy and Boolean Logic Gates Based on DNA Nanotechnology

Logic gates, especially Boolean logic gates, are the fundamental building blocks of today's computers built with silicon-based circuitry; without them, digital information could not be processed. In a way, organisms are—living—computers, too. In that regard, the human body so far is the ultimate "wet computer"—a highly efficient, biomolecule-based information processor that relies on chemical, optical and electrical signals to operate.

Instead of logic gates made with electronic transistors, the biochemical circuits found in Nature are operated with molecules such as DNA: rather than electrons flowing in and out of transistors, DNA-based logic gates receive and produce molecules as signals.

Research related to molecular logic gates is a rapidly growing and very active field and molecular devices have become a new frontier in computing.

Researchers in Denmark have designed and synthesized self-assembled DNA complexes that sense two environmental signals and produce a fluorescent output corresponding to the operation of all six Boolean logic gates AND, NAND, OR, NOR, XOR, and XNOR.

This study could help improvements in the fields of molecular computation and intelligent drug delivery.

"The logic gate senses two different oligonucleotides as inputs and produces fluorescence resonance energy transfer (FRET) as an output signal," says Reza M. Zadegan, a the time of this work a post-doctoral researcher in the Department of Molecular Biology and Genetics at Aarhus University (now a researcher in the Nanoscale Materials and Device Group at Boise State University). "As a proof-of-principle, we demonstrated the implementation of the NOR and fuzzy logic gates in our previously reported 3D DNA origami box."

He points out that integrating logic gates with DNA origami systems opens a vast avenue to many applications in the fields of nanomedicine, biocomputing and bioelectronics.

This is one of very few works that reports production of all Boolean logic gates from a single construct.

Moreover, the team showed that the designed gates can be incorporated into larger DNA structures by implementing the logic gate in their previously reported 3D DNA origami box.[24]

Another important aspect of the work is a fuzzy logic gate system with high fidelity and reliability based on DNA.

"In addition to the molecular logic calculations, our DNA logic gates have been designed to operate with micro RNA (miRNA) as input and can thus act as biosensors for specific cancer type detection," notes Zadegan. "Combination of these cancer detection abilities and incorporation of anti-cancer drugs is possible *via* our DNA origami box, where we can imagine the origami boxes that contain a therapeutic agent—antibody, toxin, enzyme or nucleic acid—caged in."

In such a scenario, the DNA origami box would enter cells and only respond to the cancer cells that are programmed into the biosensor and subsequently deliver and release the drug to kill the tumor cells. This "smart" targeted process would only respond to a cancerous cell and thus avoid the undesired side effects of the anti-cancer drug on healthy cells.

In addition to the applications mentioned previously, fuzzy logic systems like this could help improve our understanding of both biophysical systems and natural controlling mechanisms in organisms.

As Zadegan explains, one can adapt such a system to mimic many of the natural mechanisms *in vitro*, for instance the binding of multiple miRNAs to regions of messenger RNA (mRNA) to down-regulate protein expression in a joined manner. "Our fuzzy logic gate setup in the 3D origami box was designed to respond to miR-21, miR-30c, miR-191, and miR-223. All these four miRNAs are up-regulated in colon, pancreas, or prostate cancers."

Featured scientist: Dr Reza M. Zadegan (http://nano.boisestate.edu/about-us/staff/reza-m-zadegan/)
Organization: Department of Molecular Biology and Genetics, Aarhus University (Denmark)
Relevant publication: R. Zadegan, M. Jepsen, L. Hildebrandt, V. Birkedal and J. Kjems, Construction of a Fuzzy and Boolean Logic Gates Based on DNA, *Small*, 2015, **11**(15), 1811, DOI: 10.1002/smll.201402755

6.4   Sensors for Everything, Everywhere

Nanosensors go far beyond just the effort to miniaturize sensors by reducing the size of the sensing part and/or the transducer—they take advantage of the physical or chemical phenomena that matter exhibits at the nanoscale. Therefore, a nanosensor can have—often vastly improved—performance characteristics compared to a classically modeled sensor merely reduced in size.

6.4.1   Cheap Paper-Based Gas Sensors

Great efforts have been devoted to exploiting semiconductor-based sensors, ranging from one-dimensional wires to two-dimensional arrays or films. Since monitoring needs for environmental, security, and medical purposes are growing fast, the demand for sensors that are low cost, low power consuming, have high sensitivity, and selective detection is increasing as well.

Because it is easily available, is low-cost, insulating, flexible, and portable, paper has been recognized as a particularly attractive class of supporting matrix for accommodating sensing materials. However, methods for depositing semiconductor materials with high stability and continuity on papers are still sparse.

A team of Chinese researchers has developed low-cost gas sensors by trapping single-walled carbon nanotubes (SWCNTs) in paper and demonstrated their effectiveness by testing it on ammonia.

"We developed a suction filtration method—which is commonly used in chemistry labs—to successfully deposit SWCNTs on paper," says Yapei Wang, a professor of chemistry at Renmin University of China. "Deposition of SWCNTs is just a start, and we will try other semiconductor materials as well. As the air pollution in our cities is becoming more serious, we are encouraged even more to exploit cheap sensors that can be extensively used."

By using a common suction filtration method, water-dispersible SWCNTs were sucked into porous filter paper to be assembled there. Wang notes that the porous paper was fully filled with SWCNTs, which were well distributed through the paper, affording robust sensing arrays with the same thickness as the paper.

The team then fabricated sensors with the SWCNT-impregnated paper by cutting it into small pieces and depositing gold electrodes on it.

"Detection sensitivity routinely relies on how fast the gas comes into contact with the SWCNTs and how much the SWCNT responds to the adsorbed gas molecules," explains Wang. "The former is a physical diffusion process which is affected by the size of the sensing area and surface chemistry of SWCNTs. A larger sensing area can allow more gas molecules to contact the SWCNTs. Our tests indicated that thinner arrays afforded a stronger response to ammonia gas at a concentration of 20 ppm than did the thicker arrays."

He notes that a similar rule also applies to sensing arrays with the same feature width but different line numbers—an increase of array number depressed the sensitivity of the sensor.

The paper sensors developed in this work can be integrated into a common electric circuit, performing a function similar to traditional analytical equipment that is huge and difficult to carry. Wang also stresses the environmental impact of paper chips: cellulose, the raw material for paper, is a renewable product. And by using paper instead of plastic chips, there will be less environmental impact from non-biodegradable waste.

Estimating the cost of their paper-based gas-sensing chips, including all material, equipment usage, energy and labor cost, the team estimates a total price of below $1 per sensor chip.

"We believe that our approach offers a generally applicable approach for integrating active nanomaterials into paper matrices, opening opportunities in flexible sensors and optoelectronics strategies," says Wang.

While this sensor works with ammonia, detection of other gases broadly found in polluted air is also of vital importance. This may be accomplished by stepwise implantation of multiple detectors in a paper chip.

"However," says Wang, "the challenge of synchronous gas detection and elimination still remains. It is possible for paper chips to act as an antenna to recognize the target gas, deliver the information, and then catch, even eliminate, the toxic gas. We hope this work will open the door to new, exciting applications of paper chips."

Featured scientist: Prof. Yapei Wang (http://chem.ruc.edu.cn/shizililiang/ShowArticle.asp?ArticleID=126)
Organization: Department of Chemistry, Renmin University of China, Beijing (PR China)
Relevant publication: J. Wang, X. Zhang, X. Huang, S. Wang, Q. Qian, W. Du and Y. Wang, Forced Assembly of Water-Dispersible Carbon Nanotubes Trapped in Paper for Cheap Gas Sensors, *Small*, 2013, **9**, 3759, DOI: 10.1002/smll.201300655

6.4.2 Plasmonic Smart Dust to Probe Chemical Reactions

Conventional probing methods for localized surface properties often rely on ultra-high vacuum conditions. Consequently, with approaches such as scanning tunneling microscopy, it is difficult to resolve surface changes under realistic reaction conditions. Tip-enhanced Raman spectroscopy (TERS) can investigate arbitrary substrates and more diverse reaction environments but suffers from weak Raman scattering signals. Also, the fabrication of robust, reproducible, and highly enhancing tips is still challenging.

A novel platform for the optical detection of localized chemical reactions on surfaces can help overcome these difficulties by offering a sensitive, reliable, and easy-to-implement technique to probe local chemical reactions while they occur under diverse environmental conditions.

Dubbed *plasmonic smart dust*, this all-optical platform consists of silica shell-isolated gold nanoparticles which can convert incident light into highly localized subdiffraction-limit hot-spots of the electromagnetic field and can thus optically report subtle environmental changes at their pinning sites on the probed surface.

"Our smart dust nanoparticles can be dispersed on many reactive and catalytically active surfaces to probe the local reaction kinetics in real-time and under realistic reaction conditions," says Harald Giessen, a professor at the 4th Physics Institute at the University of Stuttgart. "In our experiment, we apply this approach to the investigation of hydrogen dissociation and subsequent uptake in palladium thin-films. Our platform can resolve subtle local environmental changes, both in the dielectric properties of the palladium film during hydrogen incorporation, as well as in its shape and morphology (Figure 6.6)."

Related work on plasmonic hydrogen sensing has been done, for example, using a nanofabricated plasmonic gold antenna to enhance the changes in a neighboring palladium nanoparticle. Also, arrays of palladium disks or single gold/palladium core–shell nanoparticles have been studied.

"However, these approaches do not allow the localized detection of reaction kinetics on extended reactive or catalytic surfaces," notes Andreas Tittl, a PhD student in Giessen's group and first author of a paper on this work. "The key insight that led to the development of our sensor platform is that plasmonic smart dust particles can work as sensitive local reaction reporters, even when moving from ensemble to single-particle measurements."

The combination of straightforward sensor chip fabrication *via* drop-coating and the strongly localized response of the smart dust nanoparticles leads to a versatile, label-free, real-time, and high-resolution platform for probing local reaction kinetics.

To test the ability of their plasmonic smart dust for resolving different hydrogen contents in palladium, the researchers tracked hydrogen storage on palladium surfaces using the dust. They repeatedly exposed the sample to hydrogen concentrations ranging from 0.5% to 3% and found that different hydrogen concentrations can be clearly identified as intensity changes and spectral shifts of the plasmonic resonance with response times on the order of seconds.

"The observed fast reaction kinetics compared to bulk palladium systems are due to the shorter diffusion lengths associated with palladium thin films

Figure 6.6 Plasmonic probing of local chemical reactions. Smart dust (silica shell-isolated gold nanoparticles) is spread on a thin palladium (Pd) film to locally probe the dissociation and subsequent uptake trajectory of hydrogen in Pd. (Image: Andreas Tittl, 4th Physics Institute, University of Stuttgart.)

and nanomaterials," says Tittl. "In general, higher hydrogen concentrations lead to increased resonance intensity and spectral blueshift of the resonance position."

"One of the most interesting potential applications of our work is the possibility to produce a local reaction map by pinning multiple smart dust nanoparticles at reaction sites of interest and optically monitoring spatially distinct local chemical reactions simultaneously," notes Na Liu, group leader at the Max Planck Institute for Intelligent Systems in Stuttgart. "Especially, this approach can shed light on the influence of localized features such as cracks, facets, defects, and boundaries on the performance of reactive or catalytic materials."

Furthermore, this method can be extended to investigate a plethora of chemical reactions on surfaces, ranging from the reduction and oxidation steps in fuel cells to catalytic water splitting.

The team hopes that these exciting results will pave the way towards the development of inexpensive and high-output reaction sensors for real-world applications.

Featured scientists: Prof. Harald Giessen's research group (http://www.pi4.
uni-stuttgart.de/home): Dr Laura Na Niu's research group (http://www.
is.mpg.de/liu/)
Organizations: 4th Physics Institute, University of Stuttgart; Max Planck
Institute for Intelligent Systems, Stuttgart (Germany)
Relevant publication: A. Tittl, X. Yin, H. Giessen, X. Tian, Z. Tian and C.
Kremers, *et al.*, Plasmonic Smart Dust for Probing Local Chemical Reac-
tions, *Nano Lett.*, 2013, **13**(4), 1816, DOI: 10.1021/nl4005089

6.4.3   A Human-Like Nanobioelectronic Tongue

The concept of e-noses—electronic devices that mimic the olfactory systems
of mammals and insects—is very intriguing to researchers involved in building better, cheaper and smaller sensor devices. Less well known is the fact
that equivalent artificial sensors for taste—electronic tongues—are capable
of recognizing dissolved substances.

Conventional electronic tongues (e-tongues) utilize pattern recognition
for analysis using arrays of synthetic materials such as polymers, artificial
membranes and semiconductors, for applications in the food and beverage
industries.

"Even with current technological advances, e-tongue approaches still
cannot mimic the biological features of the human tongue with regard to
identifying elusive analytes in complex mixtures, such as food and beverage
products," explains Tai Hyun Park, a professor in the School of Chemical and
Biological Engineering at Seoul National University.

Park, together with Professor Jyongsik Jang and their collaborators, have
developed a human bitter-taste receptor as a nanobioelectronic tongue.

The nanobioelectronic tongue uses a human taste receptor as a recognition element and a conducting polymer nanotube field effect transistor
(FET) sensor as a sensor platform. Specifically, the Korean team functionalized carboxylated polypyrrole nanotubes with the human bitter taste
receptor protein hTAS2R38. They say that the fabricated device could detect
target bitter tastants with a detection limit of 1 femtomole and high selectivity (Figure 6.7).

In previous work, Park's research group had already developed bioelectronic noses using human olfactory receptors. Since olfactory receptors and
taste receptors have a similar structure, they were able to apply a similar
approach for the development of a nanobioelectronic tongue using human
taste receptor protein.

The result is a nanobioelectronic tongue that has high sensitivity and
selectivity with human tongue-like properties.

"In the case of bitter taste, our e-tongue can be used for sensing quantitatively the bitter taste of, for example, coffee, chocolate drinks, drugs and

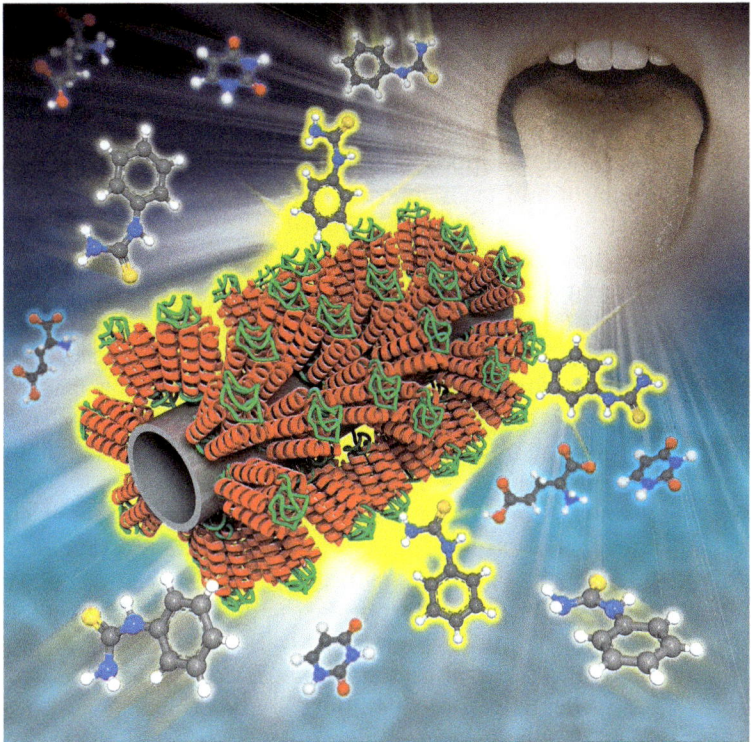

Figure 6.7 A human tongue-like nanobioelectronic tongue. Illustration of the hTAS2R38-functionalized carboxylated polypyrrole nanotube. (Image: Prof. Tai Hyun Park, Seoul National University.)

oriental medicines," says Park. "Our nanobioelectronic tongue can be used as an alternative to time-consuming and labor-intensive sensory evaluations and cell-based assays for the assessment of quality, tastant screening and basic research on the human taste system."

There are five basic types of tastes including bitter, sweet, salty, sour and umami. In order to mimic the real human tongue more closely, Park notes that the sensor should be a multiplexed array type containing all kinds of taste receptors. This will be the subject of future research.

Featured scientist: Prof. Tai Hyun Park (http://biotech.snu.ac.kr/index2.asp)
Organization: School of Chemical and Biological Engineering, Seoul National University, Seoul (Republic of Korea)
Relevant publication: H. Song, O. Kwon, S. Lee, S. Park, U. Kim and J. Jang, *et al.*, Human Taste Receptor-Functionalized Field Effect Transistor as a Human-Like Nanobioelectronic Tongue, *Nano Lett.*, 2013, **13**(1), 172, DOI: 10.1021/nl3038147

6.4.4   Electronic Sensing with Your Fingertips

Advances in materials, fabrication strategies and device designs for flexible and stretchable electronics and sensors make it possible to envision a not-too-distant future where ultra-thin, flexible circuits based on inorganic semiconductors can be wrapped and attached to any imaginable surface, including body parts and even internal organs. Robotic technologies will also benefit as it becomes possible to fabricate "electronic skin" that, for instance, could allow surgical robots to interact, in a soft contacting mode, with their surroundings through touch.

Professor John A. Rogers and his research group at the University of Illinois at Urbana-Champaign have demonstrated that they can integrate high-quality silicon and other semiconductor devices on thin, stretchable sheets, to make systems that not only match the mechanics of the epidermis, but also take on the full three-dimensional shapes of the fingertip—and, by extension, other appendages or even internal organs, such as the heart (Figure 6.8).

In previous work,[25] Rogers and his team demonstrated thin, stretchable "epidermal" sheets of electronics, suitable for integration on regions of the skin that are relatively flat, such as the forearm.

"We became interested in figuring out ways to do similar classes of devices, but with full 3D shapes, matched precisely to the body," says Rogers. "The fingertip was a good starting point to demonstrate the ideas."

Instrumented gloves have been demonstrated in the past, although not with fully functional, flexible silicon electronics. Also, the glove substrates in these cases are thick leather or cloth, very different to the ultra-thin, stretchable silicone membranes that Rogers' team uses.

To fabricate their "fingertip electronics", the researchers first formed an array of interconnected sensors and electronics on a silicon wafer in an open mesh geometry. They then lifted this device structure onto the surface of a PDMS (polydimethylsiloxane) slab *via* transfer printing. The backside of the mesh and the supporting PDMS stamp are coated with a thin layer of silicon dioxide and then pressed onto a thin elastomeric sheet. Removal of the PDMS completes the transfer.

By choosing an elastomeric closed-tube structure, specially formed to match the shapes of fingertips, the team demonstrated their technique with a set of multifunctional fingertip devices that include electrotactile electrode arrays multiplexed with silicon nanomembrane diodes, strain sensors based on silicon nanomembrane gauges, and tactile sensor arrays that use capacitors with low-modulus, elastomeric dielectrics.

"A key step in the manufacturing is a process by which we 'flip inside out' the finger-shaped membrane, to move the electronics from the outer surface, where we do the integration, to the inner surface where it can contact the skin," explains Rogers.

Apart from the demonstrated sensor arrays, these fingertip devices could also be fitted with sensors for measuring motion and temperature, with small-scale heaters as actuators for ablation and other related operations.

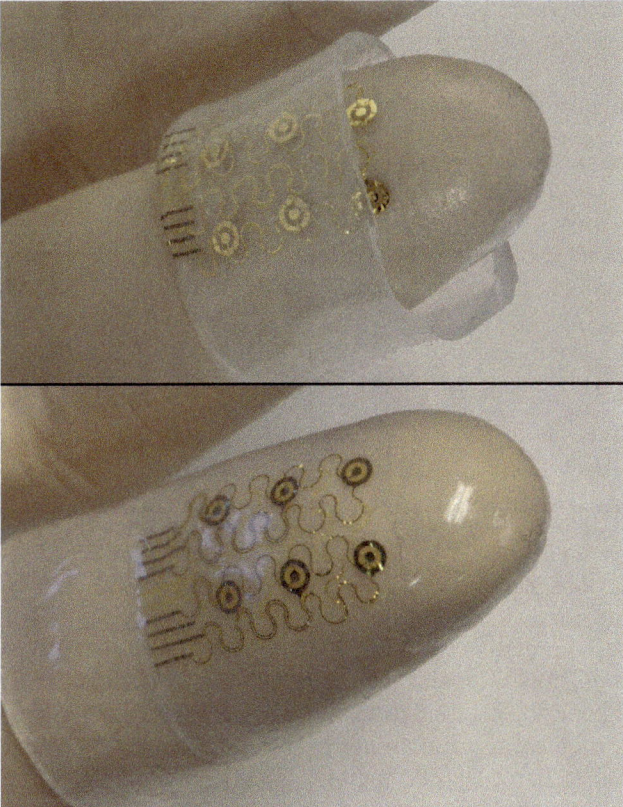

Figure 6.8   Top: turning the tube inside out relocates the array on the inner surface of the finger-tube, shown here at the midway point of this flipping process. Bottom: multiplexed array of electrotactile stimulators in a stretchable, mesh geometry on the inner surface of an elastomeric finger-tube. (Images: Rogers Group, University of Illinois at Urbana-Champaign.)

Overall, this fabrication technique solves the problem of integrating a hard, brittle material such as silicon, with a soft, stretchable membrane of silicone, in a way that affords overall stretchable mechanics but avoids fracture in the silicon.

"We use nanomembranes of silicon, structured into narrow 'serpentine' ribbons, such that buckling physics can accommodate in-plane strains without reaching fracture thresholds in the silicon," says Rogers.

The team believes that these devices can, in a general sense, increase the sensitivity of existing forms of perception through the skin of the fingertip, and also bring new types of sensing function. According to Rogers, an initial application area might be in advanced surgical gloves that improve the sense of touch, and also allow for electrophysiologial measurements, blood pressure assessments, and even potentially therapeutic function (*e.g.* radiofrequency ablation to eliminate aberrant tissues).

The researchers are already working on 3D integration of similar classes of technologies, but with expanded advanced functionality, to organs such as the heart, where the device can measure electrical activity and also stimulate as an advanced class of pacemaker with feedback control.

Featured scientists: Prof. John A. Rogers' research group (http://rogers. matse.illinois.edu/)
Organization: The Frederick Seitz Materials Research Laboratory, University of Illinois at Urbana-Champaign, Urbana, IL (USA)
Relevant publication: M. Ying, A. Bonifas, N. Lu, Y. Su, R. Li and H. Cheng, *et al.*, Silicon nanomembranes for fingertip electronics, *Nanotechnology*, 2012, **23**(34), 344004, DOI: 10.1088/0957-4484/23/34/344004

6.4.5   Electronic Skin Takes Your Temperature

As we have seen in the work discussed in the previous section, conventional, silicon-wafer-based fabrication techniques can be modified to apply electronics conformally to the heterogeneous topography of the skin.

Additionally, previous work in medical thermal imaging has shown the utility of precision temperature mapping of human skin in the study of pathological conditions that affect how the body regulates temperature.

However, medical thermal imaging suffers from the high cost of infrared cameras (more than $100 000 for high-precision models), and the cameras struggle to measure the curved or moving surfaces of living tissues.

An international multidisciplinary team including researchers at the University of Illinois at Urbana-Champaign and the National Institute of Biomedical Imaging and Bioengineering (NIBIB) has extended ideas from these previous works to enable continuous mapping of the temperature of curved, moving skin—or other tissue—to a precision better than 0.02 °C, that can be manufactured for pennies.

The team demonstrated the development of a device platform that enables high precision temperature mapping of the skin in ways that have, until now, been extremely difficult in research and impossible to implement for widespread use (Figure 6.9).

"Our findings extend previously demonstrated ideas in ultra-thin skin-mounted electronics to real, medically relevant applications that benefit from high precision thermal mapping and heating," says Professor John A. Rogers from the University of Illinois at Urbana-Champaign. "Our advancement of the field is two-fold: (1) the fabrication schemes and thermal and mechanical modeling lay a groundwork of information that will be relevant to the future development of stretchable and/or bio-integrated electronics where temperature or heat generation is a concern, and (2) the devices we have demonstrated enable a new avenue in medicine for investigating the minute temperature signals, as well as responses to heat, that relate to human physiology."

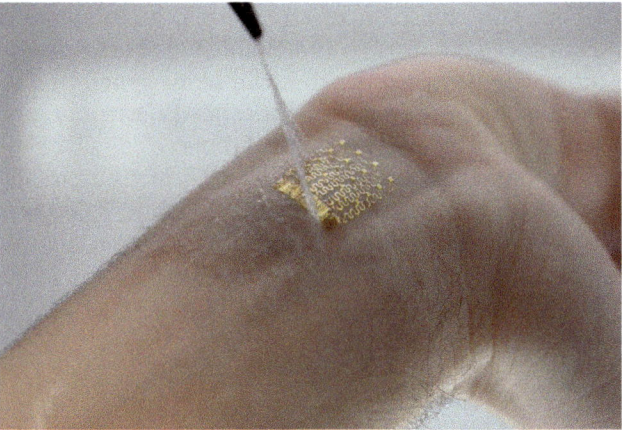

Figure 6.9   Image of a 4 × 4 sensor array after application to the skin using a water-soluble adhesive tape based on poly(vinyl alcohol). (Image: Rogers Research Group, University of Illinois at Urbana-Champaign.)

The temperature sensor array is a variation of a novel technology, originally developed in Rogers' lab, called *epidermal electronics*, consisting of ultra-thin, flexible skin-like arrays, which resemble a tattoo of a micro-circuit board.

The devices incorporate microscale temperature sensors that can simultaneously act as micro-heaters (actuators) in arrayed layouts on thin, low-modulus elastic sheets. The sensors/actuators rely on either thin serpentine features of 50-nm-thick gold layers or PIN diodes formed by patterned doping of 320-nm-thick silicon nanomembranes.

In their experiments, the researchers tested their "diagnostic skin" against measurements by an infrared camera, the gold standard in measuring localized skin temperature. In all cases, both methods delivered virtually identical results.

Rogers notes that the most direct applications of this work will be in future research on continuous, precise temperature mapping and localized heating of human skin. Examples include continuous monitoring of healing wounds, new studies in minute temperature oscillations due to various diseases, and thermally activated delivery of drugs through tissue.

"These represent a few areas of interest for us, but we hope that the technology leads to more new medical research and applications than we could hope to accomplish on our own," he says.

Currently, the biggest challenge for this form of ultra-thin skin-mounted electronics is to make them completely wireless. This requires both wireless data transmission as well as wireless power generation in the device. Various ideas have been developed to solve these problems at a basic level, but nothing yet has provided a solution that would enable the precise mapping capabilities shown here in a format that conforms to the skin and is completely wireless. This is an active area of research for Rogers' group, and others, which they hope to improve on continuously in the coming years.

Future directions in this field of stretchable, biointegrated electronics include more advanced circuitry in stretchable formats to enable more complex health monitoring, electronics that can be designed to resorb into the body after a certain amount of time, and advanced electronic surgical devices that are not limited to rigid plastic or metal tools.

Rogers points out that converting silicon-wafer-based electronics to stretchable, biocompatible formats is challenging and requires a very different approach to almost every level of device design than is typical in electronics. "However, the problems are solvable and we expect a continuous road of development moving forward."

Featured scientists: Prof. John A. Rogers' research group (http://rogers. matse.illinois.edu/)
Organization: The Frederick Seitz Materials Research Laboratory, University of Illinois at Urbana-Champaign, Urbana, IL (USA)
Relevant publication: R. Webb, A. Bonifas, A. Behnaz, Y. Zhang, K. Yu and H. Cheng, *et al.*, Ultrathin conformal devices for precise and continuous thermal characterization of human skin, *Nat. Mater.*, 2013, **12**(11), 1078, DOI: 10.1038/nmat3779

6.4.6   Nanocurve-Based Sensor Reads Facial Expressions

One of the challenges of fabricating flexible electronics has been the trade-off between a material's high flexibility and adaptability, and its conductivity. Exploring feasible methods for guiding conducting or semiconducting nanomaterials into elastomeric matrices will be key to further progress in this area.

A promising approach has been developed by scientists in China, who have come up with a facile printing strategy to assemble silver nanoparticles into micro- and nano-curve structures *via* a pillar-patterned silicon template.

The curves with various tortuosity morphologies have differential resistive strain sensitivity, which can be integrated into a multi-analysis flexible sensor to perform complex recognition of human facial expressions.

"The realization of multi-recognition facial expressions or eyeball movements based on integration analysis of silver nanoparticle-based curved array sensors is a real breakthrough for the next generation of wearable sensors, which has already been indicated in a recent review paper,"[26] says Yanlin Song, a professor at the Institute of Chemistry, Chinese Academy of Sciences, and Director of Key Laboratory of Green Printing.

Avoiding traditional methods that involve complex etching processes and toxic by-products, Song's team realized the printable fabrication of elastomeric matrices with arbitrary curve patterns and large-scale output with high resolution at the macro- and nanoscale.

"After investigating the dependence of curved nanoparticle assembly arrays on the template patterns, we can achieve different complex curves by designing the templates accordingly," explains Song.

Compared with other flexible sensors, the team's nanoparticle curve sensor with the characteristic of adjustable resistance responses to strain is a real breakthrough for wearable sensors.

The curved pattern fabrication method manipulates the nanoparticle ink to form diverse curved patterns. The scientists point out that, during the assembly process of the nanocurves, the liquid surface tension and viscosity must be well controlled, which, in combination with well-designed pillar templates, is key to yielding regularly arranged curved arrays.

To test their arrays, the researchers fabricated microelectrode sensors, which they used to measure facial expressions. They attached the sensors to different locations on the face—the upper eyelid, lower eyelid, and canthus—to investigate small strain sensing for eight types of facial expressions: anger, disgust, fear, laughter, sadness, smiling, surprise, and relaxed as a control.

"Our nanoparticle curve-based strain sensors can achieve detailed monitoring of complex micro muscle group movements," notes Song. "Actually, the most remarkable application will contribute to skin micromotion manipulation auxiliary apparatuses for paraplegics and quadriplegics."

In the next stage of their research, Song's team will use this effective and simple printable method to realize three-dimensional nanomaterial assembly. Sensors based on 3D structures will show anisotropic differential responses to external stimuli, which provide new opportunities for the fabrication of intelligent electronic skins.

Featured scientists: Prof. Yanlin Song's research group (http://159.226.64.162/web/29070/home)

Organization: Institute of Chemistry, Chinese Academy of Sciences, Beijing (PR China)

Relevant publication: M. Su, F. Li, S. Chen, Z. Huang, M. Qin and W. Li, *et al.*, Nanoparticle Based Curve Arrays for Multirecognition Flexible Electronics, *Adv. Mater.*, 2016, 28(7), 1329, DOI: 10.1002/adma.201504759

6.4.7 Selective Gas Sensing with Pristine Graphene

It has been known for some time that graphene can be used for the detection of individual gas molecules adsorbed on its surface. Back in 2007, the discoverers of graphene, Andre Geim and Kostya Novoselov, had already reported a graphene sensor that can detect just a single molecule of a toxic gas.[27] In that work, the researchers showed that gas molecules gently attach themselves to graphene without disrupting its chicken wire structure. They only add or take away electrons from graphene, which results in notable changes in its electrical conductance.

However, the extremely high sensitivity of graphene does not necessarily translate into its selectivity to various molecules. In other words, it can be detected that some molecules attached to the graphene surface change the resistivity of a graphene field-effect transistor but one cannot say what kind of molecules have attached. Scientists have therefore thought that truly selective gas sensing with graphene devices requires the functionalization of the graphene surface with some agents specific for different gas molecules.

A breakthrough in selective gas sensing with a single pristine graphene device was achieved when researchers found that chemical vapors change the noise spectra of graphene transistors.

A team of researchers from the Center for Integrated Electronics of the Rensselaer Polytechnic Institute, Nano-Device Laboratory of the University of California – Riverside (UCR), The Ioffe Physical-Technical Institute of The Russian Academy of Sciences, and the GE Global Research center demonstrated that the use of an unusual sensor parameter—low-frequency electronic noise—allows one to perform selective gas sensing for many vapors with a single device made of pristine graphene, and no functionalization of the graphene surface is required.

These results prove that graphene's capabilities for selective gas detection are better than previously thought if low-frequency electronic noise is used as an additional sensing parameter. The noise signal for each gas is reproducible, opening the way for practical, reliable and simple gas sensors made from graphene.

"It is interesting that the motivation for this work was not a desire to make a better sensor but rather a curiosity of what happens with the low-frequency noise response of graphene devices exposed to various vapors," Alexander Balandin, professor of Electrical Engineering and Founding Chair of Materials Science and Engineering at UCR, points out. "It turned out that some gases induce very clear peaks in the noise spectra, which could be used as their signatures in sensor devices."

What is so interesting here is that noise is usually considered as one of the main limiting factors for a gas sensor's operation. The fact that low-frequency noise spectra of graphene is affected in different ways by the vapors of different chemicals, demonstrates that noise can be used to discriminate between different gases.

The team shows that vapors of different chemicals produce distinguishably different effects on the low-frequency noise spectra of graphene. Specifically, they found that gases modify the noise spectra by inducing Lorentzian components with distinctive features.

The characteristic frequency of the Lorentzian noise bulges in graphene devices is different for different chemicals and varies from 10–20 Hz to 1300–1600 Hz for tetrahydrofuran and chloroform vapors, respectively. The results obtained indicate that the low-frequency noise in combination with other sensing parameters allows selective gas sensing with a single pristine graphene transistor. This proposed method of gas sensing with graphene does not require graphene surface functionalization or fabrication of an array of the devices with each tuned to a certain chemical.

Balandin points out that, while the gas signatures in the noise spectra of graphene devices are well reproducible, the physics of processes leading to the characteristic peaks in the noise spectra are not completely understood. "We plan to spend some effort on correlating the parameters of the molecules attached to the graphene surface with the energies of the peaks."

Featured scientist: Prof. Alexander A. Balandin's nano-device laboratory (http://ndl.ee.ucr.edu/)
Organization: Bourns College of Engineering, University of California – Riverside (USA)
Relevant publication: S. Rumyantsev, G. Liu, M. Shur, R. Potyrailo and A. Balandin, Selective Gas Sensing with a Single Pristine Graphene Transistor, *Nano Lett.*, 2012, 12(5), 2294, DOI: 10.1021/nl3001293

6.4.8 Detecting Single Nanoparticles and Viruses with a Smartphone

Optical imaging of nanoscale objects, whether it is based on scattering or fluorescence, is a challenging task due to a reduced detection signal-to-noise ratio and contrast at sub-wavelength dimensions. While advances in light microscopy have led to techniques that can image individual nanoparticles, these methods rely on relatively sophisticated and expensive microscopy systems.

A research group led by Aydogan Ozcan, a professor in the Electrical and Bioengineering Department at UCLA and Associate Director of the California NanoSystems Institute (CNSI), created a field-portable fluorescence microscopy platform installed on a smartphone for imaging of individual nanoparticles as well as viruses using a light-weight and compact opto-mechanical attachment to the existing camera module of the phone.

The results constitute the first time that single nanoparticles and viruses have been detected using a smartphone-based, field-portable imaging system.

"Our field-portable fluorescent imager smartphone attachment involves a compact laser diode based excitation at 445 nm that illuminates the sample plane at a high incidence angle; a long-pass thin-film interference filter; an external low numerical aperture lens; and a coarse mechanical translation stage for focusing and depth adjustment," Ozcan describes the design. "The oblique illumination light on the sample plane is by and large missed by the low numerical aperture of the external collection lens and only the scattered excitation beam needs to be blocked through the long-pass filter, creating a very efficient background rejection mechanism that is necessary to isolate the extremely weak fluorescent signal arising from individual nanoparticles or viruses."

He points out that the same imaging system is also useful for reducing the alignment sensitivity to depth of field, such that a coarse mechanical translation stage would be sufficient to focus the smartphone microscope to the sample plane even in field conditions.

The team tested the performance of their smartphone microscope by detecting individual 100 nm fluorescent particles as well as single human cytomegaloviruses that were fluorescently labeled.

"These results constitute the first time that a smartphone-based field-portable imaging platform has been able to detect single viruses or deeply subwavelength objects," says Ozcan. "We believe that the new imaging performance reached through this work would provide a complementary addition to various other smartphone-based microscopy, sensing, and diagnostics tools, which might provide new opportunities in telemedicine and point-of-care applications, among others."

Specifically, this field-portable fluorescence microscopy attachment—weighing only 186 grams—could be used for sensitive imaging of sub-wavelength objects including various bacteria and viruses, and therefore could provide a valuable platform for researchers in field settings and for conducting *e.g.*, viral load measurements and other biomedical tests even in remote and resource-limited environments.

This research together with some other smartphone-based micro-analysis techniques are being commercialized by a start-up company that Ozcan founded: Holomic LLC.

Featured scientist: Prof. Aydogan Ozcan's research group (http://innovate. ee.ucla.edu/)
Organization: Electrical and Bioengineering Department, University of California – Los Angeles, CA (USA)
Relevant publication: Q. Wei, H. Qi, W. Luo, D. Tseng, S. Ki and Z. Wan, *et al.*, Fluorescent Imaging of Single Nanoparticles and Viruses on a Smart Phone, *ACS Nano*, 2013, 7(10), 9147, DOI: 10.1021/nn4037706

6.4.9   Smartphone Nano-Biosensors for Early Detection of Tuberculosis

Tuberculosis (TB) is a widespread infectious disease caused by *M. tuberculosis*, which most commonly affects the lungs but can also attack other body parts. It is estimated that one-third of the world's population is infected with *M. tuberculosis* at least once during their lifetime and new infections continue to occur in about 1% of the population each year. More than 80% of TB infections come from underdeveloped/developing communities that lack easy access to healthcare providers.

A typical preliminary test for TB includes culturing the samples for at least 1–2 weeks in a lab, followed by examination under a fluorescence microscope.

The lack of rapid, accurate, and inexpensive point-of-care tools for detecting low amounts of *M. tuberculosis* is a critical bottleneck in early diagnosis and appropriate treatment.

Yet another smartphone-based sensor has been developed by a team of researchers from Clemson University and Sri Sathya Sai Institute of Higher Learning. It is a rapid and flexible nano-biosensor for diagnosing TB in its early stages using smartphones.

These inexpensive biosensors are fabricated by coating thin films of silver and fullerenes (spherical carbon cages, also known as buckyballs) onto a flexible cellulose acetate substrate.

In these sensors, the isotropic fluorescence from dye-stained bacteria can be coupled to the silver-coated cellulose substrate using surface plasmons. The emission from the dye could be observed by taking a picture from a smartphone camera attached to a simple diffraction grating made from compact discs.

The intensity of the emission recorded by the smartphone provides information on the density of bacteria and thus stage of the infection.

These new findings show that surface plasmons, when coupled optimally through fullerenes, boost the efficiency to levels where even a single bacterium could also be detected.

"Fluorescence is always isotropic and only 1% of emissions from the dye can be captured using traditional lab-based detectors," notes Pradyumna Mulpur, a graduate student at the Sri Sathya Sai Institute of Higher Learning, who carried out the study. "We used the concept of surface-plasmons to gain the ability to enhance the emission directly, which implies that we can detect bacteria even at low densities that exist in the early stages of the disease."

Although surface plasmons have been known for at least two decades, a major challenge in coupling them to fluorescence from dyes has been the absence of efficient spacer layers. An optimal distance between the dye and silver films that exhibit surface plasmons is necessary to achieve high signal enhancements.

"We have been looking for the best spacer layer with a controllable thickness to enhance the coupling," says Ramakrishna Podila, an assistant professor in the Department of Physics and Astronomy at Clemson University and principal investigator of this study. "We realized that graphene and fullerenes are the best choice. In fact, they not only provide optimal spacing but also protect silver films from oxidation. This enhancement allows us to use a simple camera app on a smartphone to capture vital information about the progression of the disease."

In addition to detecting TB, the team showed that these sensors could be used as forensic tools for sensing body fluids such as semen at crime scenes.

Most of the on-field diagnostic tests, such as UV-illumination, for detecting semen are presumptive. However, the plasmon-based sensors developed in this study can selectively identify semen over other body fluids such as saliva and urine.

These flexible sensors could also be used as swabs to collect samples in crime scenes. The team hopes that these sensors could be used for forensic purposes or point-of-care sensing in resource-limited settings to effectively diagnose and thereby prevent and cure new TB infections.

Featured scientists: Prof. Ramakrishna Podila's nano-bio lab (https://clem-sonnnanobio.wordpress.com/)
Organization: Clemson University, Clemson, SC (USA)
Relevant publication: P. Mulpur, S. Yadavilli, P. Mulpur, N. Kondiparthi, B. Sengupta and A. Rao, *et al.*, Flexible Ag–C_{60} nano-biosensors based on surface plasmon coupled emission for clinical and forensic applications, *Phys. Chem. Chem. Phys.*, 2015, **17**(38), 25049, DOI: 10.1039/C5CP04268B

6.4.10   One-Step Detection of Pathogens and Viruses with High Sensitivity

Early detection of food-borne pathogenic bacteria is critical for preventing disease outbreaks and preserving public health. Unfortunately, current detection techniques such as ISO method 6579, fluorescent-antibody (FA), enzyme-linked immunosorbent assay (ELISA), and polymerase chain reaction (PCR) are time-consuming, cumbersome, and have limited sensitivity. They are inadequate as they lack the ability to detect bacteria in real time.

This has led to urgent demands to develop highly efficient strategies for isolating and detecting these microorganisms in connection with food safety, medical diagnostics, water quality, and counter-terrorism.

What is required are alternative platforms for the rapid, sensitive, reliable and simple isolation and detection of *E. coli* and other pathogens.

To address this issue, researchers in China have developed a magnetic bead (MB)-based sensor that combines magnetic separation (MS) and a magnetic relaxation switch (MRS) for one-step detection of bacteria and viruses with high sensitivity and reproducibility.

"Compared to conventional assays for detection of bacteria and viruses, the MS-MRS assay is easy to operate without laborious pre-treatment and purification, and can be adapted to point-of-care tests easily," says Xingyu Jiang, a professor in chemical biology at the National Center for Nanoscience and Technology in Beijing.

Existing MRS-based approaches suffer from two problems: (1) operation is rather complex because it requires sample pretreatment before the assays; (2) many factors influence the outcome of the results and sometimes one has to know *a priori* what the sample is, for example, bacteria or viruses, in order to obtain the correct results.

The new MS-MRS assay completely resolves this issue and the user does not need to know details about the analytes before testing. This makes it highly suitable for clinical diagnosis, environmental monitoring and food safety testing.

The MS-MRS sensor developed by Jiang and his collaborators exploits the phenomenon that a small magnetic field (0.01 T, a common stationary magnet) can separate large-size magnetic beads (250 nm, abbreviated as MB250) from small-size ones (30 nm, abbreviated as MB30).

MB250 can be rapidly separated by this small magnetic field within 1 minute due to its high saturation magnetization. In comparison, MB30, which is used as the magnetic probe of MRS in this study, cannot be separated by the same magnetic field even after 60 minutes because of its low saturation magnetization.

"Based on the great difference in the separation speed of MB30 and MB250, we use the transverse relaxation time (T_2) of water surrounding the un-aggregated MB30 as the readout," explains Jiang. "The more analyte, the less remaining MB30, the higher the T_2."

He notes that the magnetic signal in the MS-MRS sensor is only related to the amount of the analyte in the samples, thus the MS-MRS improves upon two of the major limitations found in conventional MRS sensors:

Sensitivity—the sensitivity of MS-MRS improves by two orders of magnitude compared to conventional MRS for detection of salmonella in milk samples.

Stability—the inter-assay relative standard deviation (RSD) and the intra-assay RSD is lower than that of conventional MRS.

"A point-of-care testing device that employs MS-MRS is the future," says Jiang. "Having a mobile/wearable device that allows rapid readout of assays for biomarkers for diseases such as heart attack, cancer and diabetes would dramatically change healthcare diagnostics."

Featured scientist: Prof. Xingyu Jiang
Organization: National Center for Nanoscience and Technology, Beijing (PR China) (http://english.nanoctr.cas.cn/)
Relevant publication: Y. Chen, Y. Xianyu, Y. Wang, X. Zhang, R. Cha and J. Sun, *et al.*, One-Step Detection of Pathogens and Viruses: Combining Magnetic Relaxation Switching and Magnetic Separation, *ACS Nano*, 2015, **9**(3), 3184, DOI: 10.1021/acsnano.5b00240

6.4.11   A Nanosensor for One-Step Detection of Bisphenol A

Detection of very small amounts of a chemical contaminant, virus or bacteria in food systems is an important potential application of nanotechnology. The possibility of combining biology and nanosensor technology holds the potential of increased sensitivity and therefore a significantly reduced response time to sense potential problems.

"Graphene oxide (GO) has potential applications in a variety of biological fields because of its unique characteristics. In addition, due to large absorption cross-sections and the non-radioactive electronic excitation energy transfer from a fluorophore to GO, GO has been employed to construct fluorescence resonance energy transfer (FRET) biosensors," explains Professor Chuanlai Xu from the State Key Lab of Food Science & Technology, and Director, Joint Lab of Biointerface and Biodetection, Jiangnan University.

Xu and his collaborators have developed a novel, rapid, and sensitive fluorescence sensor to detect bisphenol A (BPA).

BPA is a chemical produced in large quantities for use primarily in the production of polycarbonate plastics and epoxy resins. Polycarbonate plastics have many applications including food and drink packaging, *e.g.*, water bottles, food packaging materials, impact-resistant safety equipment, and medical devices. Epoxy resins are used as lacquers to coat metal products such as food cans, bottle tops, and water supply pipes.

Some research has shown that BPA can seep into food or beverages from containers that are made with BPA. Exposure to BPA is a concern because of possible health effects of BPA on the brain, behavior and prostate gland of fetuses, infants and children.

While the actual toxicity of BPA is still debated, the direct measurement of BPA is difficult because of the weak response given by conventional electrochemical sensors, and current optical analysis methods are susceptible to the influence of interfering substances.

The novel BPA biosensor developed by the Chinese team provides a method for the rapid detection and risk assessment of BPA with high sensitivity and selectivity.

Aptamers—single-stranded oligonucleotides that can be generated for a target molecule with high affinity—are highly suitable receptors for the selective and high-proficiency detection of a wide range of molecular targets. For instance, researchers have previously shown that aptamer-functionalized graphene can detect mercury in mussels.[28]

"Our sensor is based on water-soluble and well-dispersed graphene oxide, which was used as the fluorescence quenching agent, and a specific anti-BPA aptamer labeled by FAM (FAM-ssDNA)," Xu explains. "In the absence of BPA, FAM-ssDNA can be adsorbed onto the GO surface, leading to FRET between GO and FAM-ssDNA. Subsequently, the fluorescence can be quenched quickly. Conversely, BPA can interact with FAM-ssDNA and switch its conformation to prevent the adsorption of GO, resulting in fluorescence recovery in the sensing system."

Under different concentrations of BPA, based on the target-induced conformational change of anti-BPA aptamer and the interactions between the fluorescently modified anti-BPA aptamer (FAM-ssDNA) and GO, the team's experimental results show that the intensity of the fluorescence signal was changed. They say that these results are comparable to traditional ELISA as well as other instrument-based methods, suggesting that this novel sensor might find applications in food safety testing and the monitoring of industrial production processes.

The researchers suggest that their GO-based assay offers several advantages:

- Fluorescent sensors tend to have higher sensitivity compared to most of the colorimetric sensors;

- The relationship between GO and FAM-ssDNA provides theoretical support for the experiment;
- GO can be easily chemically synthesized in large quantities. Besides, the method avoids the dual labeling of ssDNA with fluorophore and quencher units, which significantly lowers the detection cost.

Featured scientist: Prof. Chuanlai Xu
Organization: State Key Lab of Food Science & Technology, Jiangnan University (PR China)
Relevant publication: Y. Zhu, Y. Cai, L. Xu, L. Zheng, L. Wang and B. Qi, *et al.*, Building An Aptamer/Graphene Oxide FRET Biosensor for One-Step Detection of Bisphenol A, *ACS Appl. Mater. Interfaces*, 2015, 7(14), 7492, DOI: 10.1021/acsami.5b00199

6.4.12 Optical Sensor Platform Based on Nanopaper

Nanopaper, made from cellulose like regular paper, shows much lower surface roughness and much higher transparency than traditional paper. This is due to the nanoscale dimensions of the cellulose fibers used for its production. Nanocellulose is a sustainable material that can be extracted from plant cellulose pulp or synthesized by certain bacteria.

Nanocellulose has been explored in various fields, including filtration, wound dressing, as a replacement for toxic dyes in textile or security applications, as sponges to combat oil pollution, or as substrate material for flexible and transparent electronics.

Another significant area for nanopaper applications is sensors. "To date, bacterial nanopaper has been scarcely explored for optical (bio)sensing applications," says Arben Merkoçi, ICREA Research Professor and Director of the Nanobioelectronics & Biosensors Group at Institut Català de Nanociencia i Nanotecnologia. "Hence, we sought to design, fabricate, and test simple, disposable and versatile sensing platforms based on this material."

Merkoçi and his team describe various nanopaper-based nanocomposites that exhibit plasmonic or photoluminescent properties that can be modulated using different reagents.

"For the first time, we report various nanopaper-based optical sensing platforms and describe how they can be tuned, using nanomaterials, to exhibit plasmonic or photoluminescent properties that can be exploited for sensing applications," says Eden Morales-Narváez, a postdoctoral researcher in Merkoçi's group. "We also describe several nanopaper configurations for simple devices, including cuvettes, plates and spots that we printed or punched on bacterial cellulose nanopaper."

The sensing platforms include a colorimetric-based sensor based on nanopaper containing embedded silver and gold nanoparticles; a photoluminescent-based sensor, comprising CdSe@ZnS quantum dots conjugated to nanopaper (which can be photoexcited using UV-visible light); and a potential up-conversion sensing platform constructed from nanopaper functionalized with $NaYF_4:Yb^{3+}@Er^{3+}\&SiO_2$ nanoparticles (which can be photoexcited using infrared light).

Morales-Narváez explains that the proposed nanopaper-based composites can be obtained using different pathways including by exploiting the hydroxyl-containing groups of the bacterial cellulose as a reducing agent for chemical reduction of noble metal ions to metal nanoparticles; by adding bacterial cellulose as a nanonetwork to embed metallic nanoparticles during their synthesis and by producing surface carboxylic groups on the cellulose, for subsequent coupling with protein/amino-functionalized nanoparticles.

The team explored modulation of the plasmonic or photoluminescent properties of these platforms using various model biologically relevant analytes (the drug methimazole, the toxic compounds thiourea and cyanide, and iodide).

Moreover, the scientists demonstrate that bacterial cellulose nanopaper is an advantageous preconcentration platform that facilitates the analysis of small volumes of optically active materials (~4 µL).

"In all of these approaches, we took advantage of the optical transparency, porosity, hydrophilicity, and amenability to chemical modification, of bacterial cellulose nanopaper," notes Merkoçi. "We expect that these novel platforms will pave the way to new optical (bio)sensors or theranostic devices that are simple, transparent, flexible, disposable, lightweight, miniaturized and perhaps wearable."

Featured scientists: Prof. Arben Merkoçi's nanobioelectronics and biosensors group (http://www.nanobiosensors.org/)
Organization: Institut Català de Nanociència i Nanotecnologia (ICN2), Barcelona (Spain)
Relevant publication: E. Morales-Narváez, H. Golmohammadi, T. Naghdi, H. Yousefi, U. Kostiv and D. Horák, *et al.*, Nanopaper as an Optical Sensing Platform, *ACS Nano*, 2015, 9(7), 7296, DOI: 10.1021/acsnano.5b03097

6.4.13 Ultrahigh-Resolution Digital Image Sensor Achieves Pixel Size of 50 Nanometers

The digital image sensor, widely used in digital cameras and other modern vision information technologies, is a device that converts an optical image into an electronic signal. Existing digital image sensors can be divided into two categories: charge-coupled device (CCD) and complementary metal-oxide semiconductor (CMOS). The structures of CCD and CMOS are complex and each pixel contains multiple parts.

With the advance of microelectronics, the pixel size of image sensors keeps decreasing to improve the sensor resolution. Currently, the CCD pixel size and CMOS pixel size can be fabricated as small at 1.43 microns and 1.12

microns, respectively. However, it is extremely difficult to further decrease the pixel size due to the limitations of the silicon base material that is used and current sensor architecture.

The main pixel structure of current CMOS image sensors uses silicon p–n photodiodes as the core part. However, as the size of the photodiodes decreases, their performance also decreases. Active CMOS image sensors therefore employ FET amplifiers to enhance the signal coming from the p–n photodiodes. This structure increases the complexity of the device and requires more space for integrating one pixel.

To overcome the pixel size limitation of existing digital image sensors, both new materials with enormous photoelectric properties and novel device architectures are required. But the benefits will be huge. Ultra-high resolution will bring revolutionary changes in photography, telecommunication, and machine-vision. Furthermore, ultra-small pixels are a size smaller than the wavelength of visible light, and thus can serve as an effective tool to explore the interaction between light and matter.

Researchers have reported a new, ultra-high resolution nanorod digital image sensor (NDIS), which is fabricated by sandwiching vertically aligned zinc oxide nanorod arrays between orthogonal top and bottom nanostripe electrodes.

"The most important application of the NDIS is as a next-generation digital image sensor with ultra-high resolution, well beyond the limit of existing techniques," says Jinhui Song, assistant professor in the Department of Metallurgical & Materials Engineering at The University of Alabama. "Unlike existing CCD or CMOS sensors with their complexity, the NDIS has a simple structure and ultra-high resolution by taking advantage of the novel properties of nanomaterials and a unique device design."

Song notes that his team achieved two significant breakthroughs for the fabrication of ultra-high resolution digital image sensors.

Firstly, they defined and invented a new basic element of digital integrated circuit, the photo-effect transistor (PET).

"The PET has an extremely simple architecture of two terminal electrodes and it can sense photo intensity and amplify the photoelectric signal," explains Song. "Existing digital image sensor pixels, such as CMOS, have two units—a photodiode and a field-effect transistor (FET)—to realize the photo sensing function. The photodiode, which occupies most of the area of the pixel, senses photo intensity and converts it into an electrical signal. The FET amplifies the electrical signal coming from the sensor."

By contrast, he points out, the simple two-terminal PET combines both photo-sensing and electrical amplification functions in one unit. This means that one PET could function as an entire pixel, which significantly reduces the pixel complexity and enables sensor pixel miniaturization.

Secondly, as a proof of concept, the researchers fabricated an ultra-high resolution digital image sensor with a pixel size of 50 nanometers by using vertical photoelectric nanorod arrays as 3D PET pixels.

"This innovative sensor structure greatly decreases the planar area of each pixel yet still maintains their excellent optoelectronic response," says Song.

In their work, Song's team conducted a comprehensive study of the opto-electronic properties of a nanorod PET pixel with regard to size, surface functionalization, and Schottky barrier. As a result, they were able to increase the nanorod pixel on–off ratio, photoelectric response speed, and photo response spectra.

Overall, the resolution of this nanorod digital image sensor is two orders higher than that of existing CCD and CMOS digital image sensor techniques.

Going forward, the scientists plan to increase the pixel number to 100 000 in order to realize a large-scale digital image sensor with unprecedented high resolution.

Another issue that they will work on is the NDIS' light sensing response time of 0.1 s—which is considerably slower than the fastest values (10^{-8} s) that have been reported for ZnO nanomaterials.

"This might be attributed to the crystal quality of the ZnO nanorods we used and could be improved by growth optimization," Song suggests.

Furthermore, although the ZnO nanorod pixel has a strong absorption from 250 nm to 450 nm, through doping or surface modification the response spectra of ZnO nanorods could be extended to a longer wavelength to cover the entire visible spectra and even near-infrared.

Featured scientists: Prof. Jinhui Song's nanoscience group (http://nanoscience.eng.ua.edu/)
Organization: Department of Metallurgical & Materials, University of Alabama, Tuscaloosa, AL (USA)
Relevant publication: C. Jiang and J. Song, An Ultrahigh-Resolution Digital Image Sensor with Pixel Size of 50 nm by Vertical Nanorod Arrays, *Adv. Mater.*, 2015, **27**(30), 4454, DOI: 10.1002/adma.201502079

6.5   Metamaterials

Metamaterials are engineered composite materials that gain their properties from their structure rather than directly from their composition. These materials, first theorized in 1967 by the Russian physicist Victor Veselago and demonstrated in 1999 by a group led by the physicist David R. Smith at Duke University, have extraordinary properties when it comes to diverting and controlling waves, especially sound and light: for instance, they can make an object invisible, or increase the resolving power of a lens.

6.5.1   Topological Transitions in Metamaterials for More Efficient Solar Cells, Sensors, and LEDs

Physicists have uncovered a new method to manipulate light by borrowing an idea from the field of mathematical topology that deals with the properties of objects undergoing deformations, such as stretching and twisting.

They created a metamaterial that can transform from regular dielectric—a substance like glass or plastic, which does not conduct electricity—to a medium that behaves like metal (reflects) in one direction and like dielectric (transmits) in the other.

A team of scientists, led by Vinod M. Menon from the Department of Physics, Queens College, City University of New York (CUNY), has described a way to manipulate the propagation of light within a metamaterial, thereby opening the door to more efficient solar cells, ultra-high sensitive sensors and single photon sources necessary for quantum communication protocols and quantum computers. Through engineering the transmission properties of these systems, and by combining them with light emitters, one may also realize super bright LEDs that would be useful for display applications.

"The ellipsoid and hyperboloid belong to different classes of surfaces, the former being closed (bound) and the latter being open (unbound). The main idea here is the manipulation of the dielectric constants of the material along different directions in the structure that creates the unusual dispersion (hyperbolic) which in turn results in a huge increase in the optical field inside the material," explains Menon. "Optical properties of this metamaterial can be mapped onto a topological transformation of an ellipsoid into a hyperboloid.[29] The topological transition from such a bound (elliptic) to an unbound (hyperbolic) surface manifests itself in the real world as a dramatic increase of light intensity inside the material."

As the team demonstrated, this optical topological transition can be exploited to manipulate the propagation of light and control its interaction with the material.

To experimentally observe the signature of the predicted optical topological transition manifested through enhancement in spontaneous emission rate, the researchers investigated a metamaterial structure with multiple quantum dot emitters positioned on its top surface. They were able to demonstrate modifications in the light emission of these quantum dots.

The experimental part of the work was led by Queens College of CUNY, in collaboration with City College of CUNY and the theoretical work was carried out at Purdue University and the University of Alberta. Part of the experimental work was carried out at the Center for Functional Nanomaterials at Brookhaven National Laboratories.

Menon notes that a theoretical prediction of huge increases in the light intensity in anisotropic metamaterials has already been made previously.

"The motivation for this work, from my standpoint, was the potential to realize broadband control of spontaneous emission," he says. "Unlike cavities and other similar systems that rely on resonance and hence are bandwidth limited, the anisotropic metamaterial approach provides enhancement in spontaneous emission over a wide bandwidth. One could apply this in reverse to realize broadband absorbers as well."

The research team expects optical topological transition to be the basis for a number of applications of both fundamental and technological importance through use of metamaterial-based control of light–matter interactions. As a next step, they will attempt to realize such structures using other materials where the optical losses are lower.

Featured scientists: Prof. Vinod Menon's laboratory for nano and micro photonics (http://lanmp.org/)
Organization: Department of Physics, The City University of New York, NY (USA)
Relevant publication: H. Krishnamoorthy, Z. Jacob, E. Narimanov, I. Kretzschmar and V. Menon, Topological Transitions in Metamaterials, *Science*., 2012, **336**(6078), 205, DOI: 10.1126/science.1219171

6.5.2 New Cloaking Material Hides Objects Otherwise Visible to the Human Eye

Metamaterials can also be engineered with certain electromagnetic properties that allow them to act as invisibility cloaks. These materials bend all light or other electromagnetic waves around an object hidden inside a metamaterial cloak, to emerge on the other side as if they had passed through an empty volume of space.

Researchers have already been experimenting with cloaking devices for various, usually longer wavelengths such as microwave or infrared waves. Even graphene has been added to the family of cloaking materials (read more in the next section).

A team of scientists at UC Berkeley has devised an invisibility cloak material that hides objects from detection using light that is visible to humans. This work makes actual invisibility for the light seen by the human eye possible.

"The cloaking device we made in our group demonstrates that we are now able to control the flow of light in the desired paths for the whole spectrum of visible light," says Majid Gharghi, a postdoctoral fellow in Xiang Zhang's lab at UC Berkeley. "It is a proof of concept that invisibility and other optical illusion phenomena can be achieved with visible light."

He explains that invisibility cloaks usually require electromagnetic properties with extreme values that are only achievable in metallic metamaterials and have been experimentally demonstrated for cloaking at microwave frequencies.

"Because of the significant metallic loss at optical frequencies, the implementation of invisibility cloaks for visible light has been difficult," says Gharghi. "The cloak we designed uses quasi conformal mapping (QCM) and is fabricated in a silicon nitride waveguide on a specially developed nanoporous silicon oxide substrate with an exceptionally low refractive index of less than 1.25; both the waveguide and the substrate are transparent. This unique substrate increases the available index modulation and enables the implementation of transformation optics for guided visible light."

The carpet cloak works by concealing an object under the layers, and bending light waves away from the bump that the object makes, so that the cloak appears flat and smooth like a normal mirror.

According to the researchers, there are still some problems that need to be solved. To achieve the cloaking effect for the full visible range, they need to employ nanofabrication techniques to make the required metamaterials. Today, these are still difficult and time consuming processes, and therefore, the cloak that the team demonstrated is very small, 6 micrometers wide and 300 nm tall—barely enough to cloak a red blood cell.

"To cloak a large object, the difficulty and complexity of the nanofabrication processes grow proportionally, and that is the main bottleneck towards large scale invisibility," says Gharghi. "The most significant challenge will be large-scale and three-dimensional manufacturing that enables up-scaling of metamaterial fabrication techniques. Also, new materials systems need to be engineered that can provide the range of optical properties required for achieving some of the more sophisticated optical phenomena."

Besides rendering objects invisible and creating other optical illusions, the use of metamaterials to control the flow of light can have many applications in the fields of energy, medical imaging, information technology, *etc.* Examples are using transformation optics and metamaterials for better use of solar light in energy devices; for elimination of sources of noise in imaging and microscopy; and for controlling on-chip propagation of optical signals.

Featured scientist: Dr Majid Gharghi; Prof. Xiang Zhang's research group (http://xlab.me.berkeley.edu/)
Organization: Department of Mechanical Engineering, University of California Berkeley, CA (USA)
Relevant publication: M. Gharghi, C. Gladden, T. Zentgraf, Y. Liu, X. Yin and J. Valentine, *et al.*, A Carpet Cloak for Visible Light, *Nano Lett.*, 2011, **11**(7), 2825, DOI: 10.1021/nl201189z

6.5.3 The Thinnest Possible Invisibility Cloak

There has been tremendous interest in cloaking technology using metamaterials, and Andrea Alù's group in the Department of Electrical and Computer Engineering at The University of Texas at Austin, has been very active in the field, putting forward two exciting possibilities to obtain drastic scattering reduction from moderately-sized objects.

One is the concept of *plasmonic cloaking*, which the group put forward in 2005 and the following years together with Nader Engheta at the University of Pennsylvania, both theoretically and experimentally, and is based on the use of a thin metamaterial cover to suppress the scattering from a passive object.[30]

The second one is the concept of *mantle cloaking*, which is based on a simple impedance surface to achieve similar effects.[31]

In subsequent work, the team has shown that even a single layer of atoms, with the extraordinary conductivity properties of graphene, may achieve similar functionality in planar and cylindrical geometries.

"The graphene cloak idea stems from the mantle cloaking concept, which we have proposed at microwaves using frequency-selective surfaces, *i.e.*, properly patterned conducting surfaces that can tailor their effective surface impedance at will," explains Alù. "Due to the recent progress in understanding graphene's AC conductivity, we have realized that its unique features of ultra-high mobility and largely tunable Fermi level may naturally provide the required reactive properties in a single atomic layer. The effective surface impedance of graphene can be tuned in real-time, another great advantage of this graphene cloak, which makes dynamically tunable and switchable cloaking operation possible."

Once realized, this concept represents the thinnest possible cloak, operating in the terahertz (THz) spectrum.

Alù points out that the tunability properties of graphene may also realize real-time tunable cloaks for a variety of devices.

"Our results may pave the way for exciting applications in low-scattering electronic components and non-invasive sensors," he says. "There is great interest in realizing low-scattering or impedance-matched electronic components, and we believe that the use of a graphene layer may realize this effect in an ultra-thin geometry—much thinner than anti-reflection coatings or other available technology."

In addition, the operation of the graphene layers may be tuned to some degree at the desired THz frequency in real time. Alù says that this concept may realize non-invasive infrared sensors, which may detect the impinging signal avoiding scattering in any direction, extending the concept of a cloaked sensor, which the group put forward in 2009, to infrared and THz wavelengths.[32]

"This may be ground-breaking to improve the fidelity of near-field measurements," says Alù. "The sharp tunability of the graphene cloak's response may also be used to realize more efficient and thinner infrared switching devices integrated in a monolithic photonic circuitry."

Graphene's largely tunable conductivity and its ultra-thin properties may be used for a wide variety of applications—not limited to cloaking and invisibility—that scientists currently are only envisioning. Practical integration of graphene within electronic components and biasing is still at its early stages, but the potential of this technology is very promising.

"The fields of metamaterials, plasmonics and advanced materials, in which we have been active for several years, holds the promise of revolutionizing current technology," says Alù. "The recent advances in the area suggest a variety of new applications, which make working in this area very exciting."

Featured scientists: Prof. Andrea Alù's metamaterials and plasmonics research laboratory (http://users.ece.utexas.edu/~aalu/); Prof. Nader Engheta
Organization: Department of Electrical and Computer Engineering, The University of Texas at Austin, TX (USA); School of Engineering and Applied Science, University of Pennsylvania, PA (USA)
Relevant publication: P. Chen and A. Aluù, Atomically Thin Surface Cloak Using Graphene Monolayers, *ACS Nano*, 2011, 5(7), 5855, DOI: 10.1021/nn201622e

6.5.4 Novel Nanosphere Lithography to Fabricate Tunable Plasmonic Metasurfaces

In conventional nanosphere lithography (NSL), the nanosphere mono- or bilayers, which serve as the lithographic masks, can be fabricated by spin-coating, convective self-assembly or drop casting. Ultimately, the nanosphere configurations in the layers are determined by the spontaneous self-assembly process during these techniques. Therefore, the final configurations are limited to those with or close to the minimal free energy giving rise to very simple patterns. Naturally, the metal nanostructures fabricated from these masks possess simple geometries as well.

An international research team from Japan and the U.S. managed to circumvent this thermodynamical restriction by putting the monolayers in a confined environment and constructing the bilayers with sequential stacking, both of which are critical for the formation of moiré patterns.

"In this work, for the first time, we were able to construct a variety of moiré patterns from polystyrene nanosphere monolayers and subsequently employed the etched moiré patterns as masks to fabricate various novel metasurfaces," says Kai Chen, a postdoctoral researcher at the National Institute for Materials Science (NIMS) in Japan, and first author of a paper on this work. "These moiré patterns are inaccessible by conventional nanosphere self-assembly techniques due to the thermodynamical instability. We were able to overcome this restriction and fabricated bilayer moiré patterns in a layer-by-layer fashion. The metasurfaces fabricated by moiré nanosphere lithography (M-NSL) exhibited a number of complex nanostructures providing new capabilities to engineer light–matter interactions."

NSL is a scalable and versatile nanofabrication technique and has been widely used to fabricate metallic nanoparticles as well as dielectric nanostructures. M-NSL greatly extends its capabilities to fabricate complex nanostructures with tailored optical properties.

Chen recounts how he was looking for methods to construct nanosphere bilayers as lithographic masks as in conventional NSL when he noticed some very peculiar patterns on his first sample. Then he repeated this procedure and found the fascinating moiré patterns on a large scale.

"With sequential stacking or layer-by-layer technique, the top layer is separated from the bottom layer and it can move freely relative to the bottom layer," explains Chen. "With the confined environment, we significantly limited the degree of freedom of the nanospheres in the monolayers and thus some metastable configurations became stable and retained their structures. The combination of these two allows for the relative rotation of the domains in the top and bottom layers, giving rise to the moiré patterns."

This novel technique considerably extends the capability of NSL with its yield of elaborate patterns, which will provide a low-cost, high-throughput approach towards complex nanofabrication masks and novel metasurfaces, which can be used to manipulate light–matter interactions and enhance the sensitivity of molecular sensing.

Conventional NSL can only produce nanoparticles with simple shapes, such as triangles from monolayers and dots from bilayers. Combined with etching or angled deposition techniques, other irregular shapes such as crescents or rings can also be fabricated. However, the shape variations are still very limited and usually the nanoparticles on one substrate exhibit the same shape. Conventionally, more complex nanostructures with unique properties require e-beam lithography or focus ion beam lithography that is time-consuming and high-cost.

Yuebing Zheng, assistant professor of mechanical engineering and materials science & engineering at the University of Texas at Austin, who led the U.S. research team, points out that in M-NSL, a number of nanoparticles with exotic shapes can be fabricated on a single substrate. The arrays of these nanoparticles—or metasurfaces—show tunable broad surface plasmon resonances, which significantly extends the fabrication capabilities of conventional NSL. "It is expected that these metasurfaces can be employed for surface-enhanced spectroscopic studies of molecules and biological cells," he says.

"In addition, broadband resonances are also desirable for light-harvesting applications where these metasurfaces can enhance the absorption of sunlight," as pointed out by Tadaaki Nagao, group leader of Nano-system Photonics at NIMS.

Furthermore, several tunable parameters in M-NSL—such as nanosphere size, etching time and deposition conditions—allow for further engineering of the shapes of the nanoparticles.

The team expects that M-NSL will considerably improve the nanofabrication capabilities for novel metasurfaces and provide excellent platforms to engineer light–matter interactions on the nanoscale.

They are currently studying the metasurfaces fabricated by M-NSL. By comparing the experimental results with numerical simulation, they will be able to understand the different plasmon modes in the metasurfaces.

"We are also working on characterizing M-NSL by tuning the parameters in this technique, such as the size of the nanosphere, the etching time, and the metal deposition conditions," says Chen. "We anticipate that the tuning of these parameters will give rise to different moiré patterns and hence metasurfaces, enriching the family of nanostructures by M-NSL."

This means that in future, complex nanostructures can be rationally designed and then fabricated by M-NSL or other types of extensions of NSL. Complex nanopatterns and the complementary metasurface can then be engineered over large areas.

Chen cautions that, in order to achieve those goals, two sets of challenges exist: "(1) Currently, the nanosphere domain size is not controllable and is relatively small. It is always desirable to have the ability to produce large-area single-domain monolayers. (2) As for M-NSL, it is desirable to be able to control the rotation angles between the domains in top and bottom layers. Then we can rationally design and fabricate the moiré patterns and the complementary metasurfaces."

Featured scientists: Dr Kai Chen; Dr Tadaaki Nagao; Prof. Yuebing Zheng's research group (http://zheng.engr.utexas.edu/)
Organization: National Institute for Materials Science, Tsukuba (Japan) (http://www.nims.go.jp/eng/); Cockrell School of Engineering, The University of Texas at Austin, TX (USA)
Relevant publication: K. Chen, B. Rajeeva, Z. Wu, M. Rukavina, T. Dao and S. Ishii, *et al.*, Moiré Nanosphere Lithography, *ACS Nano*, 2015, 9(6), 6031, DOI: 10.1021/acsnano.5b00978

6.6 Nanotechnology Research Knows No Boundaries

Here is a motley collection of intriguing research work to round out our foray into the world of amazing new nanomaterials and nanotechnology devices and applications.

6.6.1 Superlubricity

Friction is present in numerous physical phenomena occurring at all length scales. In many cases, it is exploited to improve our life (you couldn't walk, drive or even write if there was no friction at all); but on the other hand, a reduction of friction is required to reduce mechanical energy dissipation and wear.

About one third of the world's primary energy is dissipated in mechanical friction and 80% of machinery components' failure is caused by wear. Friction and wear will also become bottlenecks for micro-/nanomechanical systems (MEMS and NEMS) featured with sliding components.

Superlubricity, a phenomenon where the friction almost vanishes between two solid surfaces, will be the key to solving these problems.

To date, most of the superlubricity observed has been realized at the nanoscale and under extreme conditions such as high vacuum or by using repulsive van der Waals forces. However, researchers have already demonstrated the practical utility of superlubricity on the microscopic scale.[33]

It was believed that macroscale superlubricity did not exist due to structural deformation of macroscale materials. Until, that is, Yingying Zhang, Fei Wei and colleagues from Tsinghua University and Peking University, reported a breakthrough in macroscale superlubricity.

The researchers demonstrated, for the first time, that superlubricity could exist in centimeter-long double-walled carbon nanotubes (DWCNTs) under ambient conditions.

"The superlubricity scale was three orders higher than the values reported before, while the shear strength was four orders lower than that reported before," says Zhang.

Double-walled carbon nanotubes—composed of two coaxial cylindrical walls with high aspect ratio—offer a good candidate model to study superlubricity. The interaction between DWCNT shells is based on van der Waals forces, and the inner shells can slide or rotate inside the outer shell.

For most DWCNTs, the interfaces between the inner and outer shells are incommensurate, which means that the two surfaces have no energetically preferred position with respect to each other, and hence they can slide relative to each other with no cost in energy.

Furthermore, for an ideal DWCNT with a partly extruded inner shell, the shear stress in the overlapped section vanishes due to the repetitive breaking and reforming of van der Waals interactions between the adjacent shells.

"Thus, only the edge section is responsible for the intershell interaction in DWCNTs during the pulling out process," explains Zhang. "The friction between the two shells of a DWCNT is independent of its length. Therefore, a centimeter-long DWCNT—on the condition that it has perfect structures and is without obvious axial curvatures—can exhibit superlubricity."

Professor Fei Wei and his team from Tsinghua University made several breakthroughs in synthesizing centimeters-long double-/triple-walled CNTs with prefect structures.[34-36] These nanotubes are perfect candidates to explore the macroscale superlubricity of DWCNTs.

In order to perform measurements for such an extensive range of length scales—from a few nanometers to a centimeter—the researchers coated the carbon nanotubes with microscale particles and used an optical microscope to analyze the frictional response.[37]

According to Zhang, the friction force measured between DWCNT shells was almost independent of the pull-out length and could be as low as 1 nanonewtons (nN). The resistant force exhibits only small fluctuations, about 2 nN, around the average value of 3 nN for a huge range of contact lengths between the extracted inner tube and its outer host; the contact length changes by six orders of magnitude, from tens of nanometers to a few millimeters.

"Our observations demonstrate that static friction at the incommensurate contact between the inner tube and its outer host is really negligible, ~10 Pascal (Pa), even for a centimeter-long contact," she says.

Professor Michael Urbakh, who reviewed[38] the findings, commented that "these encouraging results bridge the gap between what is known about friction on the microscopic and macroscopic scales. The observation of superlubricity in ultralong DWCNTs is a promising result for many practical applications, such as ultrasensitive sensors, fine positioning devices, gyroscopes, fast switches and more."

Featured scientists: Prof. Yingying Zhang (http://cnmm.tsinghua.edu.cn/en/#!/peoples/staff/yingying-zhang); Prof. Fei Wei (http://www.chemeng.tsinghua.edu.cn/scholars/weifei/index1.htm)
Organizations: Center for Nano and Micro Mechanics, Tsinghua University; Department of Chemical Engineering, Tsinghua University, Beijing (PR China)
Relevant publication: R. Zhang, Z. Ning, Y. Zhang, Q. Zheng, Q. Chen and H. Xie, *et al.*, Superlubricity in centimetres-long double-walled carbon nanotubes under ambient conditions, *Nat. Nanotechnol.*, 2013, 8(12), 912, DOI: 10.1038/nnano.2013.217

6.6.2 Microfluidics Without Channels and Troughs

In a seminal paper in *Science* in 1992, scientists described a device that was capable of causing drops of water placed on it to move uphill.[39]

However, as it turned out in subsequent research, drops of water are notoriously difficult to move from where they lie, unless they are large enough to be moved by gravity, as happens with rain drops on window panes. For instance, two research reports[40,41] showed that water was one of the fluids where drops stubbornly refused to cooperate and move when cleverly engineered surfaces were activated to produce motion of fluidic droplets.

In the absence of a microtube or a channel—which are required by most microfluidic devices—it usually is not possible to apply the pressure needed to induce liquid movement. An alternative approach, developed by researchers in Italy, is to pattern a gradient on a surface, which allows a droplet to move in order to minimize its free energy.

This novel method is simple and easily scalable to many drops and/or bigger drops. The work also provides a deep theoretical understanding of the phenomenon, and provides guidelines to improve the fabrication technique and improve the performance of this simple device.

"In our work, a drop of water is driven by surface forces against gravity and more importantly, in this case, against two competing phenomena: adhesion that would want the drop to stick to the surface where it is deposited and wetting that would want it to spread over it," says Francesco Zerbetto, a professor in the Department of Chemistry at the University of Bologna.

"The power behind the self-propelling of these droplets is a simple trick: it's chemistry working at the nanoscale," adds Massimiliano Cavallini from the Institute of Nanostructured Materials, CNR Bologna, who, together with Zerbetto, led the work.

Since liquid droplets move on horizontal surfaces only in the presence of a surface gradient—and the gradient must be sufficiently small to avoid both full wetting of the surface and pinning of the drop—the challenge is to properly prepare the wettability of the surface.

"Two of the main approaches to modify the wettability of a surface leverage the introduction of chemical heterogeneities on the surface and modification of surface topography by the creation of patterned or textured micro- or nanoscale features," explains Zerbetto. "In our work, we used the immersion method to prepare gradients of alkanethiol self-assembled monolayers (SAMs) in combination with a silicon surface functionalized by a highly hydrophobic molecule (TPOS). Optimization of the procedure allows us to demonstrate droplet motion."

Depositing a droplet of water on the functionalized surface, the researchers observed directional movement of water droplets along the gradient without any external stimulus for contact angle slopes greater than $10°$ mm^{-1}. The drop displacement was 2.8 ± 0.1 mm in the direction of the more hydrophilic area.

The ultimate goal of this research is to create a real microfluidic device—without the need of channels—where droplets can move and transport payloads and/or reagents along precise trajectories. To date, spontaneous movement without any external help is achieved using thermal gradients, sonication, *etc.*, and is limited to a few millimeters. The challenge is to increase this distance to centimeters.

Delivery is also one of the underlying issues at the basis of the work. Exploitation of the protocol proposed and demonstrated by Zerbetto and his team can deliver small quantities of water to specific locations, even when there are forces opposing such delivery, as in the case of hydrophobic conditions.

Practical applications of this work are self-cleaning surfaces where drops of water carry away pollutants that are deposited on them. Imagine solar panels, lenses and mirrors that stay clean because self-propelled water drops scrub off the dirt.

A self-propelled drop of water can also be the carrier of drugs or medication. "One could also think of situations where the molecule to be delivered is highly unstable, but can be produced by reacting two molecules that are brought together by means of two drops moving towards each other," says Zerbetto.

So far, the scientists have demonstrated that their approach is able to produce directional drop movement. This result was obtained on a surface. In future applications, they will provide more complicated paths to span complex trajectories.

"As is often the case, make it bigger, make it cheaper, make it faster, and make a lot of it is where we will go next," Zerbetto describes the team's goals. "In particular, making it bio-compatible and demonstrating that the whole thing can be inserted in a biological system will be one of our aims."

Featured scientists: Prof. Francesco Zerbetto; Massimiliano Cavallini
Organizations: Department of Chemistry, University of Bologna (http://www.chimica.unibo.it/it); Institute of Nanostructured Materials, CNR Bologna (http://www.ismn.cnr.it/index.php?lang=en) (Italy)
Relevant publication: F. Lugli, G. Fioravanti, D. Pattini, L. Pasquali, M. Montecchi and D. Gentili, *et al.*, And Yet it Moves! Microfluidics Without Channels and Troughs, *Adv. Funct. Mater.*, 2013, **23**(44), 5543, DOI: 10.1002/adfm.201300913

6.6.3 Truly Blond—Hair As a Nanoreactor to Synthesize Gold Nanoparticles

In Nature, numerous inorganic materials are synthesized by living organisms. These bioinorganic materials can be extremely complex both in structure and function, and also exhibit exquisite hierarchical ordering from nanometer to macroscopic length scales, which has not even remotely been achieved in laboratory-based syntheses. Inorganic materials in the form of hard tissues are an integral part of most multicellular biological systems.

The possibility of using such microorganisms and plants in the deliberate synthesis of nanomaterials is a recent phenomenon and scientists are now exploring the use of biological organisms and materials to literally grow nanomaterials.[42]

In a novel approach, a team of scientists from CNRS synthesized nanoparticles in hair. The purpose was to try to describe some of the chemical reactions occurring inside the hair shaft, in the so-called amorphous matrix surrounding intermediate filaments made of keratin proteins. This matrix can be seen as a set of nanoreactors.

"Our study shows that the composition and the structure of keratins induce the synthesis of nanoparticles with a homogeneous size in the cuticle (2.5 nm) and in the cortex (about 1.5 nm)," says Dr Philippe Walter, director of the Laboratoire d'archéologie moléculaire et structurale, LAMS, CNRS UMR 8220. "This is due to their environment, *i.e.* to the binding with sulfur from cysteine amino acids."

Walter adds that the idea to use gold nanoparticles to dye wool was already published and patented by the group of James H. Johnston in New Zealand with different sized nanoparticles.

Previously, Walter and his team used these nanoreactors to form PbS quantum dots[43] as well as HgS nanoparticles.[44]

In this subsequent work by the French team, the synthesis occurs *in situ*, inside the shaft. They were able to achieve a variation of color, from gold/blonde, to brown and dark brown.

"We observe the self-organization of the nanoparticles along the fiber axis due to the structure of intermediate filaments of keratin molecules," explains

Walter. "We started with the idea of trying with gold salts to use another way to check the chemical properties of keratins and of their supramolecular assemblies. Our results highlight the possible use of these chemical conditions to produce gold nanoparticles from gold salts in a one-pot nucleation, growth, morphogenesis, and passivation of extremely small gold nanoparticles."

He notes that even multiple washings of treated hairs did not affect the color of the hair as the colored gold nanoparticles are buried and stabilized inside the hair by the keratin structure.

This work has been part of a multidisciplinary project: it aims at understanding the cause of the degradation of keratin-based materials—hair and stratum corneum of mummies; feathers; archaeological and ethnographic artifacts—and what role the supramolecular organization of hair plays at scales ranging from nanometers to micrometers.

Because the gold nanoparticles formed inside the hair are fluorescent under UV light, researchers can now consider a number of applications.

"For instance, we can look at the quenching of this fluorescence after an uptake of metallic cations by the hair shaft—for example mercury and arsenic—and develop new sensors for toxic elements in water," says Walter. "We can also extract the nanoparticles and use the hair as a template for the synthesis of nanoparticles with a narrow distribution of size."

Going forward, the team would like to use the synthesis of the nanoparticles to check the preservation of the structure of archaeological hair.

"One question we would like to answer is if the sizes of the nanoparticles are the same after several thousands of years of alteration," says Walter. "Here, the synthesis of nanoparticles can be considered as a way to characterize the chemical modifications in the fiber."

Featured scientist: Dr Philippe Walter
Organization: Laboratoire d'archéologie moléculaire et structurale, Paris (France) (http://www.umr-lams.fr/)
Relevant publication: S. Haveli, P. Walter, G. Patriarche, J. Ayache, J. Castaing and E. Van Elslande, *et al.*, Hair Fiber as a Nanoreactor in Controlled Synthesis of Fluorescent Gold Nanoparticles, *Nano Lett.*, 2012, **12**(12), 6212, DOI: 10.1021/nl303107w

6.6.4 A Virus-Sized Laser

Reducing the size of photonic and electronic elements is critical for ultra-fast data processing and ultra-dense information storage. The miniaturization of a key, workhorse optical instrument—the laser—is no exception. Coherent light sources at the nanometer scale are important not only for exploring phenomena in small dimensions but also for realizing optical devices with sizes that can beat the diffraction limit of light.

Researchers at Northwestern University have found a way to manufacture single laser devices that are the size of a virus particle and that operate at room temperature. They show that subdiffraction nanoresonators based on metallic bowties, when coupled to a gain material, can generate coherent and directional light emission.

"The reason that we can fabricate nano-lasers with sizes smaller than that allowed by diffraction is because we made the lasing cavity out of metal nanoparticle dimers—structures with a 3D 'bowtie' shape," explains Teri W. Odom, Board of Lady Managers of the Columbian Exposition Professor of Chemistry and Professor of Materials Science and Engineering at Northwestern University. "These metal nanostructures—nanoparticle dimers with a gap size of tens of nanometers—support localized surface plasmons, collective oscillations of electrons, which have no fundamental size limits when it comes to confining light."

According to Odom and her team, the use of the bowtie geometry has two significant benefits over previous work on plasmon lasers: (1) the bowtie structure provides a well-defined, electromagnetic hot spot in a nano-sized volume because of an antenna effect; and (2) the individual structure has only minimal metal "losses" because of its discrete geometry.

"Surprisingly," says Odom, "we also found that when arranged in an array, the 3D bowtie resonators could emit light at specific angles according to the lattice parameters."

She explains that these bowtie resonators are an ideal system to localize and enhance electric fields within nanoscale volumes. "The primary advantage of a bowtie geometry over a symmetric, single nanoparticle as an optical resonator is their high Purcell factor. With an increased Purcell factor because of a reduced mode volume, the threshold condition to initiate lasing in bowtie nanoparticles can be readily satisfied, even at ambient temperatures."

In their experiments, the team shows that plasmonic lasing can occur at room temperature near localized surface plasmon resonance wavelengths when strong electric field localizations are present in the gap.

Odom also points out that this bowtie design is highly defect-tolerant, which is important for practical and scalable applications.

"Ultimately," she says, "the 3D resonator design can be interfaced with other gain materials, such as highly oriented molecules or single quantum dots, to improve the coupling efficiency for designing low-threshold nano-lasers or new platforms for studying quantum plasmon effects."

Featured scientist: Prof. Teri W. Odom's research group (http://chemgroups. northwestern.edu/odom/)

Organization: Department of Chemistry, Northwestern University, Evanston, IL (USA)

Relevant publication: J. Suh, C. Kim, W. Zhou, M. Huntington, D. Co and M. Wasielewski, *et al.*, Plasmonic Bowtie Nanolaser Arrays, *Nano Lett.*, 2012, **12**(11), 5769, DOI: 10.1021/nl303086r

6.6.5   High-Resolution Holograms with Nanoscale Pixels

The size of pixels is one of the key limiting features in state-of-the-art holographic displays systems. Holography is a technique that enables a light field—the product of laser light scattered off objects—to be recorded and later reconstructed when the original light field is no longer present. Holography can be thought of as somewhat similar to sound recording, whereby a sound field created by vibrating matter such as musical instruments or vocal cords is encoded in such a way that it can be reproduced later, without the presence of the original vibrating matter.

The resolution and field of view in these holographic systems are dictated by the size of the pixel, *i.e.* the smallest light scattering element.

"Light scattered from individual pixels interferes spatially to produce the projected images/diffraction patterns in holograms," says Dr Haider Butt, at the time of this research a post-doctoral researcher at the University of Cambridge's Centre of Molecular Materials for Photonics and Electronics (he now has his own nanophotonics research group at the University of Birmingham). "The pixel size controls both the resolution and field of view of these projected images. Smaller pixels allow the diffraction of light at larger angles, increasing the field of view."

Basically, this means that the smaller the pixel, the higher the resolution of the hologram. To address the limitations of current holographic systems due to their pixel size, a research team led by CMMPE's Timothy Wilkinson and Gehan Amaratunga, a professor at the University's Centre for Advanced Photonics and Electronics, set out to use nanostructures as the smallest possible light-scattering elements for producing holograms.

In their work, the scientists harnessed the extraordinary conductive and light scattering abilities of nanotubes and patterned an array of multi-walled carbon nanotubes (MWCNTs) to produce a high-resolution hologram.

"The fascinating aspect about our research is that we have demonstrated the utilization of carbon nanotubes as the smallest possible pixel for diffracting light in a highly controlled manner to produce a two-dimensional image," says Butt. "Our objective was to control the angular diffraction from the nanotube arrays by fabricating them in the form of a holographic pattern. Such an array of sub-wavelength nanotubes can act as an intensity hologram (grid of apertures) towards the incident light, producing a diffraction pattern in the far field. In this manner we can replicate the mechanism of an intensity spatial light modulator while using the world's smallest pixel defined by a carbon nanotube."

To demonstrate the capabilities of this approach, the team calculated the holographic patterns necessary to display the word "Cambridge". They performed these calculations based on the principle of diffraction optics assuming the MWCNTs acted as diffracting elements.

Based on these calculations, they then fabricated a highly ordered array of vertically aligned MWCNTs on a silicon substrate. By shining a green (532 nm) laser perpendicular to the plane of the nanotube array, a pattern with a clear "Cambridge" image was obtained on a semitransparent hemispherical screen with a radius of 15 cm.

These results pave the way towards the utilization of nanostructures for producing 3D holographic display systems with a wide field of view and high resolution.

"The high-resolution aspect makes these holograms very sensitive to the changes in materials properties, position and direction of incident light," Butt points out. "Based on this, a new class of highly sensitive holographic sensors can be developed that could sense distance, motion, tilt, density of biological materials, and wavefront of incident light."

One of the challenges for the research team is to decrease the fabrication cost of these holograms. In order to explore alternative materials, they are planning to try zinc oxide nanowires to achieve the same effects as carbon nanotubes.

Another challenge is to investigate movement in the projections. Currently, these nanoscale pixels can only render static holograms. "We will look at different techniques such as combining our nanotube pixels with the liquid crystals found in flat-screen technology to create fluid displays—possibly leading to changeable pictures and even razor-sharp holographic video," says Butt.

Featured scientist: Dr Haider Butt (http://www.birmingham.ac.uk/staff/profiles/mechanical/butt-haider.aspx)
Organizations: Centre of Molecular Materials for Photonics and Electronics (http://www-g.eng.cam.ac.uk/CMMPE/); Centre for Advanced Photonics and Electronics http://www-cape.eng.cam.ac.uk/; University of Cambridge (UK)
Relevant publication: H. Butt, Y. Montelongo, T. Butler, R. Rajesekharan, Q. Dai and S. Shiva-Reddy, *et al.* Carbon Nanotube Based High Resolution Holograms, *Adv. Mater.*, 2012, **24**(44), OP331, DOI: 10.1002/adma.201202593

6.6.6 Exploring the Complexity of Nanomaterial/Neural Interfaces

Carbon nanotubes, like the nervous cells of our brain, are excellent electrical signal conductors and can form intimate mechanical contacts with cellular membranes, thereby establishing a functional link to neuronal structures. There is a growing body of research on using nanomaterials in neural engineering.

For instance, researchers have explored the impact of carbon nanotube scaffolds on multilayered neuronal networks. Up to now, all known effects of carbon nanotubes on neurons—namely their reported ability to potentiate neuronal signaling and synapses—have been described in bi-dimensional cultured networks where nanotube/neuron hybrids were developed on a monolayer of dissociated brain cells.

In their work, a team of scientists in Italy, led by professors Maurizio Prato and Laura Ballerini, used slices from the spinal cords of mice to model multilayer-tissue complexity. They interfaced these spinal segments to multi-walled carbon nanotube (MWCNT) scaffolds for weeks at a time to see whether and how the interactions at the monolayer level are translated to multilayered nerve tissues.

The researchers report the efficacy of MWCNT/neuron interactions to generate signals, which are then translated into activity modifications.

"Our findings are entirely new in that we propose that nanomaterial/cell interactions can be instructive for network behavior and for signaling even at a distance from the interface," explains Ballerini. "This means that neurons are capable of sensing a nanomaterial and translate this interface into improved activity within a larger area of influence in spinal explants. This issue has been unexplored up to know."

According to the team, interfacing spinal cord explants to purified carbon nanotubes over a longer period (weeks) induces two major effects: first, the number and length of neuronal fibers outgrowing the spinal segment increases, associated with changes in growth cone activity and in fiber elastomechanical properties. And, secondly, after weeks of MWCNT interfacing, neurons located at as far as five cell layers from the substrate display an increased efficacy in synaptic responses—which could represent either an improvement or a pathological behavior—presumably mediated by ongoing plasticity driven by the neuron/MWCNT hybrids.

"We propose that these two effects rely on direct and indirect MWCNT interactions," says Ballerini. "The first being mediated by direct adhesion of outgrowing fibers to the nanostructured carbon substrate; the second by alterations in the activity of neuronal layers interfaced to the substrate, able to influence remote, although synaptically coupled, neuronal ensembles (Figure 6.10)."

The team says that they cannot entirely exclude that the effects on remote neurons are mediated by small amounts of detached MWCNTs, internalized by cells within the spinal explants, and not detected by routine TEM analysis. However, some of their observations seem to rule out this possibility.

Going forward, the major challenge is to unravel the signal and transduction pathways involved in the detected influence of carbon nanotubes to neuronal networks.

Ballerini notes that clarifying this issue will allow the future design of scaffolds and interfaces that, by means of their physical and chemical structure, replicate instructive features of the extracellular micro-environment at the nanoscale.

"Ongoing efforts in regenerative medicine require the development of synthetic extracellular scaffolds able to provide unique micro-environments to tissue-specific cell types," she says. "Micro- and nanoscale techniques employed to recreate interactions between cells and tissue-engineering scaffolds offer great promise in the fabrication of biological tissue constructs. These interactions are widely accepted as being essential to promote tissue

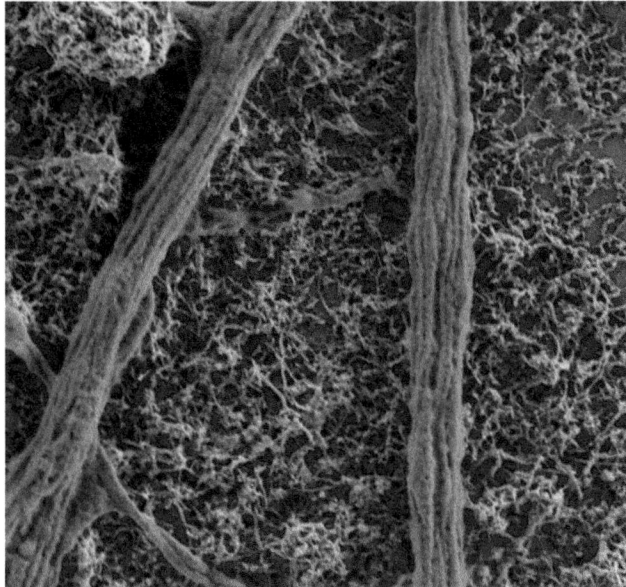

Figure 6.10   Modified (by D. Scaini, Prato–Ballerini research group) scanning electron microscopy image of spinal explant peripheral neuronal fiber on a MWCNT substrate. (Image: Prato–Ballerini research group.)

and to maintain tissue function in tissue repairing processes. Thus, exploiting physical and chemical features at the nanoscale may improve next generation transplantable devices for tissue implants."

Prato and Ballerini point out that, ultimately, nanomaterial-based scaffolds will allow the investigation of the ability of multilayered nervous tissue in translating adhesive interactions into network activity in regions relatively far from the interface itself, providing relevant information for the scientific community dealing with neuronal interfaces and carbon nanotubes.

They note that this is important because it exploits the design of artificial micro- and nanoscale devices that cooperate with neuronal network activity, thereby creating hybrid structures able to cross the barriers between artificial devices and neurons.

Featured scientists: Prof. Maurizio Prato's carbon nanotechnology group (http://www2.units.it/pratoweb/); Prof. Laura Ballerini (http://phdneurobiology.sissa.it/eng/faculty/associated/laura-ballerini.aspx)
Organizations: University of Trieste; SISSA, Trieste (Italy)
Relevant publication: A. Fabbro, A. Villari, J. Laishram, D. Scaini, F. Toma and A. Turco, *et al.*, Spinal Cord Explants Use Carbon Nanotube Interfaces To Enhance Neurite Outgrowth and To Fortify Synaptic Inputs, *ACS Nano*, 2012, **6**(3), 2041, DOI: 10.1021/nn203519r

6.6.7 Skin-Inspired Haptic Memory Devices

Our sense of touch connects us to the world around us and is an integral part of how we experience things, both physically and emotionally. Human skin is a sensitive detector of both pressure and temperature, and efforts to develop similar sensors for electronics are widespread. The exquisite sensation functions of natural skin have inspired the rapid advancements of skin-like sensing devices, especially tactile sensors, for electronic skin applications.

When skin gets stimulated by external impulses, it sends the resulting sensory information through afferent neurons to the brain to form haptic (from the Greek word *Haphe*, pertaining to the sense of touch) memory. This allows us to retain the impressions of the stimuli.

Current research on tactile sensors is mostly focused on the improvement of sensitivity and multi-functionality to emulate the function of natural skin. However, natural skin can sense external pressure and helps to form haptic memory, while current flexible tactile sensors for electronic skin can only perform sensing functions.

This functionality gap between state-of-the-art tactile sensing devices and natural skin inspired a team of researchers in Singapore to develop haptic memory devices that integrate sensor and memory functions.

The scientists came up with a strategy to realize the detection and retention of tactile sensation by haptic memory devices for the mimicry of the human sensory memory.

"Our work successfully addresses the mimicry of haptic memory by the preparation of haptic memory arrays, where applied pressure could modulate the memory states of the devices," says Xiaodong Chen, an associate professor in the School of Materials Science & Engineering at Nanyang Technological University. "Our haptic memory device not only could detect external pressure *via* resistive pressure sensors, but also 'memorize' it in the form of resistance states."

The researchers fabricated the sensitive layer of the resistive pressure sensor from microstructured PDMS film embedded with silver nanowires. Their resistive switching memory device takes advantages of typical metal–insulator–metal architecture with SiO_2 serving as a switching layer, providing fast switching time, high endurance, as well as nonvolatile memory.

The pressure information can be retained in the devices for a long time due to the nonvolatile performance memory cells used. Moreover, the external pressure distribution could be perceived by the introduction of device arrays.

According to the team, the advantage of their device lies in the rational design and integration of resistive tactile sensors and resistance switching memory cells.

"Each haptic memory device can be attributed to the integration of a resistive tactile sensor and a resistive switching memory device, where the resistance states in a memory cell can be electrically reconfigured by applied pressure on a sensor," explains Chen.

He adds that such a resistive switching memory device is an excellent candidate for integration with pressure sensors to achieve haptic memory—it could be utilized to emulate the memory functionality of the brain by constructing artificial neuromorphic networks as it behaves similarly to synapses among neurons in storing analogue value.

Information is stored and operated in resistive switching memory devices because a cell can be electrically configured between a high-resistance state and a low-resistance state, corresponding to an OFF (0) and an ON (1) state, respectively.

Similarly, a resistive pressure sensor reflects the loading or unloading of an external force by changes in resistance.

"The similarity in the resistive switching memory and resistive tactile sensing devices inspired us to build haptic memory devices to realize the nonvolatile memory of applied pressure by programming memory cells with the resistance change in tactile sensors," notes Chen.

These haptic memory arrays could be applied in humanoid robotics and human–machine interfaces to pre-process the sensation information at the contact points so as to reduce the amount of information needed to be transferred to processors and free the latter for more intelligent tasks.

The haptic memory arrays could also be integrated with prosthetics to provide feedback regarding external stimuli for rehabilitation applications.

Although state-of-the-art sensing devices have demonstrated great capabilities in responding to physical quantities, such as chemicals, gas, sound, light and pressure, there still exists a huge gap between technical sensing systems and biological systems (Figure 6.11).

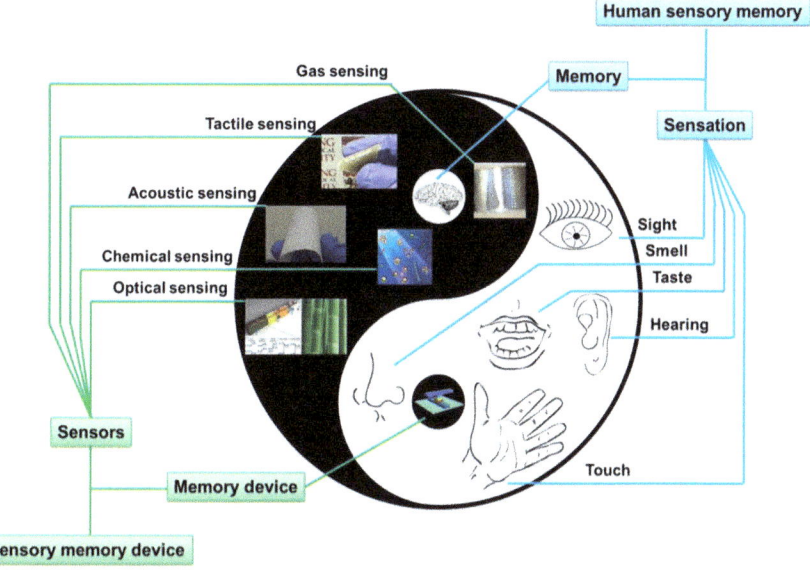

Figure 6.11 Schematic illustration of integrating different sensors with memory devices for the mimicry of human sensory memory. (Image: the Xiaodong Chen group, Nanyang Technological University.)

"The human body has five sensation modalities: touch, sight, hearing, taste and smell; these senses and their memory are not isolated but often work together," says Chen. "This has inspired us to work towards integrating memory devices with multiple sensing units to build highly integrated systems for the mimicry of human sensory memory."

Featured scientists: Prof. Xiaodong Chen's research group (http://www. ntu.edu.sg/home/chenxd/)
Organizations: School of Materials Science & Engineering at Nanyang Technological University (Singapore)
Relevant publication: B. Zhu, H. Wang, Y. Liu, D. Qi, Z. Liu and H. Wang, *et al.*, Skin-Inspired Haptic Memory Arrays with an Electrically Reconfigurable Architecture, *Adv. Mater.*, 2016, 28(8), 1559, DOI: 10.1002/adma.201504754

6.6.8   Light-Emitting Nanofibers Shine the Way for Optoelectronic Textiles

OLEDs—organic light-emitting diodes—are full of promise for a range of practical applications. OLED technology is based on the phenomenon that certain organic materials emit light when fed by an electric current and it is already used in electronic device displays in mobile phones, MP3 players, digital cameras, and increasingly also some larger TV screens.

OLEDs in fiber form could lead to revolutionary applications by integrating optical and optoelectronic devices into textiles. Combined with nano-electronic devices, we might one day see flexible optical sensors and display screens woven into shirts and other garments. You could literally wear a future generation of your smartphone or tablet on your sleeves; including the solar panels to power them.

This vision came one step closer thanks to researchers at Iowa State University, who have built a one-dimensional electroluminescent device into a single self-supporting fiber at the submicrometer scale.

"Currently, electronic light-emitting fibers use a conventional fabrication route, starting on an existing optical fiber several tens of micrometers in diameter, followed by sequential deposition of multiple thin-films," explains Liang Dong, assistant professor in the Department of Electrical and Computer Engineering at Iowa State. "The most bulky and generally only structural but not functional part of a device is the substrate. If a device did not have a substrate, then it would become much more lightweight and flexible."

Dong and his team noted that there had been various research reports on removing substrates by using chemical methods after devices were formed. But this post-processing was nontrivial and there was a risk of destroying device structures by chemicals.

Recently, however, researchers have employed electrospinning—a simple, inexpensive, and effective method for producing fibers—to develop light-emitting nanofibers with a relatively high throughput. In this process, the devices are formed by spinning a polymer nanofiber embedding a ruthenium-based ionic transition-metal complex (iTMC) onto an array of microfabricated interdigitated electrodes.

Dong's group has reported their successful effort to develop novel iTMC-based electroluminescent fibers (TELFs) using co-electrospinning.

These TELFs are electronic fabric light-emitting devices in micro/nanofiber form. The device structure is unique in that it does not need a substrate and all device components are integrated into a single small fiber.

TELFs are manufactured in two simple steps: encapsulating a conducting liquid core into a light-emitting sheath material by co-electrospinning to fabricate simultaneously two major functional layers into a single nanofiber in air; and then coating the device surface with a transparent conducting film (the electrode) by evaporation. This way, the researchers completely eliminated using a substrate for TELFs.

TELFs are probably one of the most compact electronic light-emitting fibers developed so far. Uniquely, fabrication of TELFs does not only not rely on any substrate but the process starts in air and finishes spinning major thin-films into single fibers at the moment the fibers fall onto a collector. A wide variety of materials such as clothing, plastic, steel, silicon, glass, aluminium foil, and even paper can be used as a device collector. Thus, TELFs have very high and virtually limitless substrate compatibility.

Since electrospinning is being used in many textile applications, these TELFs could be directly integrated into textile products.

"Besides general applications for lighting and displays, we are interested in integrating TELFs into bioanalytical lab-on-a-chip devices where tiny light sources are needed," says Dong. "Another potential application is light coupling. Coupling external light into nanostructures is a very challenging task in nanophotonics. The TELF technology could provide a possible solution for generating light directly inside nanofibers, without introducing external light into nanofibers."

Featured scientist: Prof. Liang Dong (http://www-archive.ece.iastate.edu/who-we-are/faculty-and-staff/faculty-new/index/detail/abc/293.html)
Organization: Department of Electrical and Computer Engineering, Iowa State University, Ames, IA (USA)
Relevant publication: H. Yang, C. Lightner and L. Dong, Light-Emitting Coaxial Nanofibers, *ACS Nano*, 2012, 6(1), 622, DOI: 10.1021/nn204055t

6.6.9 Protecting Satellite Electronics with Reinforced Carbon Nanotube Films

Nanotechnology will play an important role in future space missions. Nanosensors, dramatically improved high-performance materials, or highly efficient propulsion systems are but a few examples.

One particularly important issue is the protection of satellites from electrostatic discharge (ESD). In space, the external insulating surfaces of a spacecraft accumulate electrostatic charge as a result of exposure to space plasma, including a high flux of charged particles especially at geosynchronous earth orbit (GEO). If that charge accumulation suddenly discharges, it may damage the electronics of the spacecraft.

The space industry, therefore, has a strong requirement to develop a flexible ESD protection layer for the exterior cover of satellites.

A study conducted by researchers at Tel Aviv University together with scientists from the space environment department at Soreq NRC, explores carbon nanotube–polyimide (CNT–PI) composite materials as a flexible alternative for the currently used indium tin oxide (ITO) coating, which is brittle and suffers from severe degradation of electrical conductance due to fracture of the coating upon bending.

"We developed electrically conducting and flexible CNT–PI films specifically for space applications using polymer solution infiltration into CVD-grown entangled CNT sheets with cup-stacked nanostructure," says Yael Hanein, a professor at Tel Aviv University and Director of the university's Center for Nanoscience and Nanotechnology. "This fabrication process prevents CNT agglomeration and degradation of the CNT properties that are common in dispersion-based processes."

"We specifically explored the electrical conducting mechanism of CNT–PI composites, given that we sought a simple method to control CNT distribution within a polymer matrix, while protecting the CNT properties," adds Nurit Atar, a PhD candidate in Hanein's group. "We found that the conductivity of the CNT sheet was preserved in spite of the insulative PI infiltration. This implies that the electrical current was enabled through the original entangled CNT network that was not interrupted by the insulative PI. This proves that the polymer solution did not penetrate into the interface at the CNT junctions and so the original continuum of ohmic contacts between adjacent CNTs was preserved."

CNT–PI composites were produced before by dispersion of CNT powder in polymer matrices. Since CNTs are insoluble and tend to agglomerate in bundles, sonication and functionalization are commonly used to improve homogeneity. These incorporation techniques often result in severe degradation of the original CNT properties (*e.g.*, electrical, thermal, and mechanical characteristics).

The preparation method used by the Israeli team is based on PI infiltration into CVD-grown CNT sheets, enabling preservation of the original CNT sheet conductivity with no degradation related to the insulating PI matrix.

Hence, higher electrical conductivity can be easily achieved, *e.g.*, by controlling the CNT growth process (CVD) to form denser CNT sheets with higher conductivities.

"Another advantage of the technique presented in our work is the compatibility with patterning of CNTs along the composites, which is not facilitated by the dispersion technique," Atar points out.

As electrically conducting films, the CNT–PI composites can prevent electrostatic charge accumulation on the exterior of satellites. Particularly, the researchers found that their CNT–PI films are durable in space environment hazards such as high vacuum, thermal cycling, and ionizing radiation.

The team is working on improving the stability of the CNT–PI film to atomic oxygen, which is dominant at a low Earth orbit space environment. The challenge is to introduce inorganic nanoparticles into the polymer matrix to create a self-passivation layer when exposed to atomic oxygen.

Featured scientists: Prof. Yael Hanein's research group (http://nano.tau.ac.il/hanein/)
Organization: Center for Nanoscience and Nanotechnology, Tel Aviv University (Israel) (http://nano.tau.ac.il/)
Relevant publication: N. Atar, E. Grossman, I. Gouzman, A. Bolker and Y. Hanein, *ACS Appl. Mater. Interfaces*, 2014, **6**(22), 20400, DOI: 10.1021/am505811g

6.6.10   A Nanoscale Color Filter

For the past decade, researchers have searched for robust, inorganic color filters that can replace traditional organic dye-based filters for better stability, lifetimes, performance, and amenability to miniaturization.

Some popular solutions have come in the form of metallic gratings, where the filtering capability is based on the excitation and interference of surface plasmon polaritons. Due to the nature of their operational principle, these types of color filters provide the best spectral-selectivity when the grating includes a large number of grating elements spaced regularly apart from one another.

Since more gratings means a larger size, achieving truly nanoscale color filters with reliable spectral-selectivity can be a challenging task with plasmonic grating-based filters.

There has been tremendous interest in trying to scale down micron-sized photonic components to the nanoscale, so that they can become compatible with nanoscale electronic components. One of the key components to photonics is the optical filter.

Work by a research team in Korea, realizing an inorganic filter that can operate with a single element, represents an important step toward nanoscale color filters. The team devised a simple design in which light can be filtered and tuned over wavelengths through the use of a single nanoscale element in the form of a zinc oxide (ZnO) nanorod integrated with a silver cavity.

These results describe the smallest color filter to date. This work also ushers in possibilities of realizing filters at even smaller sizes.

"Our work takes advantage of a recent development in the field of nanophotonics, namely metal-shell induced scattering cancellation a.k.a. metal-shell induced transparency," says Jerome K. Hyun, an assistant professor in the Department of Chemistry and Nano Science at Ewha Womans University.

Hyun and colleagues from Professor Gyu-chul Yi's team at Seoul National University, fabricated a device that consists of a ZnO nanowire sandwiched between two silver films.

"At the filter wavelength, the metal effectively acts as an imperfect cloak to the nanowire and makes the nanowire appear more transparent than its surroundings," explains Hyun. "This allows the light at this particular wavelength to transmit through the nanowire more than it would through the neighboring silver film. We can choose the color with excellent spectral selectivity by simply changing the diameter of the nanowire."

He points out that this work is an application of the theory developed by professors Andrea Alù at UT Austin and Nader Engheta at the University of Pennsylvania.[45]

The inspiration for the team's work came from the realization that metal coatings offer intriguing ways of manipulating the scattering and absorption properties of small nanostructures.

Small nanostructures already provide interesting optical characteristics not seen in the bulk material. The metal component adds another degree of freedom to modifying their properties.

"We had also been interested in the filtering properties of gratings and had been investigating new types of strategies for manipulating their transmission characteristics," recounts Hyun. "These two themes helped drive the project forward toward realization of a new model of optical filters."

According to the researchers, a remaining challenge is to increase the absolute transmission efficiency. Because this nanoscale device is several times smaller than the smallest probe size achievable with optical lenses, it is difficult to achieve transmission efficiencies greater than 60% with the current measurement setup.

"However, we found that by adding more of our filters, we can increase the transmission efficiency while maintaining the excellent spectral-selectivity," notes Hyun. "We now are working in this direction."

Future work would also address challenges in the construction of these nanofilters over a large scale, aimed at display applications, and increasing the absolute transmission efficiency using different core materials.

These nanoscale filters could serve as optical bandpass filters in nanoscale communication, for instance to clean up the signal sent from a local emitter such as a single molecule.

Another specific application is ultra-high-resolution color filters. By arranging the nanowires into three types of arrays, each consisting of nanowires of a different diameter, it is possible to build a RGB color filter but at sizes that are substantially smaller than traditional color pixels and plasmonic grating based pixels.

"I believe our work is noteworthy because it provides new possibilities for color filters with practical implications in ultra-high-resolution displays and nanoscale communication, nanoscale bio-imaging *etc.*," Hyun concludes.

Featured scientist: Prof. Jerome K. Hyun
Organization: Department of Chemistry and Nanoscience, Ewha Womans University, Seoul (Republic of Korea) (http://my.ewha.ac.kr/chem/en/)
Relevant publication: J. Hyun, T. Kang, H. Baek, D. Kim and G. Yi, Nanoscale Single-Element Color Filters, *Nano Lett.*, 2015, **15**(9), 5938, DOI: 10.1021/acs.nanolett.5b02049

6.6.11   Self-Healing Hybrid Gel System

Previously reported conductive self-healing materials usually need large amounts of inorganic conducting fillers and their self-healing behaviors are only activated under specific external stimuli, such as heat, light or pH.

Some supramolecular gels with self-healing properties have been developed in the past. However, they are seldom adopted in practical applications due to weak mechanical strength and lack of high conductivity. In the meantime, nanostructured conductive hydrogels have been developed and demonstrated to be an ideal candidate to construct conducting networks in hybrid gel materials, as well as acting as a robust matrix to support other components.

Researchers at The University of Texas at Austin together with collaborators at Texas State University developed a hybrid gel that is composed of a conductive polymer and a metal–ligand supramolecule; the novel gel exhibits attractive properties associated with both conventional polymers, such as ease of synthesis and processing, and great self-healing performance at room temperature without any stimuli.

The most important result of this work is the development of a supergel material, which shows exciting features of high conductivity, appealing mechanical and electrical self-healing properties without any external stimuli, and enhanced mechanical strength and flexibility.

"Owing to these exciting features, our designed hybrid gel could greatly extend the application of self-healing materials to flexible and printable electronics, artificial skins, durable medical devices, even in energy devices," says Guihua Yu, an assistant professor of materials science at The University of Texas at Austin.

To demonstrate the potential for practical applications, Yu and his team fabricated thin-films of the hybrid gel on flexible substrates to test their self-healing electrical properties. They showed that the high conductivity of the hybrid gel can be maintained after extensive bending and stretching tests due to their good self-healing property.

Furthermore, they demonstrated with a simple electrical circuit made of the hybrid gel that, after being cut, it only takes about one minute for the circuit to self-heal and recover its original conductivity.

"In our work, we found that this supramolecular gel showed very appealing self-healing properties without any stimuli because it could dynamically assemble or disassemble, associate or dissociate at room temperature due to the moderate bond energy of metal–ligand bonds and non-covalent interactions among supramolecules," Yu explains. "The conductive polymer gel serving as 'host' exhibited a hierarchically porous structure providing 3D interconnected paths for electron transport and an ideal interface."

This work opens a field in which supramolecular gels could be used for numerous practical applications. Supramolecular gels with decent self-healing properties could be hybridized with conventional polymer gels and the resulting hybrid gel materials could be used for various applications that require high conductivity, a room-temperature self-healing property, and decent mechanical properties.

For example, the hybrid gel could be used to construct conducting circuits in flexible electronics, which would be self-repaired under physical damage. The gel can also be adopted in implantable biosensors to act as a flexible electrode, ensuring the durability of these devices. In addition, the gel material could be used for the fabrication of artificial skins. Last but not least, it can be used in many energy devices too, for example, as a functional binder material for advanced battery electrodes in high-density Li-ion batteries or other energy storage systems.

"Fundamentally, in terms of synthetic strategies, our work synergizes the sol–gel transition property of supramolecular gels and the hierarchically porous nanostructure of conductive polymer gels," notes Yu. "The 'guest to host' strategy adopted in this work could be an important and fundamental way to design and synthesize a new class of functional polymeric materials."

Although the mechanical strength of this novel hybrid gel is enhanced when compared to those of pure supramolecular gel or PPy aerogel, it is still relatively weak and the stretchability is also relatively poor. Yu's team believes this problem could be solved by replacing the bonding metal ions in supramolecules with a higher bonding energy and tuning the structure of conductive PPy gel through other synthetic methods that potentially can lead to better structure-derived elasticity.

The team is also investigating the fundamental mechanisms for the gelation and self-healing property of their supramolecular gel.

"As scientists, we would like to understand more clearly the fundamental mechanisms of self-healing properties of supramolecular gels, including the influence of different metal ions, the effects of molecules' geometry, and the interactions between supramolecule and different solvents," says Yu. "With this deeper understanding, better materials can be developed."

"In general, supramolecules consist of large molecular subunits, and given their size and structure, the assembly of these large subunits is held together by relatively weak 'non-covalent' interactions, giving rise to dynamic reversibility of assembly/disassembly to self-repair," he elaborates. "We are trying to understand the role of the cubic cage structure of our supramolecule in the processes of gelation and self-healing."

Featured scientists: Prof. Guihua Yu's research group (http://yugroup. me.utexas.edu/)
Organization: Materials Science at The University of Texas at Austin (USA)
Relevant publication: Y. Shi, M. Wang, C. Ma, Y. Wang, X. Li and G. Yu, A Conductive Self-Healing Hybrid Gel Enabled by Metal–Ligand Supramolecule and Nanostructured Conductive Polymer, *Nano Lett.*, 2015, **15**(9), 6276, DOI: 10.1021/acs.nanolett.5b03069

6.6.12 Nanowire Structures Lead to White-Light and AC-Operated LEDs

Compared to conventional inefficient incandescent and fluorescent lighting technologies, LED light bulbs can, in principle, operate at an efficiency level of 100%. Current LED lighting technology, however, is not even close to reaching this limit.

This is due to several factors. First, current LED lamps still rely on the use of phosphors to down-convert blue light into green and red light. Associated with this down-conversion process is an energy loss of approximately 30% or more.

Second, the performance of GaN-based LEDs has been limited by the inefficient current conduction of p-GaN, which typically has a resistance of about 100 times higher than that of n-GaN, leading to poor current spreading, reduced efficiency, and efficiency droop (LEDs operate most efficiently at low currents of tens of milliamps; however, if the current increases, efficiency tails off in a phenomenon known as "efficiency droop").

Third, unlike conventional light bulbs, LEDs are low-voltage devices and cannot operate on an alternating current voltage. As a consequence, an electrical circuit is required to convert AC power to low-voltage DC power (typically 2–4 V). Such a driver adds a significant level of complexity, cost, and efficiency loss to the LED devices and systems.

By and large, these problems can be solved by employing tunnel junction integration into current nanowire LED structures.

Doing exactly that, researchers at McGill University in Montreal have developed tunnel junction nanowire LEDs that can eliminate the use of resistive p-GaN contact layers, leading to reduced voltage loss and enhanced hole injection. Moreover, by using tunnel junction interconnects, they demonstrated multiple-active-region (MAR) nanowire LEDs with significantly enhanced light intensity.

"We have also realized AC operated nanowire LEDs on a silicon platform, which operate efficiently in both polarities (positive and negative) of applied voltage," says Sharif Sadaf, a post-doctoral researcher in Zetian Mi's MBE group at McGill, and first author of a paper on this work. "Compared to the current quantum well LEDs, the demonstrated tunnel junction nanowire LED technology enables phosphor-free white emission and reduced efficiency droop. It offers extreme flexibility in the operation voltage and can completely eliminate the use of an AC/DC converter required in conventional LED lighting technologies, thereby leading to reduced cost and further enhanced efficiency."

p-GaN contact resistance has been a long-standing bottleneck in improving the performance of conventional nitride-based LEDs. Researchers have therefore been looking for fundamental design modifications.

"Polarization engineered tunnel junctions offer the unique opportunity to eliminate the p-GaN contact resistance problem by replacing resistive p-GaN with n-GaN," explains Sadaf. "Moreover, by employing the tunnel junction scheme, we demonstrated a multiple-active region nanowire LED that can potentially circumvent the 'efficiency droop' problem."

He notes that, in general, stacking multiple quantum wells/dots in planar structures is not a suitable route to realize low current, high voltage operation since it also significantly increases the densities of defects and dislocations.

"These issues can be fundamentally addressed in tunnel junction nanowire LED structures, as demonstrated in our work," says Sadaf. "Moreover, such MAR tunnel junction nanowire LEDs can be designed to operate in a broad wavelength range, leading to phosphor-free white light emission."

A unique result of this work is the substantially improved light intensity compared to single active region LEDs.

The scientists attribute this striking improvement to carrier regeneration at each tunnel junction and uniform low mobility hole injection in each active region.

This work has the potential to go a long way toward solving low-power LED applications. According to Sadaf, it is feasible to integrate more active regions into the nanowire structures to achieve higher light output power from a single chip.

It is also possible to integrate different wavelength color in a single nanowire to obtain white light emission.

More importantly, having demonstrated the AC power operation of their devices, this could give lighting devices great flexibility for usage in household applications.

In principle, this concept can be further extended to other semi-polar or non-polar III–V nanowire-based applications. Already, the team is exploring other nanowire-based optoelectronics devices such as lasers, UV-LEDs, and photodetectors.

Featured scientists: Dr Sharif Sadaf; Prof. Zetian Mi's research group (https://sites.google.com/site/zetianmi/Home)
Organization: Department of Electrical & Computer Engineering, McGill University, Montreal (Canada)
Relevant publication: S. Sadaf, Y. Ra, H. Nguyen, M. Djavid and Z. Mi, Alternating-Current InGaN/GaN Tunnel Junction Nanowire White-Light Emitting Diodes, *Nano Lett.*, 2015, **15**(10), 6696, DOI: 10.1021/acs.nanolett.5b02515

6.6.13 Spiders Inspire Better Adhesives for High-Humidity Environments

Scientists have long been fascinated by spider silk—a unique biopolymer that combines mechanical strength and elasticity to make it one of the toughest materials known to man. In addition, the silk threads are coated with an adhesive, which ranks among the strongest biological glues.

Spiders are among the most diverse species on the planet. Currently, there are about 45 000 known species of spiders living in a variety of habitats and environments. About 1/6th of these species use webs to catch prey.

The capture silk used in these webs consists of axial fibers coated with glue droplets at regular intervals. The spider glue has a unique property: its adhesion is humidity responsive; for some species, the adhesion keeps on increasing up to 100% relative humidity.

This is unlike synthetic adhesives that fail under humid conditions. From a polymer science perspective, researchers are interested in understanding the principle behind humidity responsive adhesion of spider glue to create adhesives that work in high humidity conditions.

"Adhesion in a high humidity environment is a fundamental challenge for synthetic and natural adhesives; yet, some spider species that are active in highly humid environments use glue that is the stickiest in almost 100% humidity conditions," says Ali Dhinojwala, H. A. Morton professor in the Department of Polymer Science at the University of Akron. "We find that the spider glue from five different species, living in diverse habitats, is maximally adhesive at the humidity where the spider hunts for prey. This is intuitive but beautiful to observe in data."

Previously, Dhinojwala's group had worked on understanding the mechanism of spider glue adhesion at a particular humidity.[46] In subsequent work, the team discusses the mechanism of humidity responsive adhesion of spider glue.

"We observed that the glue extensibility increased dramatically with an increase in humidity," says Gaurav Amarpuri, a PhD student in Dhinojwala's group and first author of a paper on this work. "We used high speed imaging to quantify the spreading of glue and further used the spreading power law to measure the glue viscosity (Figure 6.12)."

The scientists found that glue viscosity changes over five orders of magnitude with a change in humidity from 30–90% relative humidity. No other system changes viscosity so dramatically with relative humidity. This change is similar to glue changing from the consistency of peanut butter to the consistency of olive oil. However, at the humidity of maximum adhesion, the glue of these five diverse species had very similar viscosity in the range of $10^5–10^6$ cP (centipoise).

"This shows that there exists an optimum viscosity of maximum adhesion, and different spider species tune their glue viscosity to achieve this maximum adhesion at very different humidities," Amarpuri points out. "We found that the spider glue adhesion was maximum at the humidity the spider glue

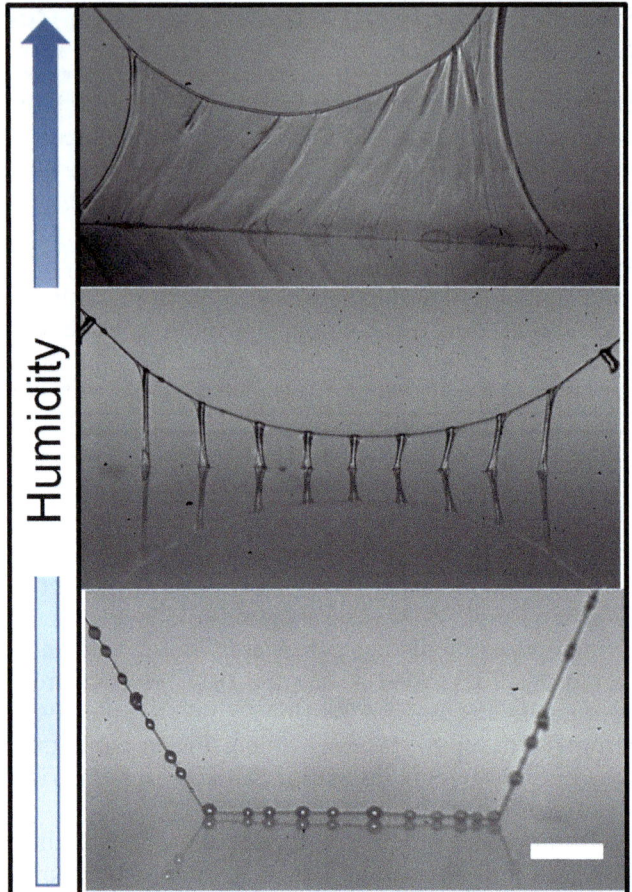

Figure 6.12   *Larinioides cornutus* capture threads peeled from a glass substrate, under low, medium, and high humidity conditions. Note the increase in glue extensibility with humidity. At medium humidity, the glue droplets form a suspension bridge structure where adhesion is at a maximum. Scale bar is 100 μm. (Image: Yizhou Chen, University of Akron.)

is supposed to function. For example, the *Argiope* species is active during the day in open fields, and its glue was stickiest at 30–50%, while the *Tetragnatha* species is active at night above streams of water, and its adhesion kept on increasing up to 100% relative humidity."

This observation poses a lot of questions for the scientists—the most obvious one being: how do these different species modulate their glue adhesion performance?

"Spider glue consists of proteins and salts," explains Dhinojwala. "Salts are low molecular weight organic compounds, which are hygroscopic. The interaction between salts and proteins causes the viscosity to change dramatically with humidity."

"Investigating the protein structure of the capture glue in different spider species is challenging due to limitation in sample size and availability," he adds. "But the salts vary significantly in ratio and composition in species. We are trying to understand the role of salts and proteins in modulating viscosity and hence adhesion."

This study highlights viscosity modulation as a potent tool that various organisms use to derive different functionalities. The mechanism of how spider species are able to adapt to their local environment and modulate viscosity will help in the development of next generation tunable adhesives.

Further efforts in quantification of bulk rheological changes in biological materials may result in a new class of smart materials.

Featured scientists: Prof. Ali Dhinojwala's research group (http://blogs. uakron.edu/dhinojwala/)
Organization: Department of Polymer Science, University of Akron, OH (USA)
Relevant publication: G. Amarpuri, C. Zhang, C. Diaz, B. Opell, T. Blackledge and A. Dhinojwala, Spiders Tune Glue Viscosity to Maximize Adhesion, *ACS Nano*, 2015, **9**(11), 11472, DOI: 10.1021/acsnano.5b05658

6.6.14   Studying Phase Transformations of a Single Nanoparticle at the Atomic Level

Catalytic nanoparticles play a crucial role in accelerating chemical reactions by offering their active sites and surfaces. Fine-tuning surface structure during synthesis and phase transformations can enhance their catalytic activity and durability by manifolds. However, control and feedback on the synthesis/transformation studies demand characterization techniques that can provide structural and compositional information on the atomic level.

"Recent developments in electron microscopy pertaining to its instrumentation (aberration correctors, ultra-fast electron energy loss spectrometers, monochromators) and the innovation with respect to specimen holders (*in situ* liquid cell, *in situ* heating, tomography) have benefitted the study of catalyst nanoparticles enormously," explains Sagar Prabhudev, a PhD student in Professor Gianluigi Botton's microscopy of nanoscale materials research group at McMaster University. "As a leap forward in studying the phase transformation of nanoparticles, we have, for the first time, coupled advanced electron microscopy techniques such as high angle annular dark field imaging (HAADF) and electron energy loss spectroscopy (EELS), with *in situ* heating ability."

By exploiting this setup, the research team studied the disorder-to-order transformation of platinum–iron (Pt–Fe) alloy nanoparticles, a material that is of great interest for fuel cell electrocatalysis as well as ultra-high density information storage (Figure 6.13).

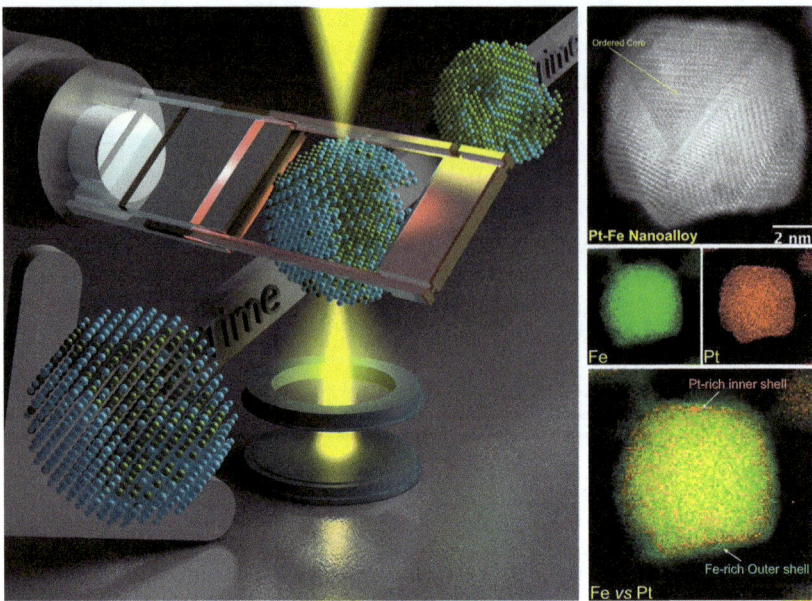

Figure 6.13   Surface segregation of Fe in Pt–Fe alloy nanoparticles. Left: schematic illustration of the study carried out in the work. One single nanoparticle is being annealed here, from initial (top right) to final (bottom left), the red-hot holder shows the high temperature/heating, the time arrow shows dynamic measurements. The yellow beam represents an electron probe on the atomic level shining on a transition-state particle, and scattered electrons are collected using various detectors. Right: gray-scale image is the atomic-resolution STEM-HAADF image of a heat-treated Pt–Fe alloy nanoparticle. Colored images are EELS compositional maps, green (Fe), red (Pt) and composite (Fe *versus* Pt). (Image: microscopy of nanoscale materials research group, McMaster University.)

The team's work is highly relevant against the backdrop of fuel cell research, specifically polymer electrolyte membrane fuel cells (PEMFC)—a highly promising energy conversion system that can convert chemical energy into electrical energy with minimal or zero emission of greenhouse gases. The cathodic reaction—called the oxygen reduction reaction (ORR)—in a PEMFC is notoriously sluggish, requiring efficient catalysts to accelerate the kinetics. Traditionally, platinum has been used in this regard, but despite its exceptional catalytic activity, it is now widely accepted to be an unrealistic choice due to depleting stocks and exceedingly high costs.

In an ongoing attempt to reduce the mass loading of Pt, there is a large body of research that suggests that a nanoscale alloying of Pt with transition metals (such as Fe) is a more viable option.

"In this regard, there is growing interest in optimizing the structure–property–performance relationship of Pt–Fe alloy nanoparticles towards fuel cell catalysis," notes Prabhudev. "One prominent approach to obtain catalytically superior Pt–Fe alloy structures is by promoting a diffusion of Pt and Fe atoms through thermal annealing (or, simply put, heating). Depending on where the atoms end up after heating, different structures result and consequently, their properties can be optimized."

"For instance," he points out, "platinum atoms on the surface—in the form of a shell—are great for catalysis but an iron shell is not. So, in order to enhance their performance, it was important for us to first understand the atomic rearrangement process taking place during heating."

By integrating atomic-resolution imaging and *in situ* heating with spectroscopy techniques, the scientists captured the interplay between various nanoscale phenomena such as surface-segregation, atomic-ordering, Ostwald ripening and coalescence. They point out that these insights were hitherto inaccessible with traditional analytical methods.

Owing to instrumental limitations, nanoparticle phase transformations have traditionally been carried out *ex situ, i.e.* heating a batch of nanoparticles in a furnace and then examining them under the microscope to see what changes did occur.

However, given the extremely small size of nanoparticles, it was difficult to track the same particle. This meant that analytical work was carried out on size-wise similar—but not identical—particles, often resulting in misleading findings.

According to Prabhudev, this work could have broad implications in various areas:

Advanced electron microscopy—the analytical methods described in this work have broadened the scope of electron microscopy towards characterizing the dynamic behavior of nanoparticles. More generally, these techniques are widely applicable to studying phase transformations of a broad range of nanomaterials. Future efforts could be expected in the direction of gas-phase annealing in environmental TEMs; optimizing electron microscopy to minimize beam-induced damage; and high-temperature drifts.

Nanoscale phase transformations—tracking the same nanoparticle is arguably the best approach to maximize insights obtained on the dynamic behavior of nanoparticles during heat-treatment. Although the interplay between surface-segregation and atomic-ordering has been illustrated in this work, there can be interplay amongst all these processes, and future efforts towards elucidating them would be of great value. Especially in the case of polydispersed samples, although time–cost intensive, experiments on a range of particle sizes would be ideal to build a 3D data cube of size *versus* composition *versus* temperature.

Fuel cell catalysis—in contrast to earlier reports suggesting Pt segregation, the team's findings reveal surface-segregation of Fe and the formation of a Fe-rich shell. However, from an ORR electrocatalysis viewpoint, it is not desirable to have Fe-rich shells. Consequently, future efforts could be directed towards performing a similar study in relation to other catalytic nanoalloy systems such as (Pt–Co, Pt–Ni, Pt–Cu)—and if the transition metal (Fe, Co, Ni, Cu) segregation is inevitable, then employ routine surface-cleaning techniques such as voltammetric dealloying as a post-treatment after annealing.

Interestingly, numerous recent studies have shown that it is possible to also achieve Pt-shell structures (ideal for catalysis) with a controlled voltammetric dealloying procedure.

Featured scientists: Sagar Prabhudev; Prof. Gianluigi Botton's microscopy of nanoscale materials research group (https://sites.google.com/site/mnmrgccem/home)
Organization: Department of Materials Science and Engineering, McMaster University, Hamilton, ON (Canada)
Relevant publication: S. Prabhudev, M. Bugnet, G. Zhu, C. Bock and G. Botton, Surface Segregation of Fe in Pt–Fe Alloy Nanoparticles: Its Precedence and Effect on the Ordered-Phase Evolution during Thermal Annealing, *ChemCatChem*, 2015, 7(22), 3655, DOI: 10.1002/cctc.201500380

References

1. N. Perur, M. Yahara, T. Kamei and N. Tamaoki, A non-nucleoside triphosphate for powering kinesin-microtubule motility with photo-tunable velocity, *Chem. Commun.*, 2013, 49(85), 9935, DOI: 10.1039/C3CC45933K.
2. W. Wang, *et al.*, Rational Synthesis of Helically Coiled Carbon Nanowires and Nanotubes Through the Use of Tin and Indium Catalysts, *Adv. Mater.*, 2008, 20, 179, DOI: 10.1002/adma.200701143.
3. see the original 2006 paper: P. W. K. Rothemund, Folding DNA to Create Nanoscale Shapes and Patterns, *Nature*, 2006, 440, 297, DOI: 10.1038/nature04586.
4. G. Loget and A. Kuhn, Electric Field-Induced Chemical Locomotion of Conducting Objects, *Nat. Commun.*, 2011, 2, 535, DOI: 10.1038/ncomms1550.
5. W. Gao, A. Uygun and J. Wang, Hydrogen-Bubble-Propelled Zinc-Based Microrockets in Strongly Acidic Media, *J. Am. Chem. Soc.*, 2012, 134(2), 897, DOI: 10.1021/ja210874s.
6. Watch a the micromotor in action in this video: https://youtu.be/O0Dxn3_24fY.
7. A free download available on the FDA's website: http://www.fda.gov/Food/FoodborneIllnessContaminants/CausesOfIllnessBadBugBook/.
8. W. Gao, S. Sattayasamitsathit, J. Orozco and J. Wang, Highly Efficient Catalytic Microengines: Template Electrosynthesis of Polyaniline/Platinum Microtubes, *J. Am. Chem. Soc.*, 2011, 133(31), 11862, DOI: 10.1021/ja203773g.
9. S. Campuzano, D. Kagan, J. Orozco and J. Wang, Motion-driven sensing and biosensing using electrochemically propelled nanomotors, *Analyst*, 2011, 136(22), 4621, DOI: 10.1039/C1AN15599G.
10. Y. Shi, M. Wang, C. Ma, Y. Wang, X. Li and G. Yu, A Conductive Self-Healing Hybrid Gel Enabled by Metal–Ligand Supramolecule and Nanostructured Conductive Polymer, *Nano Lett.*, 2015, 15(9), 6276, DOI: 10.1021/acs.nanolett.5b03069.

11. Y. Huang, Y. Huang, M. Zhu, W. Meng, Z. Pei and C. Liu, *et al.*, Magnetic-Assisted, Self-Healable, Yarn-Based Supercapacitor, *ACS Nano*, 2015, **9**(6), 6242, DOI: 10.1021/acsnano.5b01602.

12. This video shows how originally dispersed swimming nanomotors gradually accumulate and aggregate into a surface crack: https://youtu.be/kV1eNc1wXKM.

13. C. Zhong, Y. Deng, A. Roudsari, A. Kapetanovic, M. Anantram and M. Rolandi, A polysaccharide bioprotonic field-effect transistor, *Nat. Commun.*, 2011, **2**, 476, DOI: 10.1038/ncomms1489.

14. You can watch an amazing video of a camouflaging octopus here: https://youtu.be/PmDTtkZlMwM.

15. M. Ryou, J. Kim, I. Lee, S. Kim, Y. Jeong and S. Hong, *et al.*, Mussel-Inspired Adhesive Binders for High-Performance Silicon Nanoparticle Anodes in Lithium-Ion Batteries, *Adv. Mater.*, 2013, **25**(11), 1571, DOI: 10.1002/adma.201203981.

16. D. García-Hernández, S. Iglesias-Groth, J. Acosta-Pulido, A. Manchado, P. García-Lario and L. Stanghellini, *et al.*, The Formation of Fullerenes: Clues from New C_{60}, C_{70}, and (Possible) Planar C_{24} Detections in Magellanic Cloud Planetary Nebulae, *Astrophys. J.*, 2011, **737**(2), L30, DOI: 10.1088/2041-8205/737/2/L30.

17. Q. Li, L. Xia, Z. Zhang and M. Zhang, Ultraviolet Extinction and Visible Transparency by Ivy Nanoparticles, *Nanoscale Res. Lett.*, 2010, **5**(9), 1487, DOI: 10.1007/s11671-010-9666-2.

18. M. Zhang, M. Liu, H. Prest and S. Fischer, Nanoparticles Secreted from Ivy Rootlets for Surface Climbing, *Nano Lett.*, 2008, **8**(5), 1277, DOI: 10.1021/nl0725704.

19. S. Baluschev, T. Miteva, V. Yakutkin, G. Nelles, A. Yasuda and G. Wegner, Up-Conversion Fluorescence: Noncoherent Excitation by Sunlight, *Phys. Rev. Lett.*, 2006, **97**(14), 143903, DOI: 10.1103/PhysRevLett.97.143903.

20. J. Greffet, Nanoantennas for Light Emission, *Science*, 2005, **308**(5728), 1561, DOI: 10.1126/science.1113355.

21. M. Busson, B. Rolly, B. Stout, N. Bonod, E. Larquet and A. Polman, *et al.*, Optical and Topological Characterization of Gold Nanoparticle Dimers Linked by a Single DNA Double Strand, *Nano Lett.*, 2011, **11**(11), 5060, DOI: 10.1021/nl2032052.

22. A. Curto, G. Volpe, T. Taminiau, M. Kreuzer, R. Quidant and N. van Hulst, Unidirectional Emission of a Quantum Dot Coupled to a Nanoantenna, *Science*, 2010, **329**(5994), 930, DOI: 10.1126/science.1191922.

23. M. Marini, L. Piantanida, R. Musetti, A. Bek, M. Dong and F. Besenbacher, *et al.*, A Revertible, Autonomous, Self-Assembled DNA-Origami Nanoactuator, *Nano Lett.*, 2011, **11**(12), 5449, DOI: 10.1021/nl203217m.

24. R. Zadegan, M. Jepsen, K. Thomsen, A. Okholm, D. Schaffert and E. Andersen, *et al.*, Construction of a 4 Zeptoliters Switchable 3D DNA Box Origami, *ACS Nano*, 2012, **6**(11), 10050, DOI: 10.1021/nn303767b.

25. D. Kim, N. Lu, R. Ma, Y. Kim, R. Kim and S. Wang, *et al.*, Epidermal Electronics, *Science*, 2011, **333**(6044), 838, DOI: 10.1126/science.1206157.

26. X. Wang, L. Dong, H. Zhang, R. Yu, C. Pan and Z. Wang, Recent Progress in Electronic Skin, *Adv. Sci.*, 2015, **2**, 1500169, DOI: 10.1002/advs.201500169.

27. F. Schedin, A. Geim, S. Morozov, E. Hill, P. Blake and M. Katsnelson, *et al.*, Detection of individual gas molecules adsorbed on graphene, *Nat. Mater.*, 2007, **6**(9), 652, DOI: 10.1038/nmat1967.

28. J. An, S. Park, O. Kwon, J. Bae and J. Jang, High-Performance Flexible Graphene Aptasensor for Mercury Detection in Mussels, *ACS Nano*, 2013, **7**(12), 10563, DOI: 10.1021/nn402702w.

29. Watch an animation of the optical topological transition occurring in metamaterials: https://youtu.be/FG6Obwddca4.

30. See for instance: A. Alù and N. Engheta, Achieving transparency with plasmonic and metamaterial coatings, *Phys. Rev. E: Stat., Nonlinear, Soft Matter Phys.*, 2005, **72**(1), 016623, DOI: 10.1103/PhysRevE.72.016623.

31. A. Alù, Mantle cloak: Invisibility induced by a surface, *Phys. Rev. B*, 2009, **80**(24), 245115, DOI: 10.1103/PhysRevB.80.245115.

32. A. Alù and N. Engheta, Cloaking a Sensor, *Phys. Rev. Lett.*, 2009, **102**(23), 233901, DOI: 10.1103/PhysRevLett.102.233901.

33. J. Yang, Z. Liu, F. Grey, Z. Xu, X. Li and Y. Liu, *et al.*, Observation of High-Speed Microscale Superlubricity in Graphite, *Phys. Rev. Lett.*, 2013, **110**(25), 255504, DOI: 10.1103/PhysRevLett.110.255504.

34. Q. Wen, W. Qian, J. Nie, A. Cao, G. Ning and Y. Wang, *et al.*, 100 mm Long, Semiconducting Triple-Walled Carbon Nanotubes, *Adv. Mater.*, 2010, **22**(16), 1867, DOI: 10.1002/adma.200902746.

35. R. Zhang, Q. Wen, W. Qian, D. S. Su, Q. Zhang and F. Wei, Superstrong Ultralong Carbon Nanotubes for Mechanical Energy Storage, *Adv. Mater.*, 2011, **23**(17), 3387, DOI: 10.1002/adma.201100344.

36. Q. Wen, R. Zhang, W. Qian, Y. Wang, P. Tan and J. Nie, *et al.*, Growing 20 cm Long DWNTs/TWNTs at a Rapid Growth Rate of 80–90 μm/s, *Chem. Mater.*, 2010, **22**(4), 1294, DOI: 10.1021/cm903866z.

37. R. Zhang, Y. Zhang, Q. Zhang, H. Xie, H. Wang and J. Nie, *et al.*, Optical visualization of individual ultralong carbon nanotubes by chemical vapour deposition of titanium dioxide nanoparticles, *Nat. Commun.*, 2013, **4**, 1727, DOI: 10.1038/ncomms2736.

38. M. Urbakh, Friction: Towards macroscale superlubricity, *Nat. Nanotechnol.*, 2013, **8**(12), 893, DOI: 10.1038/nnano.2013.244.

39. M. Chaudhury and G. Whitesides, How to Make Water Run Uphill, *Science*, 1992, **256**(5063), 1539, DOI: 10.1126/science.256.5063.1539.

40. J. Berná, D. Leigh, M. Lubomska, S. Mendoza, E. Pérez and P. Rudolf, *et al.*, Macroscopic transport by synthetic molecular machines, *Nat. Mater.*, 2005, **4**(9), 704, DOI: 10.1038/nmat1455.

41. K. Ichimura, Light-Driven Motion of Liquids on a Photoresponsive Surface, *Science*, 2000, **288**(5471), 1624, DOI: 10.1126/science.288.5471.1624.

42. See for instance: J. Huang, Q. Li, D. Sun, Y. Lu, Y. Su and X. Yang, *et al.*, Biosynthesis of silver and gold nanoparticles by novel sundried Cinnamomum camphora leaf, *Nanotechnology*, 2007, **18**(10), 105104, DOI: 10.1088/0957-4484/18/10/105104.

43. P. Walter, E. Welcomme, P. Hallégot, N. Zaluzec, C. Deeb and J. Castaing, *et al.*, Early Use of PbS Nanotechnology for an Ancient Hair Dyeing Formula, *Nano Lett.*, 2006, **6**(10), 2215, DOI: 10.1021/nl061493u.

44. G. Patriarche, P. Walter, E. Van Elslande, J. Ayache and J. Castaing, Characteristics of HgS nanoparticles formed in hair by a chemical reaction, *Philos. Mag.*, 2012, **93**(1–3), 137, DOI: 10.1080/14786435.2012.674225.

45. A. Alù and N. Engheta, Achieving transparency with plasmonic and metamaterial coatings, *Phys. Rev. E*, 2005, **72**(1), 016623, DOI: 10.1103/PhysRevE.72.016623.

46. G. Amarpuri, V. Chaurasia, D. Jain, T. Blackledge and A. Dhinojwala, Ubiquitous distribution of salts and proteins in spider glue enhances spider silk adhesion, *Sci. Rep.*, 2015, **5**, 9030, DOI: 10.1038/srep09030.

Nanotechnology to the Rescue—Environmental Applications

There is a general perception that nanotechnologies will have a significant impact on developing "green" and "clean" technologies with considerable environmental benefits. The best examples are the use of nanotechnology in areas ranging from water treatment to energy breakthroughs and hydrogen applications.

Conflicting with this positive message is the growing body of research that raises questions about the potentially negative effects of engineered nanoparticles on human health and the environment. This area includes the actual processes of manufacturing nanomaterials and the environmental footprint they create, in absolute terms and in comparison with existing industrial manufacturing processes.

7.1 A Simple Test Kit for the Detection of Nanoparticles

The use of engineered nanoparticles in commercial products and materials continues at an increasing rate and it appears that no industry can do without them. Nanoparticles are widely added to cosmetics, textiles, catalysts, pharmaceuticals, polishing media, paints, sensors, magnetic fluids, lubricants, foodstuffs, coatings, water treatment and plastics—and that is just a random and incomplete list of applications.

Due to the widespread use of nanoparticles, there is concern from the scientific community and environmental groups regarding the health and

Nanotechnology: The Future is Tiny
By Michael Berger
© Michael Berger 2016
Published by the Royal Society of Chemistry, www.rsc.org

environmental impact of nanoparticles. The current lack of knowledge concerning the environmental fate and transport of nanoparticles is partly due to the absence of simple and affordable methodologies for their detection and characterization in complex samples—*e.g.*, waste waters, lakes, oceans, foods, biological fluids, and tissues.

Current analytical methods to detect and characterize nanoparticles require highly trained personnel to operate specialized equipment, expensive instrumentation, time-consuming sample preparation, and fixation that may disturb the natural state. Most importantly, these methods do not work well in environmental and biological samples.

Meeting the need for a reliable, sensitive, and accurate methodology for the detection of nanoparticles in complex samples, using low-cost and portable instrumentation, scientists at the University of Washington (UW) have developed a novel technique to quickly screen for the presence and reactivity of nanoparticles in commercial, environmental, and biological samples.

The researchers' ultimate goal is to develop a sensitive, simple, affordable, and portable alternative testing protocol that will determine the surface catalytic activity of nanoparticles, which ultimately screens for nanoparticles in complex matrices.

A team led by Jonathan D. Posner, Bryan T. McMinn Endowed Associate Professor at UW, developed a colorimetric assay—similar to a swimming pool test kit—that tests for the presence or absence of nanoparticles in relevant biological and environmental samples with sufficient sensitivity of parts per billion concentration levels (Figure 7.1).

"Surface reactivity of nanoparticles is a key emerging property related to potential toxicity of materials with living organisms," says Posner. "We leverage the surface catalytic redox properties of nanoparticles to provide a simple colorimetric detection assay."

He notes that there are three potential applications of this catalytic assay: (1) monitoring known nanoparticles in complex media over time during lab or field studies; (2) industrial hygiene settings where known nanoparticles are being used; and (3) screening for the presence of unknown surface reactive nanomaterials that would be complemented later by more advanced analytical techniques for identification.

In their experiments, the group tested several commercially available nanoparticles and showed that their assay functions in a wide range of complex matrices and does not require elaborate sample preparation, advanced instrumentation or highly trained personnel.

"We have selected a dye-reductant system that meets several key requirements," explains Charlie Corredor, a PhD graduate research assistant in Posner's group, and first author of a paper on this work. "This system can be detected colorimetrically (dye has different colors in its oxidized and reduced forms); consists of a dye whose reduction is thermodynamically, but not kinetically, favorable; exhibits minimal human toxic effects; uses materials that are not flammable and are stable at room temperature; and is inexpensive for future potential use as a commercially available screening assay for nanoparticles."

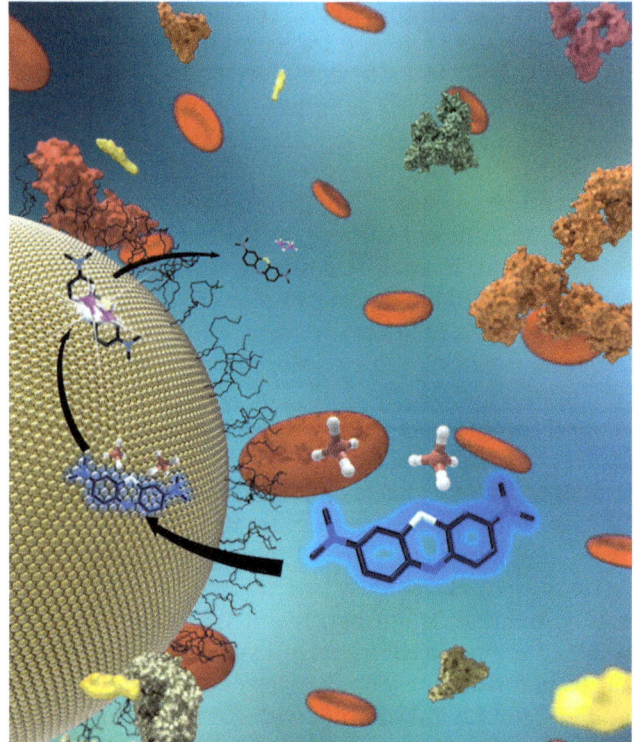

Figure 7.1   Schematic of reaction between nanoparticle surface reactivity and colorimetric assay components in complex matrices. (Image: Posner Research Group, University of Washington.)

The group's future work will focus on the development of a colorimetric assay that will be specific to a nanoparticle's characteristics, such as composition, size, concentration, and reactivity. This assay could be integrated into portable devices as well as testing equipment for commercial nanoenabled products during their life cycle.

"We are also interested in correlating nanoparticles' catalytic activity measured with our assay with cellular toxicity assays," concludes Posner. "Overall, this work represents a step toward reliable, sensitive, and accurate methodologies for the detection of nanoparticles in environmental and biological relevant samples, using low-cost and portable instrumentation."

Featured scientist: Prof. Jonathan D. Posner's research group (http://posner.uw.edu/)

Organization: Department of Mechanical Engineering, University of Washington, Seattle, WA (USA)

Relevant publication: C. Corredor, M. Borysiak, J. Wolfer, P. Westerhoff and J. Posner, Colorimetric Detection of Catalytic Reactivity of Nanoparticles in Complex Matrices, *Environ. Sci. Technol.*, 2015, **49**(6), 3611, DOI: 10.1021/es504350j

7.2 Low-Cost Nanotechnology Water Filter

The effluents of the tannery, paint, paper and textile industries containing different types of dyes are often discharged untreated into water bodies. The no. 1 polluter (after agriculture) of clean water is the textile industry, one of the most chemically intensive industries on the planet. The World Bank estimates[1] that 17 to 20 percent of industrial water pollution comes from textile dyeing and the finishing treatment given to fabric. Some 72 toxic chemicals have been identified in water solely from textile dyeing, 30 of which cannot be removed. This causes a serious environmental threat to aquatic and human life. Moreover, water treatment plants are very prone to fouling due to microorganism growth in the contaminated water, resulting in higher energy consumption and operating cost.

The development of sustainable, robust, energy-efficient and cost-effective water purification technologies is a challenging task. Conventional practices adopted for water purification—which can be classified into physical, chemical and biological methods—suffer from certain limitations such as high cost, low adsorption capacity, and generation of toxic sludge.

These technologies—which include coagulation, flocculation, reverse osmosis, membrane separation, oxidation and ozonation, adsorption—are expensive or inadequate to remove dye. Adsorption with activated carbons, which is a cheap and effective method, has been demonstrated to remove dye from wastewater. But this approach is not suitable for industrial wastewater treatment because activated carbon can only be used once and then it is commonly disposed of in landfills. Moreover, removal of pathogens from treated water requires additional processes such as chlorination, ozonation, *etc.*, which increase the cost of treatment.

A possible solution to tackle this problem has been demonstrated by scientists in India. They developed nanotechnology-based water purification using a nano-silica-silver composite material as an antifouling, antimicrobial and dye adsorptive material. With this process, pathogenic bacteria and dye present in contaminated water can be treated simultaneously without using any chemicals, high temperature, pressure or electricity.

"We synthesized a nanosilica supported silver nanocomposite material through ecofriendly protein mediated reduction of nano-silica bound silver ions," explains Dr Sujoy Das from the CSIR-Central Leather Research Institute in Chennai, India. "The proteins extracted from *Rhizopus oryzae*—a zygomycete fungus—served both as a reducing and a protecting agent for the silver nanoparticles and prevented their oxidation under environmental conditions. The result is a low-cost, highly effective nanomaterial for sustainable water purification."

The simple, low-temperature bio-synthesis fabrication process—it does not require any elaborate or expensive equipment—works without any chemicals for the reduction of silver ions and subsequent production of silver nanoparticles, thus minimizing the environmental load of toxic chemicals during the fabrication of this nanocomposite material. The coating of proteins on the nanoparticles' surface prevents the leaching of silver ions—which in itself could be a source of water contamination—and provides long-term stability of the nanocomposite.

The research team notes that the as-synthesized nanocomposite demonstrates very high dye removal capacities and exhibits antimicrobial and antifouling properties. The nanocomposite removes the dyes at wide pH, temperature and dye concentration in solution. Moreover, the nanocomposite kills the microorganisms frequently present in the contaminated water.

"Most importantly," says Das, "the silver nanocomposite very efficiently removes dyes and pathogenic microorganisms from water bodies in a single-step operation. In addition, the nanocomposite material could be regenerated after treatment of dye bearing wastewater and the regenerated nanocomposite could be stored and reused for several more cycles."

Nanosilica has a very high surface-to-volume ratio and contains a large number of surface hydroxyl groups, which provide an electrostatic binding energy for dye molecules on its surface. Combining this with the antibacterial activity of the silver nanoparticles on the surface of nano-silica results in a synergistic effect of the nanocomposite, which is responsible for high removal of dyes and microorganisms from contaminated water.

The nanocomposite also prevented attachment of floating microorganisms and inhibited the formation of biofilms on its surface. This makes it possible to use it for prolonged times in contaminated water.

"We believe that the long-term antibacterial, antifouling and high dye adsorption properties of our functional nanomaterial are exceptionally promising for the development of high-efficiency and low cost water purification technologies," concludes Das.

After completing their tests, the team is planning to develop a filter for industrial wastewater treatment in a larger volume. They also intend to make potable water filters so that people can use it for domestic water purification.

Featured scientists: Dr Sujoy Das' bioinspired nanomaterials laboratory (http://sujoyclri.webs.com/)
Organization: Central Leather Research Institute, Chennai (India)
Relevant publication: S. Das, M. Khan, T. Parandhaman, F. Laffir, A. Guha and G. Sekaran, *et al.*, Nano-silica fabricated with silver nanoparticles: antifouling adsorbent for efficient dye removal, effective water disinfection and biofouling control, *Nanoscale*, 2013, 5(12), 5549, DOI: 10.1039/C3NR00856H

7.3   Carbon Nanotube Ponytail Cleanser

One of the most heavily used materials for water purification filters is carbon, usually in the form of activated carbon—*i.e.* carbon treated with oxygen to open up microscopic pores. This high degree of porosity and the resulting high surface area make this material ideal, among other things, for removing pollutants that are attracted to carbon (such as volatile organic compounds, pesticides and benzene), from water.

One of the problems with activated carbon is the disposal of adsorbed contaminants along with the adsorbent. Another concern is that its pores are often blocked during adsorption. By contrast, carbon nanotubes' (CNTs) open structures offer easy, undisrupted access to reactive sites located on the nanotubes' outer surface. That is why researchers see CNTs as an attractive potential substitutes for activated carbon.

"An issue with using CNTs for water purification of course is the fact that unbounded nanotubes would pose health risks to humans and the ecosystem because they are difficult to separate from treated water," explains Chong-zheng Na, at the time of this work an assistant professor in the Department of Civil and Environmental Engineering and Earth Sciences at the University of Notre Dame (now an associate professor at Texas Tech University).

Na says that the challenge of improving CNTs' separability has been tackled by other researchers in the past. "The motivation is to eliminate an important concern regarding the use of CNTs for water purification, namely how the CNTs will be collected after treatment. From an application point of view, a high efficiency of recollection is important for cost saving reasons. From a health point of view, a high efficiency of recollection eliminates concerns of potential harms that loose CNTs may do to humans and ecosystems."

Na, postdoc fellow Haitao Wang, and graduate student Hanyu Ma, have demonstrated that individual CNTs can be integrated into micrometer-sized colloidal particles without using a heavy or bulky particulate support.

Organizing individual nanotubes into hierarchical structures represents a new strategy to scale up nanomaterials for macroscopic engineering applications.

The researchers grew CNT arrays hundreds of micrometers in length on nanometer-thin mineral discs with a negligible mass and volume, a structure they termed carbon nanotube ponytails (CNPs). The layered double oxide discs are slightly magnetic. This magnetization is sufficiently weak to prevent CNPs from aggregating under self-attraction but strong enough to be utilized for separation.

"Compared to individual CNTs, CNPs can be more effectively separated from water using gravitational sedimentation, magnetic attraction, and membrane filtration while having the ability to perform adsorption, disinfection, and catalytic degradation of contaminants in water," explains Na (Figure 7.2).

To address the CNT separability issue, researchers previously did the same thing as Na's team, *i.e.* increase the overall size of CNTs. For example, CNTs were fixed on inactive colloidal particles. What makes the CNT ponytails special, though, is that they contain little supporting material.

An interesting aspect from a material synthesis point of view is that this work demonstrates the preparation of CNT colloidal particles nearly free of support without involving exotic procedures—*i.e.*, CNPs are made using the

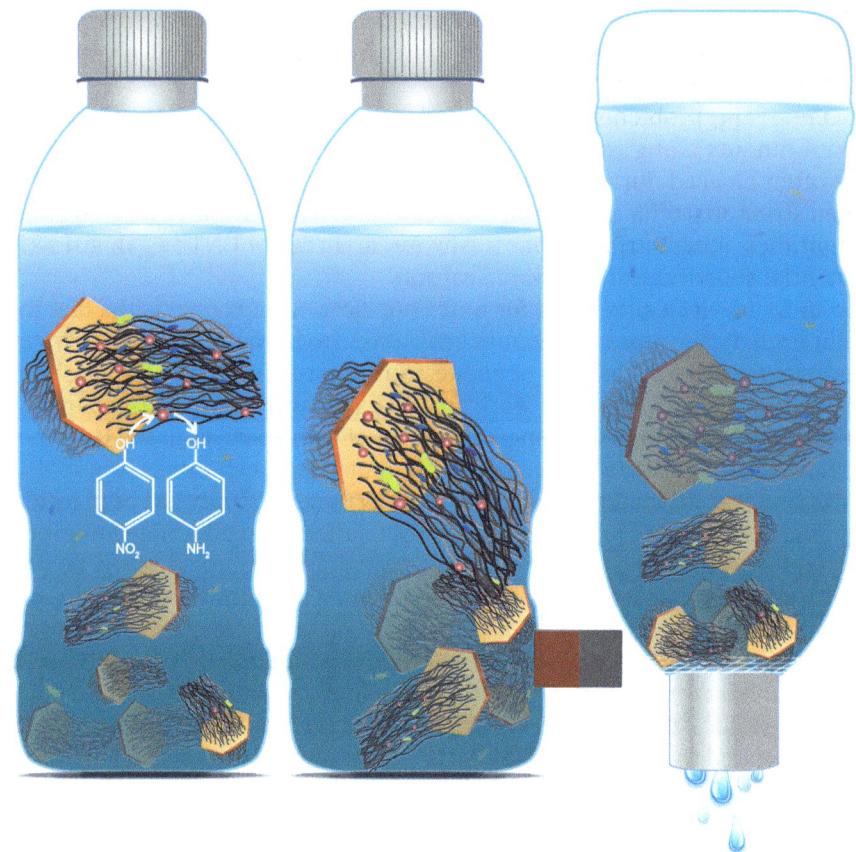

Figure 7.2   Schematic of water purification with carbon nanotube ponytails. (Image: Xia Zhao and Na Lab, University of Notre Dame.)

same standard chemical vapor deposition method for CNT growth using a slightly different catalyst material.

"The elimination of extensive use of support transforms CNT colloids from a composite material containing only a minor mass/volume fraction of the active component (*i.e.*, CNTs) to a material almost exclusively made of active CNTs," says Na. "Obvious advantages include saving energy for transportation and saving space in use."

In their water purification experiments, the researchers showed that their CNPs are as good as CNTs for removing contaminants from water as adsorbents, disinfectants, and catalyst supports.

For water purification, processes using CNPs might be helpful in situations where commonly used water treatment infrastructures are not available, for instance, providing clean water in developing countries or remote rural areas. In addition, CNPs can be used in industrial processes—*e.g.*, as catalyst supports—and emergency response, *e.g.*, adsorbents for chemical spills.

Going forward, the team would like to further reduce the cost of making multifunctional and recollectable hierarchical materials for water purification so that the treatments can become more affordable.

"Obviously," says Na, "this would be tied to the advancement of CNT synthesis if we keep using CNTs as the building material. Alternatively, we are looking into materials that are less expensive than CNTs. Our study with CNPs has provided insights for what properties we should be looking for in potential alternative materials."

Given the recent advances in nanotechnology, the water community has been searching for ways to incorporate nanomaterials into treatment processes. There is, however, a tremendous dimensional disparity between the nano world and the macro world of water. No one is interested in nanoliters of water. When people talk about water, they talk in units such as million gallons per day.

The challenge is how to connect these two worlds.

"What we did in making CNPs has done precisely that," says Na. "By integrating nanotubes into colloidal particles, we helped CNTs climb from the nanometer rung to the micrometer rung on the dimension ladder. By incorporating magnetism into CNPs, we elevated CNTs even further to the centimeter scale."

"In our future research, we will keep looking for better ways to connect nanomaterials to the macro world that we live in—in technical terms, we will be researching for new strategies to scale up nanoscale properties for water purification," he concludes.

Featured scientists: Prof. Chongzheng Na's research group (http://www.chongzhengna.com/)
Organization: Department of Civil and Environmental Engineering and Earth Sciences, University of Notre Dame, IN (USA)
Relevant publication: H. Wang, H. Ma, W. Zheng, D. An and C. Na, Multifunctional and Recollectable Carbon Nanotube Ponytails for Water Purification, *ACS Appl. Mater. Interfaces*, 2014, **6**(12), 9426, DOI: 10.1021/am501810f

7.4 Just Shake It! A Simple Way to Remove Nanomaterial Pollutants from Water

As the production and use of nanotechnology-based products in our daily life is rapidly growing, the risks of environmental pollution due to nanomaterials are increasing as well.

The spreading of nanomaterials in manufacturing, product application, and waste management processes will eventually lead to some degree of contamination of water. While "conventional" water contaminants can be cleaned up by state-of-the-art technologies with filtration and condensation processes, this is not the case for nanoscale pollutants.

The size of nanomaterials is commonly below 100 nm—thousands of times smaller than regular water contaminants and much smaller than the micron-sized or larger pores used in conventional water filters. Simply reducing the filter pore size is not a solution either since that will cause clogging of the filtration line.

"We have demonstrated that water contaminated with nanomaterials can be cleaned up by a 'hand shaking' approach that can be performed even in a kitchen," says Yoke Khin Yap, a professor in the Department of Physics at Michigan Technological University. "Our approach is simple and universal, and can be used for many one-dimensional (1D) and two-dimensional (2D) nanomaterials including nanotubes, nanowires, graphene, and nanosheets. Therefore, our approach would support continued development of nanotechnology by reducing the risk of water contamination."

Yap and his team recently found that microscopic fluid interface dynamics could be utilized to extract and remove nanomaterials from contaminated water.

This novel technique is applicable to the extraction of 1D and 2D nanomaterials from water with an efficiency of almost 100%. The approach involves emulsification of the contaminated water with oil—or other organic solvents—by hand shaking. The nanomaterials will then be captured in the oil phase. They can be removed after the oil and water are separated, *i.e.* the water is condensed.

The extraction of nanomaterials demonstrated here is based on the capillary force generated at the interface of oil (or organic solvent) droplets in water during the emulsification process.

According to the researchers, the extraction of functionalized nanomaterials is dependent on the concentration of surfactants used. If the concentration is sufficiently low, then the extraction is possible with high efficiency.

"Our attempts to extract spherical particles confirmed that the shape of the particles has an effect on the extraction mechanism," Yap points out.

He also notes that, in principle, the nanomaterials can be reused once the organic solvent/oil has been evaporated.

While in general, this simple approach will be applicable to extract many other particulates or solids in water, the extraction of zero dimensional (0D) materials such as quantum dots is still low in efficiency. The team hopes to overcome this issue in a future study.

The idea for this technique goes back several years to when the team devised a nanofilter to separate water and oil and published a well-cited paper on the results.[2]

"We thought that the superoleophilicity of nanotubes will allow us to extract nanotubes from contaminated water by the technique reported in our present paper," says Yap. "Indeed, it works fine but we also found that the same technique works as well for nanowires, which are not superoleophilic. That prompted us to find the explanation for our data and the capillary force was the answer."

Featured scientists: Prof. Yoke Khin Yap's research group (http://www.phy. mtu.edu/yap/)

Organization: Department of Physics, Michigan Technological University, Houghton, MI (USA)

Relevant publication: B. Tiwari, D. Zhang, D. Winslow, C. Lee, B. Hao and Y. Yap, A Simple and Universal Technique To Extract One- and Two-Dimensional Nanomaterials from Contaminated Water, *ACS Appl. Mater. Interfaces*, 2015, 7(47), 26108, DOI: 10.1021/acsami.5b07542

7.5 The Challenge of Testing Nanomaterial Ecotoxicity in Aquatic Environments

Aquatic ecotoxicity test methods, which are routinely applied to testing of nanomaterials, were originally developed for water-soluble chemicals. Nanomaterials are fundamentally different from many "conventional" chemicals as they often have limited or no solubility at all and are potentially released to the environment in a particulate form (*e.g.* carbon nanotubes). Only limited nano-specific guidance on ecotoxicity testing is currently available. These guidelines exhibit a number of specific shortcomings mainly related to characterization, exposure preparation, quantification and monitoring concentrations, and dose-metrics.

Scientists in Denmark have contributed to the progress in algae testing of nanomaterials. This work also aids in the development of additional guidance as it adds to the understanding of pros and cons of different techniques for biomass quantification.

"The motivation for carrying out this research was partly related to large variations in algal growth inhibition test results for nanoparticles—both in our own lab but also comparing results reported in the literature," says Nanna Hartmann, who carried out this work as part of her PhD project[3] at the Department of Environmental Engineering at Technical University of Denmark (DTU). "One explanation could be that the behavior of nanoparticles changes with media properties, over time and due to interactions with organisms. At the same time the nanoparticles have an effect on the organisms. Hence, we wanted to understand better how biomass quantification methods were influenced by these dynamic interactions and try to identify the most optimal technique (Figure 7.3)."

Although several papers exist on the toxicity of nanomaterials to algae, this is an issue that has previously been investigated and described in depth.[4]

If algal biomass can be correctly and accurately quantified, and the influence of artefacts eliminated, then the results of the algal growth inhibition tests will be much more reliable, comparable and applicable to risk assessment.

To compare differences in algal toxicity of different nanoparticles, a first step is to make sure that observed effects are actually toxic effects and not influenced by the biomass quantification technique.

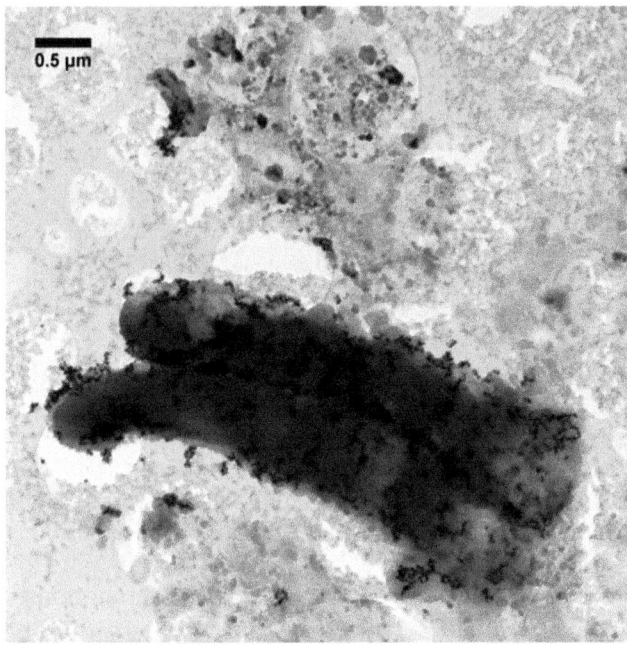

Figure 7.3   Algal cells after 24 hours' exposure to gold nanoparticles. (Image: Christian Engelbrekt, DTU.)

In their study, the team, led by Anders Baun, a professor at DTU who heads the nanotechnology and risk group, and Professor Jens Ulstrup from DTU's nanoscale chemistry group, did an in-depth investigation of the method that is used for testing effects of chemicals on algal growth and its applicability to testing of nanomaterial toxicity.

"Specifically, we evaluated three techniques which are routinely used for quantifying algal biomass, namely: coulter counting; cell counting by use of a haemocytometer; and measuring the fluorescence of algal pigment extracts," explains Hartmann. "Based on our results, we found that the fluorometric method was most suitable as it allows for a physical separation of the nanoparticles and the biomass surrogate (pigment). This technique needs to be combined with visual inspections of the algal cells in a microscope. This provides additional information about the exposure of the cell, interactions between particles and cells, and it allows for detection of physiological effects."

She notes that the fluorescence methods need some further optimization to minimize particle interference on the fluorescence measurements.

The team's results also highlight another important issue: as the test system is dynamic, both quantitative and qualitative changes in exposure should be monitored over time. Otherwise, time-dependent changes in toxicity cannot retrospectively be linked to particle behavior and characteristics.

The researchers point out that there are still some challenges in the test designs that have to be solved. For example, a better separation of particles and algal pigment extracts will decrease particle background interference. Also, there are other measuring techniques—such as *e.g.* flow cytometry—that could be evaluated and compared to the techniques investigated in this study.

Featured scientist: Dr Nanna Hartmann
Organization: Department of Environmental Engineering, Technical University of Denmark, Kgs. Lyngby (http://www.env.dtu.dk/english)
Relevant publication: N. Hartmann, C. Engelbrekt, J. Zhang, J. Ulstrup, K. Kusk and A. Baun, The challenges of testing metal and metal oxide nanoparticles in algal bioassays: titanium dioxide and gold nanoparticles as case studies, *Nanotoxicology*, 2012, 7(6), 1082, DOI: 10.3109/17435390.2012.710657

7.6 Water Quality Testing with Artificial "Microfish"

Advances in micro- and nanoscale engineering have led to various mobile devices that either can move on solids or swim in fluids. Researchers are applying various strategies to design nanoscale propulsion systems by either using or copying biological systems such as the flagellar motors of bacteria or by employing various chemical reactions. Different practical micromotor applications, ranging from drug delivery to target isolation and environmental remediation, have already been reported.

Yet, there are no reports on a nanomachine-based toxicity assay approach, analogous to the use of live aquatic organisms for testing the quality of our water resources (one of the oldest and most widely used approaches for water quality testing relies on changes in the swimming behavior and lifetime expectancy of live fish in the presence of toxic substances).

Until now. Professor Joseph Wang's team at the University of California - San Diego (UCSD), took advantage of recently developed biocatalytic—enzyme-powered—tubular microengines that offer efficient propulsion in various real-life media and on well-established enzyme inhibition processes.

The researchers describe a new, simple and cost-effective strategy for water-quality testing based on changes in the propulsion behavior and lifetime of artificial biocatalytic microswimmers in the presence of aquatic pollutants, in a manner analogous to changes in the swimming behavior and survival of fish used for toxicity testing.

"Our novel micromotor toxicity testing concept mimics live-fish water testing and relies on the toxin-induced inhibition of the enzyme catalase, responsible for the biocatalytic bubble propulsion of tubular microengines," Wang explains the results. "The basis for this exciting development is the influence of the toxin upon the activity of the enzyme powering the micromotor, and hence upon the propulsion efficiency and survival time of such artificial biocatalytic microswimmers."

The novel microswimmer water-toxicity assay strategy addresses drawbacks associated with fish-toxicity assays, including standardization and reproducibility problems, and major ethical concerns associated with live-fish toxicity bioassays.

The researchers fabricated their artificial microfish by a template-based electrodeposition of PEDOT/gold bilayer microtubes. These have a defined conical microtube configuration with a length of ~8 μm, outer diameters of 2.0 μm and 1.2 μm, along with inner openings of 1.6 μm and 0.8 μm.

The microengine-based water toxicity testing developed by the UCSD team relies on changes in the propulsion behavior associated with the inhibition of the catalase biocatalytic layer by common pollutants.

The biocatalytic decomposition of the hydrogen peroxide fuel at the inner enzymatic layer of the microtube generates the oxygen bubble thrust at the tail and leads to an efficient swimming motion of the microfish. However, if this device encounters chemical stress—for instance heavy metals, pesticides or herbicides—the biocatalytic activity of catalase is inhibited and results in a lower bubble frequency and an impaired locomotion of the artificial microfish.

Wang notes that the optical tracking of the movement of the artificial microfish offers a simple and direct real-time visualization of these toxin-induced changes in their swimming performance.

"Unlike earlier motion-based biosensing applications involving the attachment of bioreceptors, no additional surface functionalization is required for realizing the microfish-based water-quality testing," he says. "In addition, the composition of our enzyme-powered polymer-based microengines makes these microfish highly compatible and environmentally friendly."

Featured scientists: Prof. Joseph Wang's laboratory for nanobioelectronics (http://joewang.ucsd.edu/)
Organization: University California San Diego, CA (USA)
Relevant publication: J. Orozco, V. García-Gradilla, M. D'Agostino, W. Gao, A. Cortés and J. Wang, Artificial Enzyme-Powered Microfish for Water-Quality Testing, *ACS Nano*, 2013, 7(1), 818, DOI: 10.1021/nn305372n

7.7   Microscale Garbage Trucks

The construction of artificial micro- and nanomotors is a high priority in the nanotechnology field owing to their great potential for diverse potential applications, ranging from targeted drug delivery, on-chip diagnostics and biosensing, to pumping of fluids at the microscale and environmental remediation. Particular attention has been given to self-propelled chemically powered micro/nanoscale motors, such as catalytic nanowires, microtube engines, and spherical Janus microparticles.

Researchers in Germany have reported the first example of micromotors for the active degradation of organic pollutants in solution. The novelty of this work lies in the synergy between the internal and external functionality of the micromotors.

"Previously, some groups tried to demonstrate the use of catalytic nanomotors for biomedical applications—including ours—on-chip biosensors and capture of bio species," says Dr Samuel Sánchez, Group Leader, Smart Nano-Bio-Systems, Max Planck Institute for Intelligent Systems in Stuttgart, Germany. "However, the toxicity of the fuel employed still limits their real applications. We imagined that environmental applications might be another field to explore, where the use of hydrogen peroxide is not controversial."

In that direction, Wang's group at the University of California San Diego have already reported the removal, not degrading, of oil droplets from solution.[5] Sánchez and his collaborators went one step beyond that and demonstrated the total removal of contaminants using micromotors. Indeed, the hydrogen peroxide is used for self-propulsion and remediation when it interacts with the outer layer of the micromachine.

"We have demonstrated the ability of self-propelled micromotors to oxidize organic pollutants in aqueous solutions through a Fenton process," explains Sánchez. "The combination of mixing and releasing iron ions in liquids results in a rate of removal of a model pollutant (rhodamine 6G) *ca.* 12 times higher than when the Fenton oxidation process is carried out with nonpropelling metallic iron tubes."

The research team from Sanchez' group, the Leibniz Institute for Solid State and Materials Research Dresden, and the Chemnitz University of Technology, demonstrates that micromotors boost the Fenton oxidation process (read more about Fenton reactions at the bottom of this section) without applying external energy, and complete degradation of organic pollutants is achieved.

Sánchez notes that, if desired, the micromotors can be easily recovered using a magnet once the water purification process has been completed and the excess hydrogen peroxide can be easily decomposed to pure water and oxygen under visible light.

The team fabricated their tubular bubble-propelled micromotors containing small amounts of metallic iron (from 20 to 200 nm layer thickness) as an outer layer and platinum as an inner layer.[6] The mechanism of degradation is based on Fenton reactions relying on spontaneous acidic corrosion of the iron metal surface of the micromotors in the presence of hydrogen peroxide, which acts both as a reagent for the Fenton reaction and as the main fuel to propel the micromotors. Moreover, the ability of self-propelled, tubular micromotors to improve mixing results is a synergetic effect that enhances water remediation without applying external energy.

This work can pave the way for the use of multifunctional micromotors for environmental applications where the use of hydrogen peroxide is not a major drawback but a co-reagent.

Sánchez adds that the high efficiency of the oxidation of organic pollutants achieved by the Fe/Pt catalytic micromotors reported in this work is of importance for the design of new and faster water treatments, such as the decontamination of organic compounds in wastewaters and industrial effluents.

The aim of this study was to fabricate an autonomous microscopic cleaning system that works without an external energy input in a much faster and convenient way. The micromotors offer this ability to move the catalyst around without external actuation or addition of catalysts (iron salts) to achieve water remediation, removal of organic dyes, *etc.*

However, as the researchers point out, this is an application especially for microscale environments. "Unfortunately, it is clear that we would not use the micromotors in a large reactor vessel to clean huge amounts of water," says Sánchez.

Nevertheless, the high efficiency of the oxidation of organic pollutants achieved by the Fe/Pt catalytic micromotors is of importance for the design of new and faster water treatments, such as the decontamination of organic compounds in wastewaters and industrial effluents.

"We have proven that the usefulness of the micromotors lies not solely in their capacity to move, but to exploit their motion using their external surface to enhance useful catalytic reactions," says Sánchez. "This work could open a new research line towards coupling a variety of catalytic reactions in self-propelled devices where the presence of hydrogen peroxide is not a disadvantage. In future, we expect that a rich variety of contaminants can be cleaned."

7.7.1 About Fenton Reactions

The Fenton method is one of the most popular advanced oxidation processes for the degradation of organic pollutants, utilizing the hydroxyl radical (OH$^\bullet$) as its main oxidizing agent. The generation of OH$^\bullet$ in the Fenton method occurs by reaction of H_2O_2 in the presence of Fe(II). However, one disadvantage of these processes is that Fe ions in solution must be removed after the treatment to meet regulations for drinking water. In order to diminish and, in the best scenario, solve the problems caused by the presence of Fe ions in treated effluents and decrease the costs of recovery, the use of heterogeneous Fenton catalysts is a promising strategy that could allow for the degradation of pollutants by Fenton processes without the requirement of dissolved iron salts. The micromotors fabricated in this work can be included as a new type of heterogeneous Fenton catalyst. With this method, the remaining iron in the solution is one to three orders of magnitude lower than in conventional Fenton processes.

Featured scientists: Dr Samuel Sánchez's lab-in-a-tube and nanorobotics group (http://www.is.mpg.de/de/sanchez)
Organization: Max Planck Institute for Intelligent Systems, Stuttgart (Germany)
Relevant publication: L. Soler, V. Magdanz, V. Fomin, S. Sanchez and O. Schmidt, Self-Propelled Micromotors for Cleaning Polluted Water, *ACS Nano*, 2013, 7(11), 9611, DOI: 10.1021/nn405075d

7.8 Nanomaterials that Capture Nerve Agents

Protection against nerve agents—tabun, sarin, soman, VX and others —is a major terrorism concern of security experts. Nerve agents, which attack the nervous system of the human body, are clear and colorless or slightly colored liquids and may have no odor or a faint, sweetish smell. They evaporate at various rates and are denser than air. Current methods to detect nerve agents include surface acoustic wave sensors; conducting polymer arrays; vector machines; and the most simple: color change paper sensors. Most of these systems have certain limitations including low sensitivity and slow response times.

Nanoporous materials can capture highly toxic nerve agent vapors by physical adsorption. Unfortunately, the broad range of toxic agents, environmental conditions and types of carbonaceous material simply do not allow laboratory testing of every possible combination. In this regard, screening of candidate carbonaceous materials for efficient capturing of highly toxic nerve agents is of undoubted interest for both military and civilian protective applications.

Although respirators with a cartridge filled with carbonaceous adsorbents (*i.e.*, activated charcoals, activated carbons, and activated carbon fibers) have been used since World War I, the impact of the internal porous structure on the protective time of respirators has been poorly understood.

"Everyone can argue that the internal structure of a nanoporous material affects the protective time of any respiratory device," says Dr Piotr Kowalczyk, at the time of this work a senior research fellow at the Nanochemistry Research Institute at Curtin University of Technology in Australia (now a senior lecturer at the School of Engineering and Information Technology, Murdoch University). "However, the fundamental question is how to optimize this structure to get longer protection against highly toxic agent molecules. Clearly, fundamental research is needed to answer this nontrivial question."

An international team, led by Kowalczyk and Alexander V. Neimark, a professor at Rutgers University, together with scientists from the physicochemistry of carbon materials research group at Nicolaus Copernicus University in Poland, is shedding new light on the selection of an optimal nanomaterial for capturing highly volatile nerve agents.

"We propose an entirely new strategy for combined experimental and computational screening of candidate carbonaceous materials for capturing of highly toxic molecules inside nanopores," says Kowalczyk.

He notes that the proposed idea is actually pretty simple: "First, we determined the distribution of nanopore sizes for selected commercialized porous materials (nitrogen porosimitry, non-local density functional theory, and integral theory of adsorption). Next we computed Henry constants for sarin and DMMP (its common stimulant) adsorbed in model slit-shaped carbon nanopores at 298 K (using the Metropolis–Ulam Monte Carlo method, and force-field calculations).

"Finally, we averaged Henry constants with experimental pore size distribution (note that normalized pore size distribution is the probability distribution). Knowing the averaged Henry constant for studied carbonaceous materials and concentration of sarin/DMMP in the bulk phase, we were able to predict the exact mass of sarin/DMMP adsorption (simply, the captured mass of toxic agent per mass of material). This allowed us to select the optimal carbonaceous nanomaterial for potential use in protection devices."

According to the team's calculations, the optimal carbonaceous adsorbent—pitch-based P7 activated carbon fiber—adsorbed ~100 µg per gram DMMP at 0.03 µg per cubic meter. Commercial norit activated carbon adsorbed only ~20 µg per gram DMMP at 0.03 µg per cubic meter.

"However, what is more important, we discovered the strong relationship between the pore size and the efficacy of nerve agent capturing *via* physisorption," Kowalczyk points out. "In general, it appears that slit-shaped carbon pores with pore sizes around ~0.5 nm are optimal for sarin and DMMP capture."

He explains this process in more detail: "Consider the virtual thin layer of glue that covers the walls of pores. If pores are very small—*i.e.*, their effective size is comparable with the size of agent molecules—the adsorbed molecules are very 'sticky' and they are 'captured by glue'—like insects on trapping tape. In wider pores, this 'virtual thin layer of glue' is not an effective capturing medium and agent molecules can move easily through the pores."

Therefore the key question is: how do the adhesive properties of the "virtual layer of glue" depend on the size of the nanopore? And here, the team discovered that this dependence is exponential. In other words, capturing of nerve agents is only very effective when the size of carbon pores is around ~0.5 nm. Shrinking or widening the slit-shaped carbon pore width decreases the efficiency of sarin and DMMP capturing significantly.

"Some of my colleagues asked me if I believe in our theoretical results," says Kowalczyk. "The great physicist Paul Dirac used to say: 'This result is too beautiful to be false; it is more important to have beauty in one's equations than to have them fit experiment'."

"And I truly believe that our theoretical results have to be correct—within the assumed model of nanopores—because they are so simple and beautiful," he concludes.

Featured scientists: Dr Piotr Kowalczyk (http://www.pkowalczyk.com/index.html); Prof. Alexander V. Neimark (http://sol.rutgers.edu/~aneimark/)
Organizations: Nanochemistry Research Institute, Curtin University of Technology, Perth (Australia) (https://nanochemistry.curtin.edu.au/); Department of Chemical and Biochemical Engineering, Rutgers University, Piscataway, NJ (USA)
Relevant publication: P. Kowalczyk, P. Gauden, A. Terzyk and A. Neimark, Screening of carbonaceous nanoporous materials for capture of nerve agents, *Phys. Chem. Chem. Phys.*, 2013, **15**(1), 291, DOI: 10.1039/C2CP43366D

7.9 Replacing Chemical Disinfectants with Engineered Water Nanostructures

The burden of food-borne diseases worldwide is huge, with serious economic and public health consequences. The food industry is in search of effective intervention methods that can be applied from "farm to fork" to ensure the safety of the food chain and be consumer and environmentally friendly at the same time.

In the food industry, chemicals are routinely used to clean and disinfect product contact surfaces as well as the outer surface of the food itself. These chemicals provide a necessary and required step to ensure that the foods produced and consumed are as free as possible from microorganisms that can cause food-borne illness.

Food activists are concerned that some of the chemicals used by the food industry for disinfection can cause health issues for consumers. A prime example is the current discussion in Europe about "American chlorine chicken". Most U. S. poultry is chilled in antimicrobial baths that can include chlorine to keep salmonella and other bacteria in check. In Europe, chlorine treatment was banned in the 1990s out of fear that it could cause cancer.

Trying to develop chemical-free disinfection techniques, researchers at the Center for Nanotechnology and Nanotoxicology at the Harvard T. Chan School of Public Health have explored the effectiveness of a nanotechnology-based intervention method for the inactivation of food-borne and spoilage microorganisms on fresh produce and on food production surfaces.

This method[7] utilizes engineered water nanostructures (EWNS) generated by electrospraying of water.

"Our EWNS possess unique properties," says Philip Demokritou, Associate Professor of Aerosol Physics and Director of the Center for Nanotechnology and Nanotoxicology. "They are 25 nm in diameter; remain airborne in indoor conditions for hours; contain reactive oxygen species (ROS); have very strong surface charge (on average 10 electrons per structure) and have the ability to interact and inactivate pathogens by destroying their membrane."

The results of the team's assessment of the efficacy of these tiny water nanodroplets in inactivating representative food-borne pathogens such as *Escherichia coli*, *Salmonella enterica* and *Listeria innocua*, on stainless steel surfaces and on tomatoes, showed significant log reductions in inactivation of select food pathogens.

"These promising results open the door for further exploration of the dynamics of this method in the battle against food-borne diseases," notes Demokritou, who led the study. "More importantly, this novel, chemical-free, cost effective and environmentally friendly intervention method holds great potential for development and application in the food industry, as a 'green' alternative to existing chemical-based inactivation methods."

He points out that nanotechnology can lead to new useful intervention approaches that can be used in the battle against food-borne diseases: "Using water in its engineered nanoscale form can be a 'game changer'—a cost effective approach that can be easily deployed at various intervention points across the 'farm to fork' line. This technology can be used in farms during packaging of fresh produce, during transportation, and in supermarkets as well as in your refrigerators at home."

Demokritou's team is further exploring the prospects of this novel approach and working on scaling up the technology, which could not only significantly improve inactivation efficiency and decrease the microbial load on the fresh produce but also extend the shelf life of produce and reduce the number of cases of food-borne illnesses on consumption.

Featured scientist: Prof. Philip Demokritou (http://www.hsph.harvard.edu/philip-demokritou/)
Organization: Center for Nanotechnology and Nanotoxicology, Harvard T. Chan School of Public Health, Boston, MA (USA) (http://www.hsph.harvard.edu/nano/)
Relevant publication: G. Pyrgiotakis, A. Vasanthakumar, Y. Gao, M. Eleftheriadou, E. Toledo and A. DeAraujo, *et al.*, Inactivation of Foodborne Microorganisms Using Engineered Water Nanostructures (EWNS), *Environ. Sci. Technol.*, 2015, **49**(6), 3737, DOI: 10.1021/es505868a

7.10   Nanotechnology Could Make Battery Recycling Economically Attractive

Batteries are an integral part of modern life—just go ahead and count the batteries that you use yourself in your watches, computers, cell phones, cameras, alarm clocks, flashlights, toys, remote controls, power tools, smoke detectors, cars, boats and so on. You'll come up with a staggering number. And chances are that your batteries are disposable, so you throw them out with your garbage when they are empty. Add to that the batteries used by industry, hospitals, public transport, the military *etc.*, and you get several billion batteries that are bought every year, a roughly $50 billion market.

Many batteries still contain heavy metals such as mercury, lead, cadmium, and nickel, which can contaminate the environment and pose a potential threat to human health when batteries are improperly disposed of. Not only do the billions upon billions of batteries in landfills pose an environmental problem, they are also a complete waste of reusable raw material.

Unfortunately, current recycling methods for many battery types, especially the small consumer-type ones, don't make sense from an economical point of view since the recycling costs exceed the recoverable metals' value. Therefore, recycling companies only take up spent batteries if someone pays for their service. In Switzerland, for instance, the purchase price of batteries up to a weight of 5 kg contains the cost for the battery's disposal, which finances the entire recycling process.[8]

The economic recycling problem is particularly serious in developing countries such as India where, so far, economic interests supersede environmental obligations. This situation makes the development of economically interesting battery recycling technologies quite an urgent issue.

"Most of the reported processes for the recovery of metals from spent batteries focuses on production of metal salts/oxides/ferrites," explains Dr Akash Deep, a scientist in the Biomolecular Electronics and Nanotechnology Division at the Central Scientific Instruments Organization in Chandigarh, India. "Steps involved in these technologies include ammoniacal or acidic leaching, precipitation, solvent extraction, and thermal treatment. Generally the recycling of lead-acid, lithium ion and rechargeable Ni–Cd batteries is given more importance. But, in reality, the relative consumption of alkaline Zn–MnO$_2$ batteries is much larger."

Deep and his team carried out research to address the recycling of consumer-type batteries. In their research report, they describe the recovery of pure zinc oxide nanoparticles from spent Zn–Mn dry alkaline batteries.

The researchers dismantled the spent batteries and leached the desired metals from the waste electrode materials. They introduced a solvent extraction step (using the acid extractant Cyanex 923) to selectively extract zinc from the powderized electrode materials.

"This commercially available extractant is cheaper than the previously tried Cyanex 272/Cyanex 301, offers clearer phase separations and better loading capacity, and is easy to regenerate," says Deep.

After extracting the zinc, the team incinerated the pure zinc loaded organic layer at 600 °C for the synthesis of high-purity zinc oxide nanoparticles.

"The recovered pure zinc oxide nanoparticles may find applications in piezoelectric transducers, gas sensors, photonic crystals, light-emitting devices, photodetectors, photodiodes, optical waveguides, transparent conductive films, varistors, and solar cells," says Deep.

Although this work focuses on the recovery of pure nanoparticles from spent alkaline Zn–MnO$_2$ batteries, the same approach may be utilized in recycling other types of spent batteries, including Ni–Cd, Li-ion, Zn–C, and Pb-acid.

"Our research team is working on the recovery of pure cadmium and lead nanoparticles and trying to develop green processes for the ultimate recovery of highly pure and efficient zinc, cadmium, and lead quantum dots from environmental waste," Deep describes the group's current research focus. "In future, we are expecting to turn different environmental waste, such as spent batteries, electronic parts, LCD and LED panels, or semiconductor waste into lucrative raw materials for the recovery of high cost nanoparticles, quantum dots, nanowires *etc.*"

Over the past two decades, the advancements in nanotechnologies have spurred the use of metallic and non-metallic nanoparticles in virtually every scientific field. There is an increasing demand for customized nanoparticles, nanowires, nanotubes, atomic layers and other similar kinds of photonically or electrically active nanotemplates.

Combine this with the environmental implications of the various nanofabrication routes and you'll see a trend where nanotechnologies actually build up quite a significant ecological footprint.

However, research such as the one done by Deep and his collaborators will appeal to environmental nanotechnologists who should see this as an opportunity to turn recycling of electronic waste into an economically beneficial proposition for industry.

Featured scientist: Dr Akash Deep
Organization: Biomolecular Electronics and Nanotechnology Division, Central Scientific Instruments Organization, Chandigarh (India) (http://www.csio.res.in/)
Relevant publication: A. Deep, K. Kumar, P. Kumar, P. Kumar, A. Sharma and B. Gupta, *et al.*, Recovery of Pure ZnO Nanoparticles from Spent Zn–MnO$_2$ Alkaline Batteries, *Environ. Sci. Technol.*, 2011, **45**(24), 10551, DOI: 10.1021/es201744t

7.11 Bioinspired Nanofur Reduces Underwater Drag of Marine Vessels

Nature often provides the blueprint when researchers are developing new technologies. Just think of George de Mestral and how, back in 1948, his observation of the structure of cockleburr seeds led to the development of Velcro®. In a similar vein, observations made on the *Salvinia* fern as well as the *Notonecta glauca* bug have now led researchers to develop a nanofur structure that significantly reduces fluid drag.

For instance, consider the large amount of fuel used by the 90 000 ocean-going cargo ships that roam the seas (international shipping uses about 300 million metric tons of fuel and it is estimated to be responsible for 3.5–4% of all climate change emissions). Most of the energy in shipping is used to overcome surface friction. An effective way to reduce frictional drag underwater could significantly reduce marine fuel consumption, making the shipping industry more efficient and environmentally friendly.

Additionally, in most applications that involve moving liquids through pipes and tubes of different sizes, a lot of the energy is used to overcome the drag the fluid experiences moving over the sidewalls. Here as well, a drag reducing coating could compensate this effect and increase efficiency in these areas.

Both the fern and the insect investigated for this research have surfaces covered by high-density hairs, which allow them to keep an air layer under water. This enables the *Notonecta glauca* bug to move nimbly and swiftly through the water by reducing the drag on its surface.

Scientists at the Institute for Microstructure Technology (IMT), Karlsruhe Institute of Technology, have developed an inexpensive, highly scalable method to produce a superhydrophobic, air retaining biomimetic surface—a "nanofur"—that shows not only high long-term stability but also high resistance against additional applied pressure. These properties enable the surface to significantly reduce the frictional drag experienced by fluids over a wide range of flow rates.

"The most exciting result of our research is that the produced nanofur can not only hold an air film under water for more than 31 days (as detailed in previous work[9]), but also shows a high stability regarding additional hydrostatic pressure," says Dr Maryna Kavalenka, a postdoc at IMT and the first author of a paper on this work. "Most other surfaces show a much shorter life time of the retained air film or withstand only lower hydrostatic pressure."

It is precisely this stable air film on the nanofur surface that enables it to significantly reduce fluid drag. In experiments that tested the performance of nanofur-coated surfaces, the measured pressure drop across the channels lined with nanofur is approximately 50% lower than in the channels lined with unstructured polymer. Lower pressure drop for the nanofur indicates the reduction in fluid drag by the material, resulting from the air layer retained on its nano- and microstructured surface.

The nanofur is produced using a hot pulling technique, which was developed at the IMT. In contrast to other methods used to produce superhydrophobic, air retaining and drag reducing surfaces, the hot pulling method is low-cost, because it uses sandblasted steel plates as molds, an inexpensive fabrication procedure and material.

In addition, the process is highly scalable and uses no additional chemicals, which could be toxic or harmful, thus making the nanofur easy to handle.

"Novel in our study is that we used a simple method to investigate the dynamic collapse of the air retention of the nanofur by analyzing pictures taken with a camera," explains Kavalenka. "Other studies investigated periodic or highly symmetric structures for air retention and drag reduction, which therefore mostly had one specific critical pressure at which the collapse of the air/water interface happened. The nanofur, on the other hand, is covered by a layer of randomly distributed high aspect ratio nano- and microhairs interspersed with microcavities, analogous to the natural model air-retaining surfaces."

Because of its highly irregular shape and surface, the nanofur does not have one critical pressure at which the water–air interface collapses. Because of this, the scientists investigated the dynamic behavior of the air–water interface in respect to the applied hydrostatic pressure.

Even though the air film held under water by the nanofur shows great long-term stability and a high resistance against hydraulic pressure, the team is investigating if it is possible to increase the resistance of the air-retaining layer against hydrostatic pressure. This could further improve the frictional drag reduction of the nanofur under water and open the door to a wide range of applications.

"The future direction of the drag reduction and air retention research of our group is mainly focused on increasing the long term stability, as well as the stability against additional hydrostatic pressure," concludes Kavalenka. "This, combined with research into the drag reduction for higher flow rates as they occur on container ships and other types of marine vessels, is going to be very interesting. This field poses the special challenge of shear forces, which might lead to a loss of the air film. Combining this with harsh environmental conditions for the drag reducing surfaces leads to many challenges that we'll have to overcome in fabricating these materials."

Featured scientist: Dr Maryna Kavalenka
Organization: Institute for Microstructure Technology (IMT), Karlsruhe Institute of Technology (Germany) (http://www.imt.kit.edu/english/)
Relevant publication: M. Kavalenka, F. Vüllers, S. Lischker, C. Zeiger, A. Hopf and M. Röhrig, *et al.*, Bioinspired Air-Retaining Nanofur for Drag Reduction, *ACS Appl. Mater. Interfaces*, 2015, 7(20), 10651, DOI: 10.1021/acsami.5b01772

7.12 Risk-Ranking Tool for Nanomaterials

Military organizations around the world, especially in the U. S., have been quicker than most to appreciate the potential of nanotechnology. More money is being spent on nanotechnology research for military applications than for any other nano-related research area.

Public reports about military nanotechnology research and development activities are full of sensors, batteries, wound care, filtration systems, smart fabrics, and lighter, stronger, heat-resistant nanocomposite materials *etc.* Naturally, nanomaterial safety has become an important issue for military organizations as well.

"Assessing the potential human health and environmental risks of engineered nanomaterials (ENMs) within the context of the applications and products in which they are incorporated continues to be an extremely challenging endeavor," says Khara D. Grieger, PhD, an environmental risk assessor and research scientist at RTI International. "Given the challenges of developing sufficient data that would be required for traditional risk assessment frameworks, risk assessors are continuing to refine their methods and techniques to perform risk assessments using combined quantitative and qualitative frameworks for ENMs, resulting in various alternatives for risk analysis."

The use of risk-ranking tools may be particularly advantageous to prioritize materials or products according to their risk potential, *i.e.*, identify the "riskiest" ENMs or nanotechnology products, for example, for further research or investigation. This may be useful especially in cases of resource and time constraints.

Based on the research conducted by the U. S. Army Center for Environmental Health Research (USACEHR) associated with the incorporation of various engineered nanomaterials into military applications and equipment (termed "army materiel"), a new analysis developed and implemented a risk-ranking tool to rank ENM–army materiel pairs based on their relative risks to soldier and civilian health under occupational scenarios.

The analysis included 133 unique applications and shows a wide range of nanomaterials used.

The scientific core of the work focuses on the development and application of a relative risk-ranking tool that ranks engineered nanomaterials as well as the applications in which they are embedded relevant to worker or soldier health.

"The development of this tool is important because it not only takes into account the physicochemical characteristics of ENMs but also the characteristics of the equipment in which they are embedded, relevant for current, real-world scenarios involving ENMs," explains Grieger. "The results from this work may be used to help prioritize additional research, such as in-depth risk evaluations or further nanotoxicological research pertaining to the highest ranked ENMs, materiel, or ENM–materiel pairs."

She adds that the fundamental methodology and risk-ranking algorithm developed in the ranking tools may be applicable to other occupational and environmental settings involving ENMs and could, therefore, easily be translated to other application scenarios.

Grieger also points out that this study is unique in that real-world ENM–army material pairs used in research or full-scale field applications were considered to perform the relative risk ranking rather than primarily hypothetical or pristine ENMs. It is also unique since a tool was developed for a specific organization or stakeholder, so it was able to be tailored to meet their needs more precisely.

"Ideally, it is recommended that future revisions of this ranking tool and other similar tools involving ENMs be able to tap into publicly accessible databases in order to be able to update the underlying data sets on environmental, health and safety implications of ENMs, potentially in real time," says Grieger. "As significant research efforts are currently focused on the collection and curation of data on ENMs and their behavior, these risk-ranking tools would benefit from accessing these publicly available databases."

Separate analyses that assess "new" levels of uncertainty if additional data were required, such as value of information methodologies, may also be beneficial when considering such updates based on current knowledge.

The scientists also recommend that increased funding is made available to support the development of decision support tools for emerging technologies, including those focused on risk ranking, which specialize in conditions of extreme uncertainty.

"Given the ever-increasing pace of nanomaterial development and incorporation into various product applications as well as the extreme challenges to develop the data required to satisfy quantitative risk assessment approaches, decision support frameworks which focus on handling extreme conditions of uncertainty in a transparent, structured, and robust manner are critically needed, particularly for decisions regarding emerging technologies in the twenty-first century," Grieger concludes.

Featured scientist: Dr Khara D. Grieger (http://www.researchgate.net/profile/Khara_Grieger)
Organization: RTI International, Durham, NC (USA)
Relevant publication: K. Grieger, J. Redmon, E. Money, M. Widder, W. van der Schalie, S. Beaulieu, *et al.*, A relative ranking approach for nano-enabled applications to improve risk-based decision making: a case study of Army materiel, *Environ. Syst. Decis.*, 2014 35(1), 42, DOI: 10.1007/s10669-014-9531-4

References

1. http://worldbank.270a.info/classification/indicator/EE.BOD.TXTL.ZS.html.
2. C. Lee, N. Johnson, J. Drelich and Y. Yap, The performance of superhydrophobic and superoleophilic carbon nanotube meshes in water–oil filtration, *Carbon*, 2011, **49**(2), 669, DOI: 10.1016/j.carbon.2010.10.016.
3. Ecotoxicity of engineered nanoparticles to freshwater organisms. (http://orbit.dtu.dk/fedora/objects/orbit:86084/datastreams/file_5598471/content).
4. N. Hartmann, F. Von der Kammer, T. Hofmann, M. Baalousha, S. Ottofuelling and A. Baun, Algal testing of titanium dioxide nanoparticles—Testing considerations, inhibitory effects and modification of cadmium bioavailability, *Toxicology*, 2009, **269**(2–3), 190, DOI: 10.1016/j.tox.2009.08.008.
5. M. Guix, J. Orozco, M. García, W. Gao, S. Sattayasamitsathit and A. Merkoçi, *et al.*, Superhydrophobic Alkanethiol-Coated Microsubmarines for Effective Removal of Oil, *ACS Nano*, 2012, **6**(5), 4445, DOI: 10.1021/nn301175b.
6. Watch the micromotor in action in this video:https://youtu.be/daOi-A6Q-JAA. A synergetic effect is achieved by taking advantage of the release of the iron ions from the outer layer of the micromotors and their active motion in the solution.
7. G. Pyrgiotakis, J. McDevitt, A. Bordini, E. Diaz, R. Molina and C. Watson, *et al.*, A chemical free, nanotechnology-based method for airborne bacterial inactivation using engineered water nanostructures, *Environ. Sci.: Nano*, 2014, **1**(1), 15, DOI: 10.1039/C3EN00007A.
8. http://www.inobat.ch/.
9. M. Röhrig, M. Mail, M. Schneider, H. Louvin, A. Hopf and T. Schimmel, *et al.*, Nanofur for Biomimetic Applications, *Adv. Mater. Interfaces*, 2014, **1**(4), 1300083, DOI: 10.1002/admi.201300083.

Subject Index

£ 66-99